D1251240

The Donor-Acceptor Approach to Molecular Interactions

The Donor-Acceptor Approach to Molecular Interactions

Viktor Gutmann

Technical University of Vienna
Vienna, Austria

PLENUM PRESS • NEW YORK AND LONDON

Library of Congress Cataloging in Publication Data

Gutmann, Viktor.
The donor–acceptor approach to molecular interactions.

Bibliography: p.
Includes index.
1. Chemical bonds. 2. Electron donor–acceptor complexes. I. Title.
QD461.G88 541'.224 77-25012
ISBN 0-306-31064-3

A Division of Plenum Publishing Corporation
227 West 17th Street, New York, N.Y. 10011

Printed in the United States of America

To

Professor Ingvar Lindqvist
Uppsala

in recognition of his pioneering contributions
for the development of the concept presented in this book

Preface

Recent developments in various areas of chemistry have been decisively influenced by the principles of structure and mechanism and by the ideas of coordination chemistry, in particular by the donor–acceptor approach. A unified view of almost all kinds of molecular forces is provided by quantum mechanics, and for practical purposes have been classified according to model assumptions, namely, dispersion, polarization, electrostatic, and short-range forces. The latter are divided into two- and three-center covalent chemical bonds, metallic bonds, and exchange-repulsion forces. This approach allows statements of principle and systematic analysis. However, quantitative predictions on concrete large systems are virtually impossible, and there are no general rules that account for structural and chemical changes due to intermolecular interactions.

Chemists are therefore left with qualitative descriptions in which the changes in electron densities are considered. Such models as the MO theory or the resonance concept unrealistically assume that the nuclei remain in fixed positions. Further difficulties are encountered in the attempted description on the "nature" of the chemical bond, e.g., the forces involved.

In order to avoid these difficulties an extension of the donor–acceptor concept, characterized by the comparison between equilibrium structures in different molecular environments, will be presented in this book. In this way, changes in the positions of the nuclei can be taken into account and the question of the nature of the molecular forces is no longer important. The approach is purely inductive in that experimental data, such as bond distances or spectroscopic properties, for a given molecule under different molecular conditions are correlated to thermodynamic and kinetic data. Thus, the conclusions drawn from quantum chemical analysis for interactions between and within the molecules are shaped into simple rules, the so-called bond-length variation rules, which can be applied to the great variety of intermolecular interactions. Using a chemically satisfying pictorial presentation, this concept emphasizes the structural variability of a molecule and coordinates hitherto unrelated facts and phenomena. It provides a rationalization and a unified re-interpretation of well-known phenomena ranging from solid-state physics to biochemistry.

A comprehensive application of molecular interaction theory to all these branches has not been possible within this text. The objective here has been to provide a unified presentation in many areas of chemistry in the broadest sense, and this has been sometimes at the expense of in-depth specialization.

To those sophisticated chemists who may regard this simple model of donor–acceptor interactions as an oversimplification, I can only say that I have found it extremely useful in explaining and predicting various phenomena. It is my hope that other chemists will be stimulated to test this concept in many diverse areas of chemistry and that students will find it useful in their attempts to understand basic chemical behavior. The text does not require any expert knowledge in a particular field. It has been written for any chemist with a sound background in basic inorganic, organic, and physical chemistry.

The present book would have been impossible without the systematic investigations on solution chemistry at the Institute of Inorganic Chemistry at the Technical University of Vienna, Austria for which the support of the Fonds zur Förderung der wissenschaftlichen Forschung in Österreich is gratefully acknowledged. I wish to thank all of my co-workers and a number of friends and colleagues all over the world for valuable discussions and encouragement. In particular, I wish to express my gratitude to Prof. J. C. Bailar, Jr., Urbana, Illinois, U.S.A., Prof. W. Beck, Munich, German Federal Republic, Prof. I. Bertini, Florence, Italy, Prof. G. Costa, Trieste, Italy, Dr. W. Gerger, Vienna, Austria, Dr. G. Gritzner, Vienna, Austria, Prof. E. Hengge, Graz, Austria, Prof. P. L. Huyskens, Leuven, Belgium, Prof. K. Issleib, Halle, German Democratic Republic, Dr. R. F. Jameson, Dundee, Great Britain, Prof. S. Kirschner, Detroit, Michigan, U.S.A., Dr. H. Mayer, Vienna, Austria, Dr. U. Mayer, Vienna, Austria, Prof. A. Maschka, Vienna, Austria, Prof. A. Meller, Göttingen, German Federal Republic, Prof. J. Michalski, Lodz, Poland, Prof. A. Neckel, Vienna, Austria, Prof. H. Noller, Vienna, Austria, Prof. H. Nöth, Munich, German Federal Republic, Prof. R. Paetzold, Jena, German Democratic Republic, Prof. A. J. Parker, Perth, Australia, Dr. E. Plattner, Vienna, Austria, Prof. A. I. Popov, East Lansing, Michigan, U.S.A., Dr. G. Resch, Vienna, Austria, Prof. E. G. Rochow, Captiva Island, Florida, U.S.A., Dr.

R. Schmid, Vienna, Austria, Dr. I. Schuster, Vienna, Austria, Prof. P. Schuster, Vienna, Austria, Prof. Z. G. Szabó, Budapest, Hungary, Prof. H. Tuppy, Vienna, Austria, and Prof. K. Utvary, Vienna, Austria.

Special thanks are due to Susanne Stitz for typing the manuscript, to Harald Schauer for drawing the figures, and to Mr. Thomas Lanigan and Mr. Robert Golden of Plenum Press for all their efforts in the production of this book.

Vienna
1978

V. Gutmann

Contents

Chapter 11. Complex Stabilities in Solution

Chapter 12. Kinetics of Substitution Reactions

Chapter 13. Kinetics of Redox Reactions of Metal Complexes

Chapter 14. Solvent and Reaction Effects on the Course of Substitution and Insertion Reactions

Chapter 15. Miscellaneous Applications

Chapter 16. Homogeneous and Heterogeneous Catalysis

Chapter 17. Biochemical Applications

Notation

a	Activity
AA	Acetamide
Ac	Acceptor
AC	Acetone
AN	Acceptor number
BF	Benzoylfluoride
BL	Butyrolactone
BN	Benzonitrile
Bu	Butyl
c	Concentration
D	Donor
d	Internuclear distance
d_a	Axial bond length
d_e	Equatorial bond length
δ	Chemical shift
δ^+	Fractional positive charge
δ^-	Fractional negative charge
DCE	Dichloroethane
DEA	Diethylacetamide
DEF	Diethylformamide
D_{het}	Heterolytic dissociation enthalpy
D_{hom}	Homolytic dissociation enthalpy (bond energy)
DMA	Dimethylacetamide
DMF	Dimethylformamide
DMSO	Dimethylsulfoxide
DN	Donor number
E	Redox potential
e	Electron
E^0	Standard redox potential
E_a	Activation energy
E_B	Electron affinity of B
$E_{1/2}$	Polarographic half-wave potential
$E_{1/2(BBCr)}$	Polarographic half-wave potential referred to *bis*-biphenylchromium(I)
EA	Electron acceptor
EC	Ethylene carbonate
ED	Electron donor
en	Ethylene diamine
EPA	Electron-pair acceptor
EPD	Electron-pair donor
ES	Ethylene sulfite
Et	Ethyl
Etac	Ethylacetate
F	Faraday constant
f	Force constant
Hal	Halide
HMPA	Hexamethylphosphoric amide
I	Ionization energy
L	Ligand
L^e	Entering ligand
L^l	Leaving ligand
L^n	Nonleaving ligand
K	Equilibrium constant
K_a	Acidity constant
K_{assoc}	Association constant
K_b	Basicity constant
K_{dissoc}	Dissociation constant
K_{form}	Formation constant of ion pairs from un-ionized species
K_{sep}	Dissociation constant of ion pairs into free ions

K_{out}	Outer-sphere complex formation constant
K_s	Solubility product
k	Rate coefficient
M	Metal
Me	Methyl
μ	Dipole moment
ν	Frequency
NB	Nitrobenzene
NM	Nitromethane
NMF	*N*-Methyl formamide
NMP	*N*-Methyl pyrolidinone
PDC	Propanediol-(1,2)-carbonate (propylene carbonate)
phen	Phenantroline
PY	Pyridine
S	Solvent
TBP	Tributylphosphate
TCEC	Tetrachloroethylene carbonate
TEP	Triethylphosphate
THF	Tetrahydrofuran
TMP	Trimethylphosphate
TMS	Tetramethylene sulfone (sulfolane)
x_A	Mole fraction of A

Natura enim simplex est (Nature is simple)
Newton (1687)

Chapter 1
Basic Considerations

1.1. Introduction

A Lewis acid is a molecule or ion whose incomplete electronic arrangement allows it to bind to another species by accepting an electron pair from that species. A Lewis base is a molecule or ion capable of donating an electron pair to a Lewis acid and resulting in the formation of a coordinate covalent bond (Lewis, 1923; Luder and Zuffanti, 1946). Although this concept for acids is much broader than that proposed by Brønsted (1923) and Lowry (1923), it was many years before it gained general acceptance.

In organic chemistry, the breakthrough of the Lewis ideas was due to Robinson's electronic interpretation of organic reactions, which was extended most effectively by Ingold (1933, 1934). The carbon chemists use the terms nucleophile for a Lewis base, which donates electrons to a carbon atom, and electrophile for a Lewis acid, which accepts electrons from carbon.

The Lewis concept was introduced to coordination chemistry by Sidgwick (1927), who used the terms electron donor or donor for a Lewis base and electron acceptor or acceptor for a Lewis acid. Thus a great number of coordinating interactions, including acid–base chemistry, substitution, displacement, atomic-, or group-transfer reactions can be described by the simple concept as exemplified at the top of page 2.

On the other hand, this concept has not been applied to redox reactions, which involve changes in oxidation numbers. The reducing agent can be considered a donor of electrons and the oxidizing agent an acceptor of electrons. Thus there are no clear border lines between Lewis-type and redox interactions (Ussanovich, 1939).

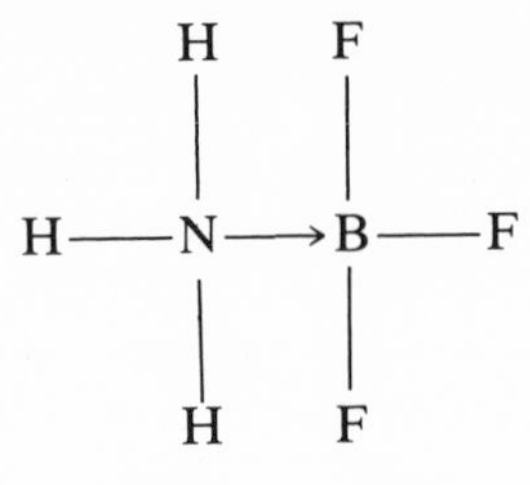

Lewis:	Base	Acid
Sidgwick:	Donor*	Acceptor
Robinson–Ingold:	Nucleophile	Electrophile

In the case of a classical acid–base reaction such as

$$\overset{-III}{N}\,\overset{+I}{H_3} + \overset{+I}{H}\,\overset{-I}{Cl} \rightleftharpoons \overset{-III}{N}\,\overset{+I}{H_4^+} + \overset{-I}{Cl^-}$$

or a classical carbon substitution such as

$$\overset{-III}{N}\,\overset{+I}{H_3} + \overset{-I}{H_3}\,\overset{+IV}{C}\,\overset{-I}{Cl} \rightleftharpoons \overset{+I}{H_3}\,\overset{-III}{N}\,\overset{+IV}{C}\,\overset{-I}{H_3^+} + \overset{-I}{Cl^-}$$

the oxidation numbers of all the atoms as shown remain unchanged and the reaction is considered as a displacement of Cl^- by H_3N at the H or the C atoms. On the other hand, the reaction

$$\overset{+IV}{S}\,\overset{-II}{O_3^{2-}} + \overset{-II}{O}\,\overset{+I}{Cl^-} \rightleftharpoons \overset{+VI}{S}\,\overset{-II}{O_4^{2-}} + \overset{-I}{Cl^-}$$

is considered a redox reaction since changes in the oxidation numbers for both sulfur and chlorine take place. Kinetics studies have been made (Basolo and Pearson, 1958) and this reaction may be thought of as a nucleophilic substitution reaction at the oxygen, with the sulfur atom acting as a donor toward the oxygen atom of the hypochlorite ion. It has been demonstrated, using ^{18}O-labeled hypochlorite ions, that the sulfate product contains the proportion of ^{18}O required by the transition state:

$$\left[\begin{array}{c} O \\ | \\ O{=}S{:}\rightarrow O{-}Cl \\ | \\ O \end{array}\right]^{3-}$$

*Note that we have distinguished between coordinate bond arrows (open arrowheads) and reaction arrows (solid arrowheads) throughout this book.

Since an oxygen-atom transfer actually takes place, it is somewhat ambiguous to consider OCl^- as containing oxygen and chlorine in the −II and +I oxidation states, respectively. The requirement of unchanged oxidation numbers in the course of this substitution reaction would be fulfilled if an oxidation number of 0 is assigned to all oxygen atoms: The oxidation numbers of both sulfur (oxidation number −II) and chlorine (oxidation number −I) would remain unchanged (Mayer and Gutmann, 1975):

$$\overset{-II}{S}O_3^{2-} + O\overset{-I}{Cl^-} \rightleftharpoons \overset{-II}{S}O_4^{2-} + \overset{-I}{Cl^-}$$

An important point, usually ignored, is that the actual chemical behavior of a molecule or ion is highly influenced by its molecular environment (Gutmann, 1971*a*). For example, water acts as a donor toward most metal ions, whereas it acts as an acceptor toward various anions; it may function as a reducing agent toward fluorine and as an oxidizing agent toward sodium. Although with excess water the reaction is expressed by the well-known equation

$$Na + \underset{\text{excess}}{H_2O} \rightleftharpoons \tfrac{1}{2}H_2 + Na^+ + OH^-$$

the reaction products are different when small amounts of water are allowed to react with excess liquid sodium (Addison, 1967), which reduces both OH ions and H atoms further to oxide and hydride ions, respectively:

$$\underset{\text{excess}}{4Na} + H_2O \rightleftharpoons Na_2O + 2NaH$$

Consequently the actual behavior of a molecule may be considered as the result of the unique matching of its electronic structure with the electronic structures of the other constituents composing the system. It is therefore necessary to consider each system as a whole rather than by its isolated constituents. This approach is provided by quantum chemistry, which becomes, however, more difficult as the system under consideration becomes larger.

It is the aim of this book to develop as an alternative an extension of the donor–acceptor approach in order to (1) include intermolecular interactions other than the coordinating type, and (2) to derive rules for the structural variability of a molecule in different molecular environments and to relate the structural differences to thermodynamic and kinetic quantities.

1.2. *Charge-Density Rearrangement Caused by an Intermolecular Interaction*

An intermolecular interaction leads to a charge-density rearrangement within the system. In the course of the reaction $A+D \longrightarrow A{\rightarrow}D$ a certain amount of charge transfer takes place from D (acting as donor) to A (acting as acceptor). However, charge transfer is not an observable quantity, since its definition depends on the particular choice of the border between the subsystems (D and A in our example) in the complex. This arbitrariness is obvious in a straightforward analysis of density difference functions. The second contribution to the charge-density rearrangement is due to the polarization effects, which cannot be separated unambiguously from the charge-transfer contributions. The calculated numbers depend on the accuracy of the individual calculations as well as on the definitions chosen in the various methods of analysis. As typical examples, the Mulliken population analysis or integration of electron densities should be mentioned. Nevertheless, general trends were found, although various kinds of analysis had been applied. For example, the results obtained by semiempirical molecular orbital calculations reveal the relative changes in fractional atomic charges involved in the formation of molecular adducts (Schuster, 1973, 1977; Schuster *et al.*, 1977). The following examples describe the significant changes in charge arrangement and bond lengths which occur when charge-transfer complexes are formed. They illustrate in an extreme form the types of changes which can be expected following any intermolecular interactions and thus indicate, for example, the types of changes that occur when solutes are dissolved in solvents.

The extent of charge transfer from an ammonia molecule to a fluorine molecule has been calculated to be 0.0483 electron (Carreira and Person, 1972). The $F_{(1)}$ atom as the acceptor atom acquires a small positive charge of 0.0181 electron, while the negative charge of 0.0644 electron is transferred to the $F_{(2)}$ atom away from the electron donor. The fractional negative charge of the N-donor atom of the ammonia molecule is increased by 0.0362 electron by withdrawing the electron charge from the terminal hydrogen atoms. Although the actual charges depend on the basis-set used, the general trends of the changes in fractional nuclear charges are well established.

The changes in charge densities are linked with changes in the relative positions of the nuclei. The internuclear distance between two given nuclei becomes longer as the heteropolarity of the bond increases (Lindqvist, 1963). The changes in fractional charge at the fluorine atoms due to coordination are associated with an increase in polarity of the F—F bond. In the example under consideration the calculated increase in the F—F bond length is 10 pm; there is also an increase in polarity for each of the N—H bonds and hence an increase in N—H bond lengths. Further examples as well as a more detailed discussion of the changes in fractional nuclear charges will be provided in Sections 3.2 and 7.2.

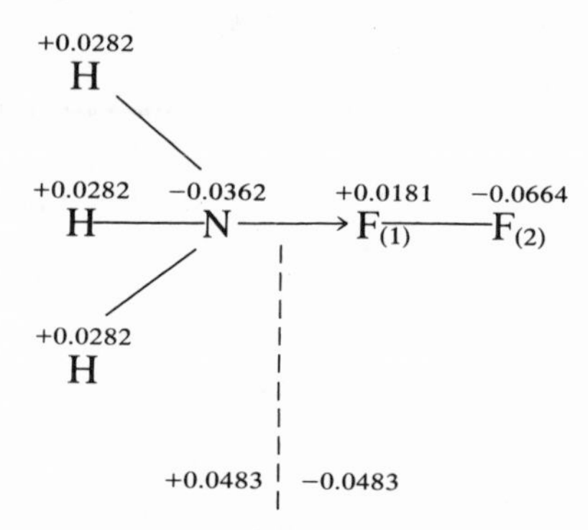

Changes in fractional nuclear charges have been derived for iodine monochloride and its 1:1 pyridine adduct from NQR-spectral analysis (Bowmaker and Hacobian, 1968). The net charge transfer of 0.26 charge units is distributed within the acceptor unit ICl to provide a small increase of 0.03 charge units in positive fractional charge at the iodine atom, which acts as an acceptor atom, while the negative fractional charge at the terminal chlorine atom is considerably increased, namely, by 0.29 charge units.

+0.32 I —— Cl −0.32 C_6H_5N ⟶ I —— Cl (I: +0.35, Cl: −0.61)

+0.26 | −0.26

Similar results are available for the complexes of 3,5-dibromopyridine with ICl, IBr, or Br_2 (Bowmaker and Hacobian, 1969) and for the pyridine complexes with iodine (Ichiba *et al.*, 1971). Qualitative information about the charge transfer involved in complex formation has been obtained from the changes in chemical NMR shifts: ^1H NMR signals in donor molecules are shifted to lower fields by complexation, indicating a decrease in electron density. Examples are the iodine and iodine-halide complexes of nitriles (Wells, 1966), sulfides (Strom *et al.*, 1967), and pyridines (Yarwood, 1967, 1970).

The structure of a molecule is therefore not rigid but rather adaptable to the environment. The *structural variability* of a given molecule in different solvents, for example, is reflected in its *chemical variability*. It is therefore most desirable to derive rules for both the structural and the chemical changes which occur as a result of intermolecular interactions such as solvation or crystal-lattice effects. The extended donor–acceptor concept is based on the comparison of equilibrium structures in different environments irrespective of the nature of the binding forces, and the structural differences are related to thermodynamic and kinetic properties.

The charge-density rearrangement due to charge transfer and polarization effects results from the interaction between an electron donor and an electron acceptor. Distinctions are sometimes made between an electron-pair donor (EPD, Lewis base, nucleophilic agent) and an electron donor (ED, reducing agent) as well as between an electron-pair acceptor (EPA, Lewis acid, electrophilic agent) and an electron acceptor (EA, oxidizing agent). Although these distinctions may be useful (Gutmann, 1971*a*), they will not be required for the approach used in this book.

The terms "donor" and "acceptor" indicate the function involved in the process of charge transfer. Coordination chemists use these terms in the more restricted sense of a Lewis base and Lewis acid, respectively, while organic chemists use the terms "nucleophilic" and "electrophilic" agent. For the sake of uniform presentation, the terms "donor" and "acceptor" will be used throughout this book.

A heteronuclear substrate $A^{\delta+}—B^{\delta-}$ will be attacked by an electron donor at the area of low electron density, while an electron acceptor will attack at the area of high electron density. Each of these interactions induces an electron drift along the bond A—B resulting from both the charge transfer from donor to acceptor and the mutual polarization interactions. These two effects result in an electron shift from A to B. The changes in electron distribution are reflected both in appropriate changes in the fractional charges and in the positions of the nuclei A and B (Gutmann and Schmid, 1974; Gutmann, 1975, 1976*a*, *b*). The increase in bond length may be represented by a full bent arrow connecting the bonded nuclei and pointing into the direction of the electron shift. This symbolism differs from the representation common in organic chemistry: In contrast to this, it emphasizes both the changes in fractional atomic charges and the changes in the positions of the nuclei.

The usefulness of these considerations lies in the fact that any *charge density pattern* is reflected in a characteristic *structural pattern*, which determines the *reactivity pattern* within the molecule: By donor attack at A the fractional negative charge at B is increased and so are both its donor and reducing properties. Acceptor attack at B leads to an increase in fractional positive charge at A and hence the acceptor and oxidizing properties of A are enhanced.

Donor$\longrightarrow A^{\delta+} \overset{\frown}{—} B^{\delta-}$ (Increase in donor and reducing properties at B)

$A^{\delta+} \overset{\frown}{—} B^{\delta-} \longrightarrow$Acceptor (Increase in acceptor and oxidizing properties at A)

Induced
increase in
A—B distance

By either of these processes the bond A—B is increased in heteropolarity, which is manifested in an increase in bond length and hence associated with a weakening of the A—B bond (Lindqvist, 1963).

These considerations, which are of greatest importance to all fields of molecular science, can be based on a few simple rules—the so-called *bond-length variation rules* (Gutmann, 1976*a*, *b*). Their formulation would not have been possible had Lindqvist not realized the need to obtain precise structural data in order to account for the bond-length variations within a molecule in different donor solvents (Lindqvist, 1963).

1.3. *The First Bond-Length Variation Rule*

The first bond-length variation rule relates the intermolecular donor–acceptor interaction to the induced intramolecular effects: The smaller the intermolecular distance D$\longrightarrow$A, the greater the induced lengthening of the adjacent intramolecular bonds both in the donor and acceptor components (Gutmann, 1975, 1976*a*, *b*):

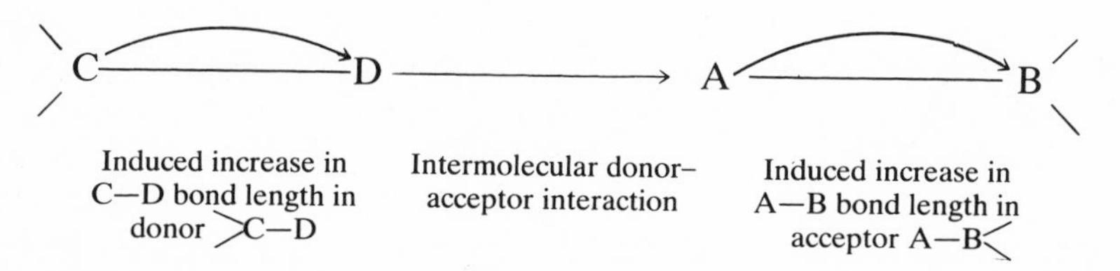

The study of induced intramolecular effects is instructive for 1 : 1 adducts, since they can be regarded as resulting from the formation of only one new bond.

The inverse relationship between intermolecular and intramolecular bond length within the acceptor component can be seen from the consideration of bond distances in complexes formed between an N-donor and boron(III) fluoride. The B—F distances increase as the N$\rightarrow$B bond distances decrease (Table 1.1). The N—B bond

Table 1.1. Bond Distances in BF_3 and Its Adducts with N-Donor Molecules

Compound	N→B distance (pm)	B—F distance (pm)
BF_3	—	130
$CH_3CN \rightarrow BF_3$	163	133
$H_3N \rightarrow BF_3$	160	138
$Me_3N \rightarrow BF_3$	158	139

distances decrease with increasing donor strength of the N-donor (Prout and Wright, 1968):

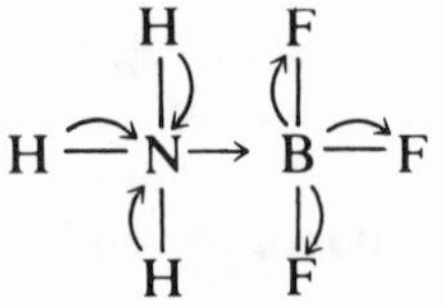

$$H_3N \rightarrow BF_3$$

The induced bond lengthening within the donor unit $>$C—D in our general example can be seen from the comparison of the N—H bond distances in ammonia (101.5 pm) and in the ammonium ion (103.1 pm).

The reader may have noted that the changes are not discussed in terms of rehybridization at the donor and acceptor atoms upon complex formation. This is because I want to develop broadly applicable bond-length variation rules within the framework of the donor–acceptor concept.

Structural data are available also for adducts of $SbCl_5$ with eight different O donors. A decrease in ΔH for adduct formation leads to shortening of the intermolecular Sb—O bond distance and, although to a smaller degree, to lengthening of the Sb—Cl bond distances (Table 1.2) (Kietaibl *et al.*, 1972). The Sb—Cl bond is greatest in the $SbCl_6^-$ ion, which has a regular octahedral configuration, while in the listed molecular adducts with O-donor molecules the octahedra are more or less distorted and the deviation from 90° of the Cl—Sb—O bond angle increases with increase in Sb—O distance (Table 1.2).

A chemical interaction may be treated as an equilibrium reaction characterized by the ΔG or the log K value. In many cases the ΔH values follow the same trends as the ΔG values and hence in the absence of the former, the ΔH values may be used in such cases. Quantitative relationships between such data and structural properties are restricted to the comparison of closely related systems. We shall describe the differences in properties of one molecule in different molecular environments; for example, the system of $SbCl_5$ as a single reference acceptor in the presence of various donor molecules or the system containing one reference donor, such as Et_3PO, in the presence of various acceptor molecules. We shall be interested in comparing the ΔG or ΔH values for the interactions of these reference molecules with a variety of donors and acceptors, respectively, and then relating ΔG and ΔH values to observed structural changes such as intermolecular and intramolecular bond lengths. This approach is in accordance with Wigner's statement: "*Physics does not describe nature; it describes regularities among events and nothing but regularities among events.*" We are seeking such regularities. A variety of surprising relationships and regularities exist once the effects of bond-length variations are realized. Differences in spectroscopic properties of a molecule in different molecular environments are related to the differences in electronic and spatial arrangements (Gutmann, 1975). For example, there is a relationship between the differences in electron-binding energies of $3d_{5/2}$ orbitals of antimony in quick-frozen solutions of $SbCl_5$ in different donor solvents and the ΔH values of complex formation with the same donors (Burger and Fluck, 1974) (Fig. 1.1).

The formation of molecular adducts from different O donors and molecular iodine has been extensively studied (Paetzold, 1968; Niendorf and Paetzold, 1973; Paetzold and Niendorf, 1974, 1975). For the interaction between donors with the same functional group and molecular iodine in solution of carbon tetrachloride, there is a linear relationship between the differences in the force constant of the X=O group and the differences in complex stability (Fig. 1.2). The appropriate changes within the acceptor iodine can be seen by comparing the differences in complex stability with the shift of the longest-wavelength absorption band of iodine, also known as the blue shift. Increase in complex stability is accompanied

Table 1.2. Internuclear Distances and Bond Angles in $>X{=}O \rightarrow SbCl_5$ Complexes

Donor	$-\Delta H_{D\cdot SbCl_5}$ (kcal·mol^{-1})	Mean distance Sb—Cl (pm)	Distance Sb←O (pm)	Cl—Sb—O angle (°)
TCEC	0.2	231	240	82.6
PhCOCl	2.3	231	219	83.8
$POCl_3$	11.7	233	217	85.8
Ph_2SO	—	235	216	86.6
Me_2SO_2	14.0	235	212	86.4
DMF	26.6	235	205	87.0
HMPA	≈38.0	236	205	88.3
Cl^-	—	247	194	90.0

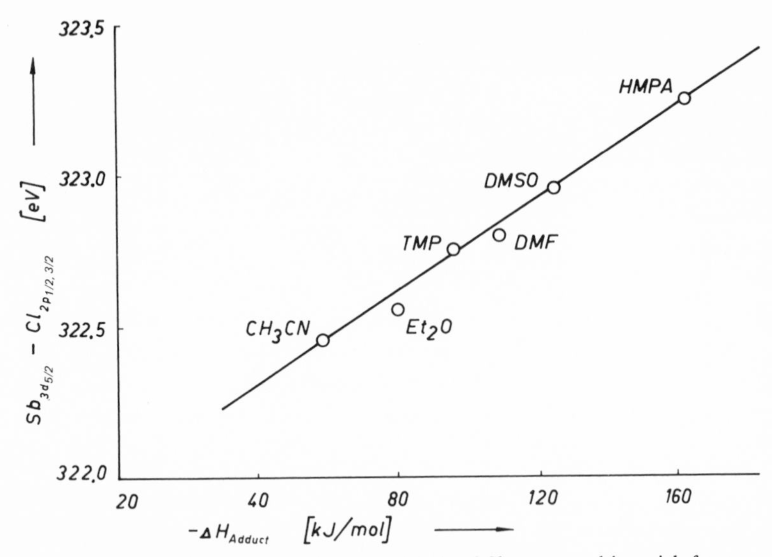

Fig. 1.1. Electron-binding energies of $3d_{5/2}$ orbitals of Sb measured in quick-frozen solutions of $SbCl_5$ with different donor solvents in $C_2H_4Cl_2$ solutions vs. the donor number of the solvent (from Burger and Fluck, 1974, courtesy of Pergamon Press).

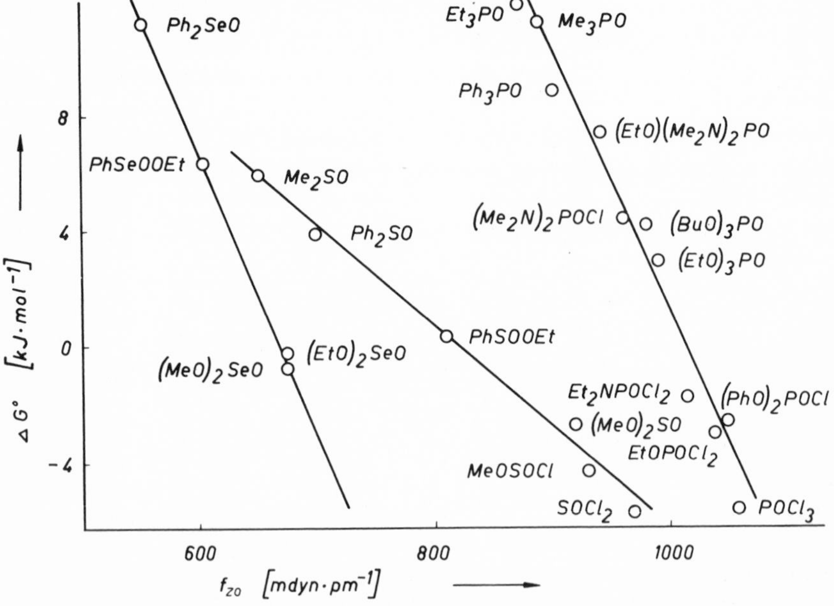

Fig. 1.2. Relationship between ΔG^0 for the complex equilibria $D + I_2 \rightleftharpoons D \cdot I_2$ in CCl_4 and the valence force constant of the coordinating group ZO of the donor molecule (from Paetzold, 1975*b*, courtesy of Butterworth and Co.).

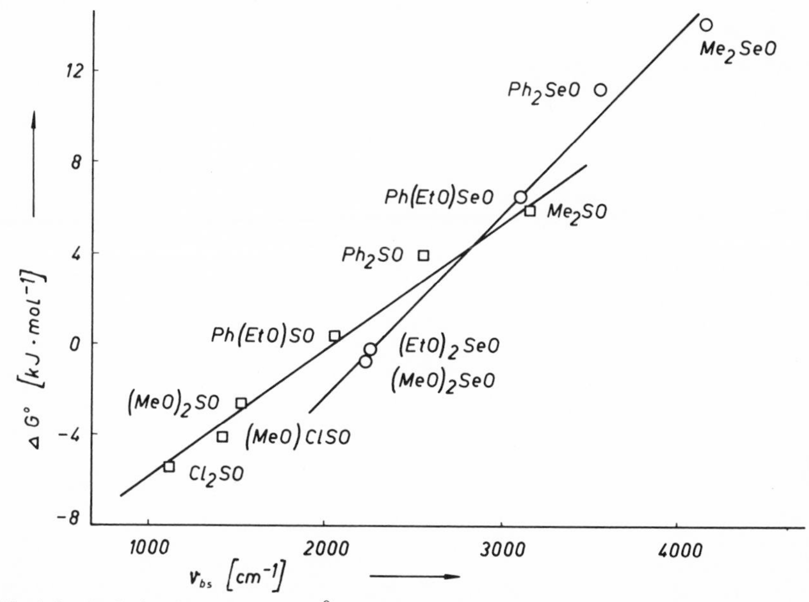

Fig. 1.3. Relationship between ΔG^0 for the complex equilibria $D + I_2 \rightleftharpoons D{\cdot}I_2$ and the blue shift ν_{bs}.

by a "blue shift," which indicates an increase in I—I bond distance (Fig. 1.3). Since the induced electronic changes in both the donor and acceptor units are related to the ΔG and ΔH values for complex formation, it follows that a relationship exists also between the induced electronic variations within the donor and acceptor components, e.g., in the oxygen-donor–iodine complexes between the differences in the force constants of the X=O bonds in the donor and the differences in blue shift of the visible I—I band in the acceptor.

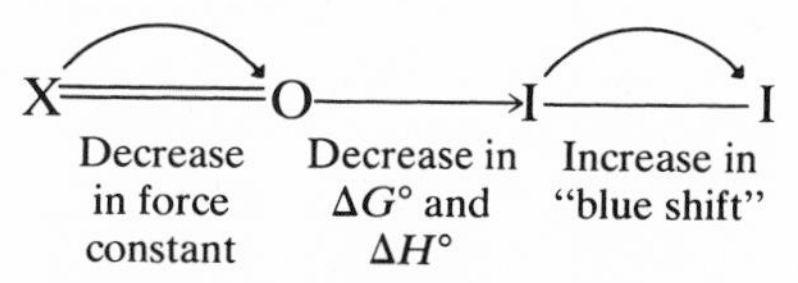

Relationships between spectroscopic data on one side and thermodynamic, thermochemical, and kinetic quantities, as well as chemical characteristics on the other, will be found throughout this book. In addition, phenomenological parameters characterizing the donor and acceptor properties, respectively, will be introduced in Sections 2.2 and 2.3.

Gilchrist and Sutton (1952) have shown that there is an inverse relationship between bond length and dissociation energy: The greater the internuclear distance, the smaller the bond energy. More quantitative relationships have been shown by Szabó and Thege (1975). Due to strong interactions the induced changes in fractional charges and in bond distances may be so great as to lead to heterolysis of a bond (Gutmann and Mayer, 1969; Gutmann, 1970, 1971*b*): Donation at A leads to stabilization of the cation A^+ while acceptance at B stabilizes the anion B^-. Heterolysis (ionization) may be considered as a displacement reaction of the anion by donor (solvent) molecules at A or as displacement of the cation by acceptor (solvent) molecules at B:

$$nD + A^{\delta+}{-}B^{\delta-} \rightleftharpoons AD_n{}^+B^-$$

$$A^{\delta+}{-}B^{\delta-} + n\,Ac \rightleftharpoons A^+BAc_n{}^-$$

Many ionizing solvents exhibit both donor *and* acceptor properties at various degrees and hence both effects may occur at the same time: The electron shift occurs in the same direction, e.g., from A to B irrespective of the attack by the donor at A or by the acceptor at B (*Push–pull effect*, Swain, 1948).

1.4. The Second Bond-Length Variation Rule

The bond-length variations are not confined to the bonds adjacent to the sites of an intermolecular interaction. Also, subsequent characteristic changes in bond lengths are induced *throughout the molecule* as the charge-density rearrangement takes place. By the second bond-length variation rule, such intramolecular changes are formulated as follows (Gutmann, 1976*b*): A σ bond is lengthened when, as a result of the interaction, the electron shift occurs from a nucleus carrying a positive fractional charge to one carrying a negative fractional charge, whereas the σ bond is shortened when the electron shift is induced in the opposite direction. Since it is impossible in many cases to decide the charge the nucleus is carrying within a molecule, it may be suggested that we use the concept of electronegativity. The bond is lengthened when the electron shift takes place from the more electropositive to the more electronegative atom in the uncomplexed species and it is shortened when the electron shift takes place from the more electronegative to the less electronegative atom.

An example is provided by the interaction of tetrachloroethylene carbonate with $SbCl_5$ (Kietaibl *et al.*, 1972): Although the ΔH value for this reaction is low (Gutmann and Wegleitner, 1976), the induced changes in bond distances in the crystalline adduct are well pronounced (the arrows indicate the direction of the electron shift; a slightly bent solid arrow denotes increase and a dashed arrow a decrease in bond distance):

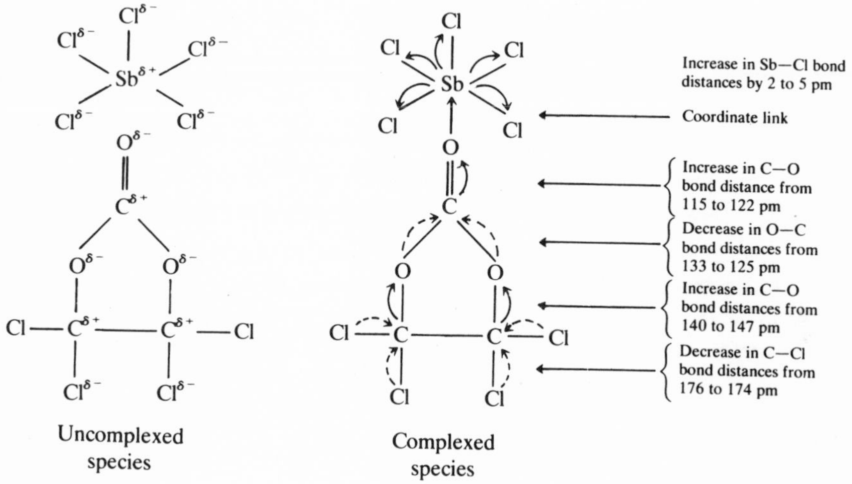

Increase in Sb—Cl bond distances by 2 to 5 pm

Coordinate link

Increase in C—O bond distance from 115 to 122 pm

Decrease in O—C bond distances from 133 to 125 pm

Increase in C—O bond distances from 140 to 147 pm

Decrease in C—Cl bond distances from 176 to 174 pm

Uncomplexed species

Complexed species

Likewise, the formation of the adduct $Cl_2SeO \rightarrow SbCl_5$ through oxygen donation to the antimony leads to an increase in the $Sb^{\delta+}$—$Cl^{\delta-}$ bond distances as well as in the $Se^{\delta+}$—$O^{\delta-}$ bond distance (Hermodsson, 1969). The increase in polarity of the $Se^{\delta+}$—$O^{\delta-}$ bond imposes an increase in positive net charge at the selenium atom, which will attract the electrons more strongly from the more electronegative chlorine atoms, so that the $Cl^{\delta-}$—$Se^{\delta+}$ bonds are shortened.

The situation is more difficult for compounds with homonuclear bonds, for example, in carbon chemistry, and it may be reasonable to assume in most of these cases alternating bond lengthening and shortening. For example, the changes in charge density at the C atoms of pyridine while forming its 1:1 complex with iodine have been deduced from ^{13}C NMR measurements (Larkindale and Simkin, 1971). The 2-, 6-, and to a greater extent the 4-carbon positions show a downfield shift with respect to the free pyridine, indicating lower electron densities than in free pyridine. On the other hand, the 3- and 5-carbon positions show an upfield shift relative to the same position in the uncomplexed donor molecule. The observed changes in the ^{13}NMR spectrum, originally attributed to a magnetic anisotropic effect, correlate approximately with the proton NMR changes in the same system. They may indicate the following alternating fractional charges in the complex relative to those in the free molecules with shortening of the bonds $C_{(5)}—C_{(6)}$ and $C_{(3)}—C_{(2)}$:

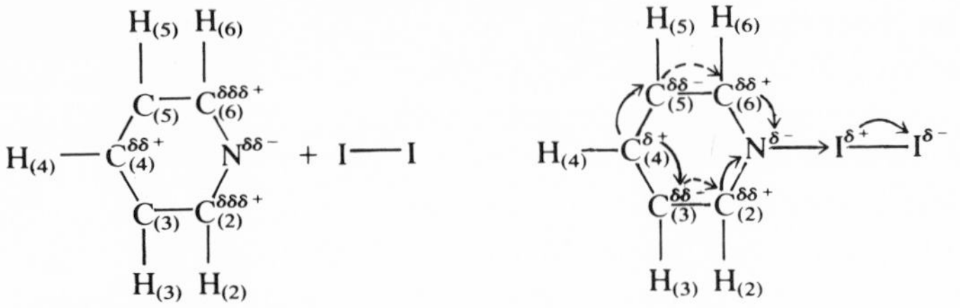

Uncomplexed species

Changes due to complexation

Such *cooperative effects* are not confined to adduct formation and indeed, they result from changes in *any molecular arrangement*, such as in a solution or in the solid state; they lead to a new structure pattern for the *whole system under consideration.*

1.5. *The Third Bond-Length Variation Rule*

A relationship between bond length and coordination number has been found for the crystalline state on geometrical grounds (Goldschmidt, 1958). The atoms or ions are regarded as rigid spheres and an electrostatic

Table 1.3. Bond Lengths M—X in Acceptor Molecules and in Their Donor–Acceptor Complexes

Acceptor	M—X (pm)	Complex ion	M—X (pm)
$CdCl_2$	223.5	$CdCl_6^{4-}$	253
SiF_4	154	SiF_6^{2-}	171
$TiCl_4$	218–221	$TiCl_6^{2-}$	235
$ZrCl_4$	233	$ZrCl_6^{2-}$	245
$GeCl_4$	208–210	$GeCl_6^{2-}$	235
GeF_4	167	GeF_6^{2-}	177
$SnBr_4$	244	$SnBr_6^{2-}$	259–264
$SnCl_4$	230–233	$SnCl_6^{2-}$	241–245
SnI_4	264	SnI_6^{2-}	285
$PbCl_4$	243	$PbCl_6^{2-}$	248–250
PF_5	154–157	PF_6^{-}	173
$SbCl_5$	231	$SbCl_6^{-}$	247
SO_2	143	SO_3^{2-}	150
SeO_2	161	SeO_3^{2-}	174
ICl	230	ICl_2^{-}	236
I_2	266	I_3^{-}	283

interpretation has been provided by Pauling (1960). These approaches do not consider the electronic effects caused by the immediate environment and hence they are limited in their application.

The bond-length variation rules of the extended donor–acceptor concept are based on the electronic effects and hence they include also the geometrical contributions. When an adduct is formed, the coordination number is increased both at the donor and at the acceptor atom and according to the first bond-length variation rule (Section 1.3) the bonds originating from these atoms are lengthened irrespective of the state of aggregation. Hence, this rule can be reformulated in terms of the changes in coordination numbers involved, and this rule may be called the *third bond-length variation rule*. As the coordination number increases, so do the lengths of the bonds originating from the coordination center (Gutmann, 1976*c*; Gutmann and Mayer, 1976). We shall make use of this rule in particular in considering phenomena at interfaces (Chapter 5).

1.6. General Aspects

The three bond-length variation rules are supplemented by rules (a) for the changes in fractional atomic charges (Section 3.2) (b) for the outer-sphere effects (Sections 7.3 and 7.4), and (c) for the relations between coordinating and redox properties (Section 8.1). In contrast to the concept of inductive and mesomeric effects within a molecule, the extended donor–acceptor approach considers not only electronic changes, but also the changes in the positions of the nuclei.

It must be emphasized that even small changes in bond polarity and hence in bond distance and bond angles—even those too small to be traced by the methods available—may be reflected in characteristic changes in chemical properties; e.g., the rearrangement of the charge pattern produces a new structural pattern which is reflected in a new "reactivity pattern" of the compound.

One of the advantages of this unified donor–acceptor approach, together with the bond-length variation rules and the empirical solvent parameters to be introduced in the following chapter, is that it offers a method of choosing conditions for desired bond polarity within a functional group by applying an appropriate molecular environment to a species to which a variety of bond polarities, bond lengths, and bond strengths can be induced.

The wide application of this extended donor–acceptor concept is due to the following advantages (Gutmann, 1976*b*):

(1) Emphasis is placed on the *observable changes* in bond properties,

such as spectroscopic data or internuclear distances that result from the charge-density rearrangement following an intermolecular interaction.

(2) The changes in such properties require *no interpretation of the bonding forces*. Changes in covalent, ionic, metallic, hydrogen, van der Waals, or other bonds are all subject to the same rules to various degrees.

(3) The concept is in no way contradicted by the *basic principles of quantum chemistry* and hence it can be applied to very weak (physical) as well as to strong (chemical) interactions, because in both cases rearrangement of the charges takes place.

(4) The approach can be applied to both homogeneous and heterogeneous interactions and indeed to interface phenomena, such as adsorption, crystallization, or heterogeneous catalysis.

Chapter 2
Empirical Parameters for Donor and Acceptor Properties

2.1. Solvent Characterization by Empirical Parameters

The concept of nucleophilic–electrophilic interactions has been applied widely to reaction mechanisms in organic chemistry for more than 40 years. On the other hand, electrolyte solutions in water and in nonaqueous solvents have been interpreted mainly by the electrostatic theory. It was shown in 1931 that perchloric acid behaves as a very weak acid in nitromethane, deviating in properties considerably from those predicted by the Debye–Hückel–Onsager theory (Wright *et al.*, 1931; Murray-Rust *et al.*, 1931). The authors rightly concluded that this might be due to the lack of donor and acceptor properties of nitromethane and suggested that these properties be considered for the interpretation of conductivity data. However, they did not pursue this point further.

Grunwald and Winstein (1948) attempted to provide a characterization of the ionizing properties of solvents by a single parameter and thus proposed the so-called Y value. This was based on a comparison of the rate coefficient for the solvolysis of *t*-butyl chloride k^{BuCl} in one solvent with the rate coefficient for solvolysis in the reference solvent mixture of 80% ethyl alcohol and 20% water k_0^{BuCl}:

$$Y = \log [K^{\text{BuCl}}/k_0^{\text{BuCl}}]$$

Kosower (1956, 1958; Kosower and Klinedinst, 1956) made an attempt to apply the covalent approach to molecular adducts, but his correct ideas were overshadowed by his introduction of the Z values, which he considered characteristic for the "solvent polarity" or the "ionizing" properties of solvents. These values are based on the UV spectra of 1-ethyl-4-carbomethoxypyridinium iodide. The charge-transfer energy of this compound in a particular solvent was considered as an empirical measure of the solvent polarity. Kosower stressed the advantage that the Z values were based on a physical process understood in great detail.

$$COOCH_3 \qquad\qquad COOCH_3$$

$$(C_5H_4N^{\oplus}\text{–}C_2H_5)\,I^{\ominus} \xrightarrow{h\nu} (C_5H_4N^{\cdot}\text{–}C_2H_5)\,I^{\cdot} + e$$

$$C_2H_5 \qquad\qquad C_2H_5$$

However, the charge-transfer energy expresses the difference between the excited and ground states of the complex and therefore is a characteristic property of the complex as a whole (in the particular environment) and not a perturbed transition of either component (Brackman, 1949). The synonymous use of the terms "solvent polarity" and "ionizing power of a solvent" has been unfortunate and misleading since there is no general relationship between solvent polarity (denoting the dipolar character of solvent molecules or more precisely the distribution and polarizability of the charges) and ionizing properties (the ability of the solvent to heterolyze covalent bonds).

The close relationship between Z values and Y values found in a number of systems has been taken as further support for the Z values, which are more convenient to measure than Y values. Unfortunately neither Z nor Y values represent effectively the donor properties of a solvent: Nearly identical Z values are assigned to solvents of vastly different donor properties but similar acceptor properties such as nitromethane, acetonitrile, and DMSO or diethyl ether and benzene. Hence numerous cases have been found where the application of the Z values failed.

Dimroth and Reichhardt's E_T values (Dimroth *et al.*, 1963; Reichhardt, 1969) are based on the solvent sensitivity of light absorption (an intramolecular charge-transfer transition) of a pyridinium phenol betain and hence may be applied to the interpretation of those solvent effects for which the Z values are applicable. A further solvent scale based on the solvatochromic comparison method has been proposed recently (Kamlet and Taft, 1976; Taft and Kamlet, 1976; Yokoyama *et al.*, 1976).

Lindqvist (1963) emphasized the changes in bond properties resulting from coordination and derived a qualitative order of solvent *donor strength* from calorimetric data on $SbCl_5$ (Lindqvist and Zackrisson, 1960). This has been expressed in a more quantitative way by the donor number concept (Gutmann and Wychera, 1966; Gutmann, 1968), which is successfully applied to ionization phenomena in aprotic solvents. More recently, another empirical parameter has been introduced to describe the acceptor properties of solvents, namely, the *acceptor number* (Mayer *et al.*, 1976). It would be extremely difficult to calculate such parameters as donor and acceptor numbers by means of quantum mechanical methods, but calculations are unnecessary because such empirical parameters can be determined by experiment.

It is indeed necessary to make a distinction between the donor properties and the acceptor properties of a solvent and to assign an empirical parameter for each of them. The application of the donor number (Section 2.2) together with the acceptor number (Section 2.3) has proved extremely useful in the interpretation and prediction of a vast amount of coordinating interactions in solution.

2.2. *The Donor Number*

For the characterization of donor properties of HB the pK_b values for the aqueous reactions $HB \rightleftharpoons H^+ + B^-$ have been applied frequently. For various monodentate ligands coordinated to the *same* metal ion, linear correlations between the pK_b and the logarithm of the stability constant have been found. For ligands which have different donor atoms, there is no such correlation. Even for ligands with the same donor atom, such as NH_3, aniline, pyridine, and imidazole, correlations are so poor as to be of no value in predicting unknown stability constants from pK_b values for complexes with cations other than H^+. This is because of the influence of specific hydration phenomena on the ionized species in water. For example, in the gas phase the base strength of amines increases in the order $NH_3 < MeNH_2 < Me_2NH < Me_3N$, while the pK_b values in water bear no relation to this, e.g., 4.75, 3.36, 3.23, and 4.20, respectively.

In order to obtain a solvent-independent representation for the donor ability of a molecule the "donor number" or "donicity" is used (Gutmann and Wychera, 1966). The donor number is defined as the molar enthalpy value for the reaction of the donor (D) with $SbCl_5$ as a reference acceptor in a 10^{-3} M solution of dichloroethane (Table 2.1):

$$D + SbCl_5 \rightleftharpoons D\cdot SbCl_5 \quad (-\Delta H_{D\cdot SbCl_5} \equiv DN)$$

For several other solvents, donor numbers have been inferred indirectly from the ^{23}Na NMR chemical shifts of $NaClO_4$ in its respective

Table 2.1. Donor Numbers (DN) for Various Solvents Obtained from Calorimetric Measurements in 10^{-3} M Solutions of Dichloroethane with $SbCl_5$ as a Reference Acceptor

Solvent	DN	Solvent	DN
1,2-Dichlorethane (DCE)	–	Ethylene carbonate (EC)	16.4
Benzene	0.1	Phenylphosphonic difluoride	16.4
Thionyl chloride	0.4	Methyl acetate	16.5
Acetyl chloride	0.7	*n*-Butyronitrile	16.6
Tetrachloroethylene carbonate (TCEC)	0.8	Acetone (AC)	17.0
		Ethylacetate	17.1
Benzoyl fluoride (BF)	2.3	Water	18.0
Benzoyl chloride	2.3	Phenylphosphoric dichloride	18.5
Nitromethane (NM)	2.7	Diethyl ether	19.2
Nitrobenzene (NB)	4.4	Tetrahydrofuran (THF)	20.0
Acetic anhydride	10.5	Diphenylphosphoric chloride	22.4
Phosphorus oxychloride	11.7	Trimethyl phosphate (TMP)	23.0
Benzonitrile (BN)	11.9	Tributyl phosphate (TBP)	23.7
Selenium oxychloride	12.2	Dimethyl formamide (DMF)	26.6
Acetonitrile	14.1	*N*-Methyl pyrolidinone (NMP)	27.3
Tetramethylenesulfone (TMS)	14.8	*N*-Dimethyl acetamide (DMA)	27.8
Dioxane	14.8	Dimethyl sulfoxide (DMSO)	29.8
Propandiol-(1,2)-carbonate (PDC)	15.1	*N*-Diethyl formamide (DEF)	30.9
		N-Diethyl acetamide (DEA)	32.2
Benzyl cyanide	15.1	Pyridine (PY)	33.1
Ethylene sulphite (ES)	15.3	Hexamethylphosphoric triamide (HMPA)	38.8
iso-Butyronitrile	15.4		
Propionitrile	16.1		

solvent (Erlich and Popov, 1971) and these donor numbers are listed in Table 2.2. However, they are only valid insofar as they have a direct linear relationship between donor number and such NMR shifts.

The molar enthalpy of 1:1 molecular adduct formation in dichloroethane is taken as an approximate measure of the energy of the coordinate bond between donor atom and the Sb atom of $SbCl_5$. The thermodynamic stability of the 1:1 $D \cdot SbCl_5$ complex is defined by the free enthalpy $\Delta G^0 = -RT \ln K$, and ΔG^0 can be used alternatively as a measure of the donor number of a solvent as long as a linear relation exists between ΔG and ΔH values. In such cases the ΔH values can be used as a guide to the relative complex stabilities. Such is the case with the donor number. Figure 2.1 shows the good relationship between ΔH values and $\log K$ values for the interactions between various donors with different donor atoms and antimony(V) chloride in dilute dichloroethane. Such linear ΔG–ΔH relationships are also found for a series of complexes with a given donor and various acceptors (Fig. 2.2). This implies the existence of a linear relationship between ΔH and ΔS values (Fig. 2.3). It is noteworthy that even water, which has unusual entropic effects as a bulk solvent, fits the linear relationship shown in Fig. 2.1 when the water is present in a dilute solution of dichloroethane.

The $-\Delta H_{D\cdot SbCl_5}$ values follow essentially the same order as the $-\Delta H$ values of the donors toward a different acceptor molecule such as iodine, phenol (Fig. 2.4), $Sn(CH_3)_3Cl$ (Fig. 2.5), or vanadyl acetylacetonate, as well as the solvation energies of various cations in the pure donor solvents (see Section 7.3). It is therefore possible to predict ΔH values for donor–acceptor interactions by making use of such empirical relationships. This is a method for the prediction of ΔH values, alternative to that by the equation of Drago and Wayland (1965).

$SbCl_5$ as a reference acceptor fulfills the following requirements:

(1) Adducts are formed in a 1:1 molar ratio with all donor molecules.

(2) Adduct formation leads to a change in the $SbCl_5$ acceptor from a bipyramidal to a distorted octahedral configuration, thus involving similar hybridization energies.

(3) $SbCl_5$ is a very strong acceptor and this allows fairly complete adduct formation when $SbCl_5$ is present in excess, even with very weak donors.

(4) The Sb—Cl bonds are not easily heterolyzed; even for interactions with strong donors ionization equilibria can be neglected.

Table 2.2. Donor Numbers Indicated by Indirect Methods

Solvent	DN
Phenylacetonitrile	14
Benzophenone	16
Methanol	19
Ethanol	20
Dimethoxyethane	20
Formamide	24
Tetramethylurea	31
Hydrazine	44
Piperidine	51
Ethylenediamine	55
Ethylamine	55
t-Butylamine	57
Ammonia (liquid)	59
Triethylamine	61

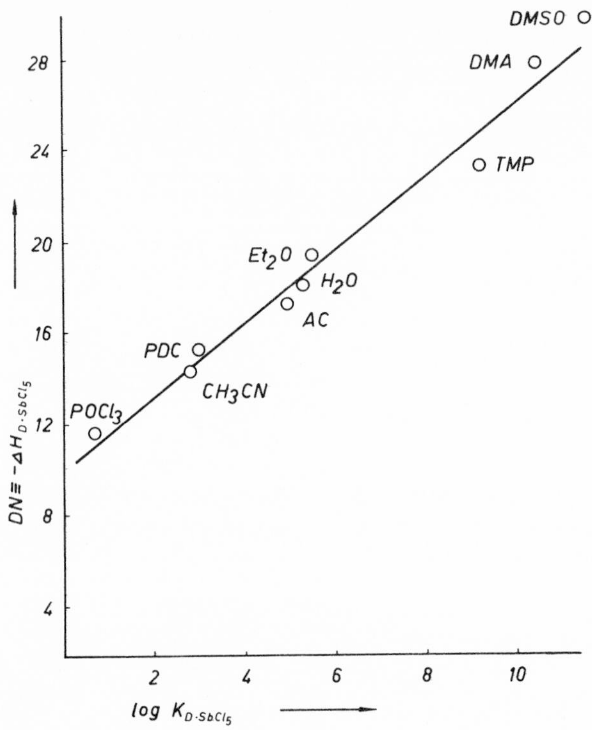

Fig. 2.1. Relationship between ΔH and $\log K$ values for the reactions of $SbCl_5$ with different donors in dilute solution of dischloroethane (from Gutmann and Wychera, 1966, courtesy of Pergamon Press).

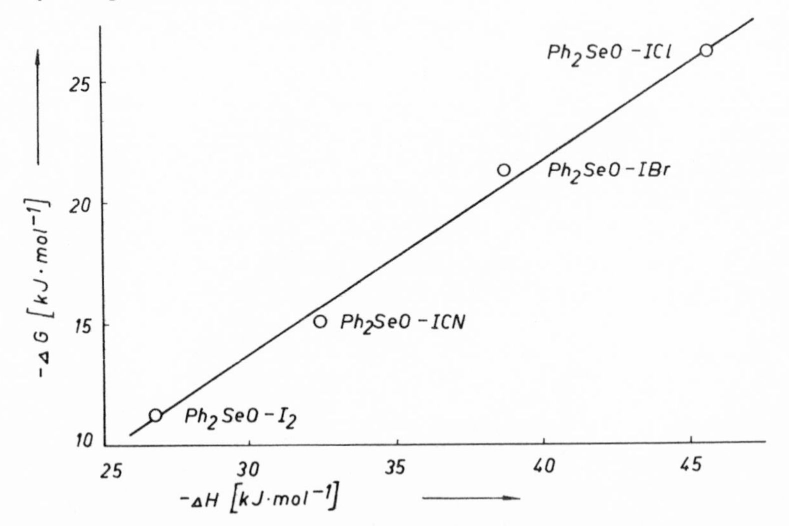

Fig. 2.2. Relationship between ΔH and ΔG values for the complex formation of Ph_2SeO with iodine and iodine halides.

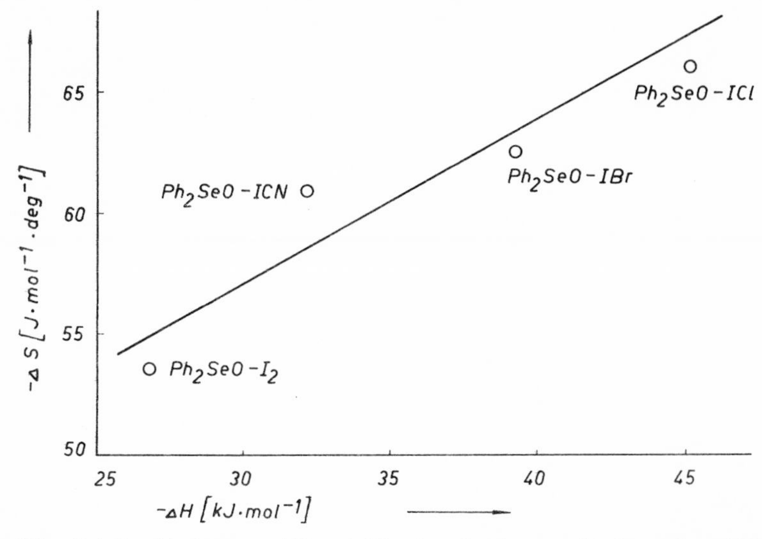

Fig. 2.3. Relationship between ΔH and ΔS values for the complex formation of Ph_2SeO with iodine and iodine halides.

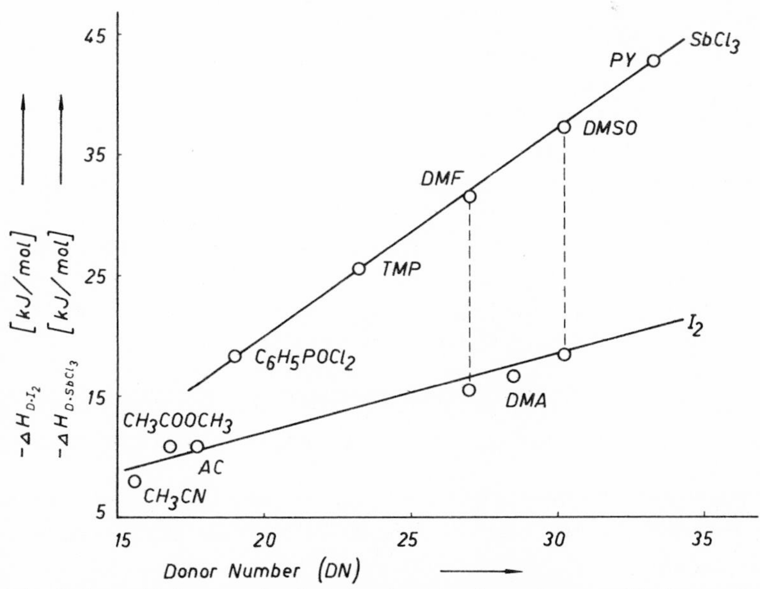

Fig. 2.4. Relationship between the donor number of various solvents and ΔH values for the complex formation of the donor solvents with iodine and with $SbCl_3$.

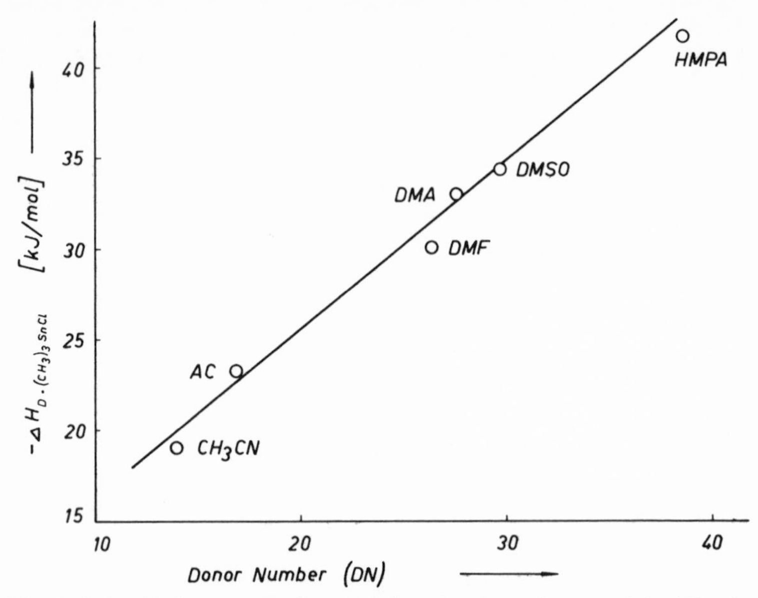

Fig. 2.5. Relationship between the donor number of various solvents and the ΔH values for complex formation of the donor solvents with $(CH_3)_3SnCl$.

Since $SbCl_5$ is a hard acceptor, the application of the donor number values is not possible for predicting interactions with soft acceptors. For example, sulfur donors (which are soft) will coordinate more strongly toward iodine (which is also soft) than the corresponding oxygen donor (which is hard) (Tamres and Searles, 1962). The reverse is found for the interaction of a pair of sulfur and oxygen donors with phenol, which is hard. It is not possible to infer the donor properties of S donors toward soft acceptors such as Ag^+, which have strong interactions, from the consideration of the ΔH values of the S donors toward $SbCl_5$, which are usually small. Nitriles behave as rather weak donor molecules toward $SbCl_5$, whereas they show a higher affinity relative to other donors toward Ag^+ and Cu^+ ions (Duschek and Gutmann, 1972).

These irregularities do not seriously restrict the applicability of the donor number concept since most of the commonly used solvents contain oxygen or nitrogen as their donor, and most solutes are rather hard or borderline cases for which the donor number concept may be applied. For example, the variations in Hg—X stretching motions of the mercury(II) halides in different hard solvents (Smith and Brill, 1976) are related to the donor numbers of the solvents (Fig. 2.6), although mercury(II) halides are often classified as moderately soft acceptors.

An example of the application of the donor number concept to secondary induced effects is provided by the relationship between the ^{29}Si chemical shift for silanols or silylamines and the donor number of the coordinating solvent molecules shown in Fig. 2.7 (Williams *et al.*, 1976):

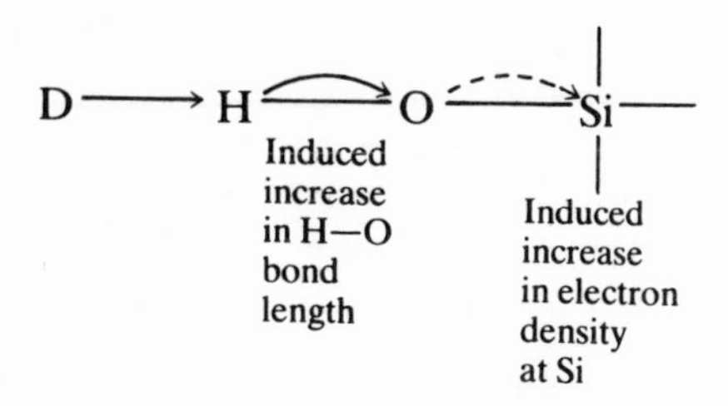

While the donor numbers of the solvents can be obtained from calorimetric measurements, it is difficult to assign donor numbers to anions in solution since their donor properties are decisively influenced by solvation: The more strongly an anion is solvated, the weaker the donor properties. Since aprotic solvents do not differ vastly in anion-solvating properties, as evidenced by comparison of their acceptor numbers (see Section 2.3), the approximate values for relative donor properties of anions

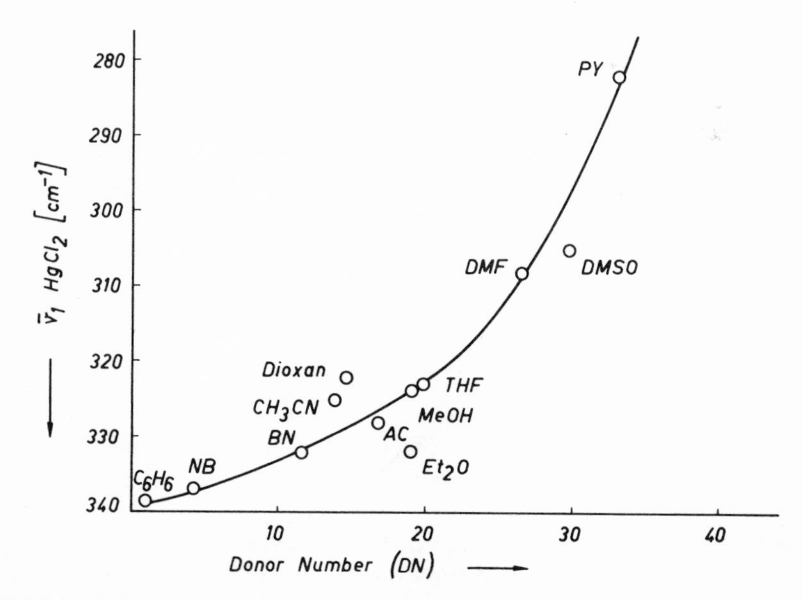

Fig. 2.6. Relationship between the donor number of various solvents and the Hg—X stretching frequencies in complexes between solvent and $HgCl_2$.

in such solvents can be obtained from equilibrium studies (Mayer and Gutmann, 1970). The ΔG values for the reactions

$$VO(acac)_2(CH_3CN) + D \rightleftharpoons VO(acac)D + CH_3CN$$

where D may be a neutral or a charged donor, have been plotted vs. the donor numbers of the neutral donor molecules, and from this plot approximate donor number values for anions were obtained (Fig. 2.8).

As can be seen from Fig. 2.8, in aprotic solvents the iodide ion is a slightly weaker ligand than acetonitrile (DN = 14.1) or propanediol carbonate (DN = 15.1). The bromide ion has donor properties that fall between trimethyl phosphate (DN = 23) and acetone (DN = 17) and the chloride ion has donor properties similar to those of dimethyl formamide

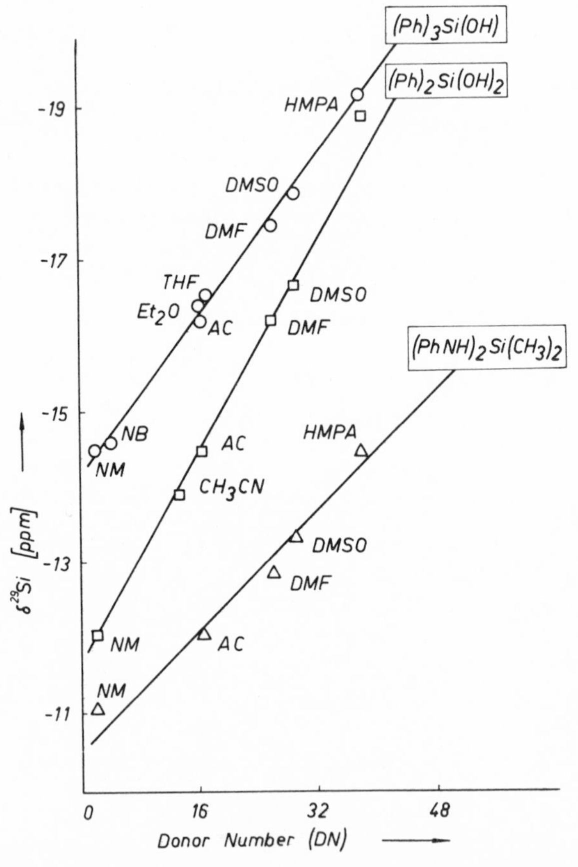

Fig. 2.7. Relationship between the ^{29}Si NMR chemical shift in silanols and silylamines in various solvents and the donor number of these solvents (data from Williams *et al.*, 1976).

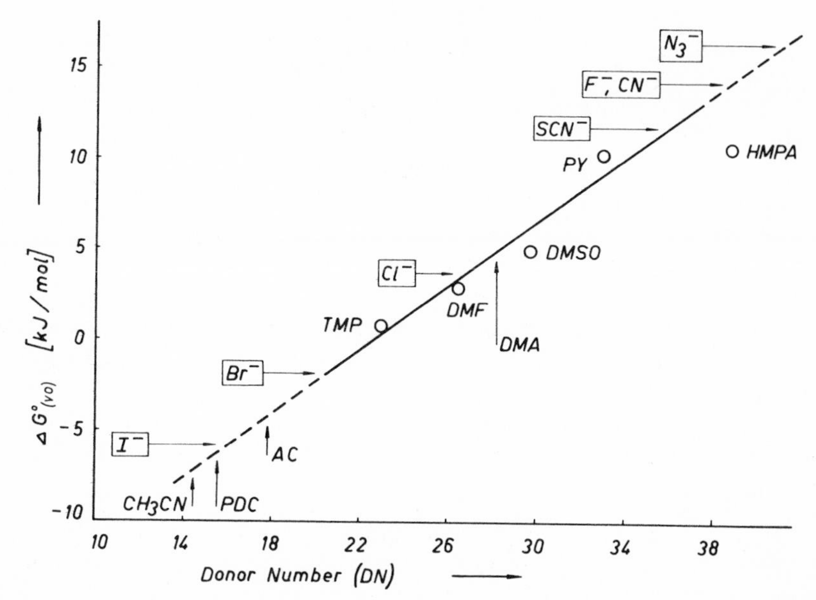

Fig. 2.8. Approximate values for donor properties of several anions obtained from ΔG^0 values for the reaction $VO(acac)_2(CH_3CN) + D \rightleftharpoons VO(acac)_2D + CH_3CN$ (from Gutmann, 1971*a*, courtesy of Springer-Verlag, Vienna, New York).

(DN = 26.6). The fluoride ion is similar in donor properties to hexamethylphosphoric triamide (DN = 38.8), while the azide ion appears to have even greater donor properties.

2.3. *The Acceptor Number*

Thermodynamic quantities which represent a measure of the acceptor properties of solvents are provided by the single-ion free energies of solvation of halide ions (Alexander *et al.*, 1972; Cox *et al.*, 1974) and by the half-wave potentials for the polarographic reduction of the hexacyanoferrate(III) ion (Gutmann *et al.*, 1976; Gritzner *et al.*, 1976*a*). An empirical parameter for the acceptor properties of solvents has been proposed recently, which is based on the ^{31}P NMR chemical shift in triethylphosphine oxide in the respective pure solvent. This approach meets the following important requirements (Mayer *et al.*, 1975):

(1) The probe nucleus is remote from the actual place of interaction, namely, the basic oxygen atom, thereby theoretically eliminating ill-definable contact contributions to the chemical-shift values.

(2) Triethylphosphine oxide is a very strong base (DN estimated ≈40). This together with the partial double-bond character of the P—O bond assures a high sensitivity of the phosphorus resonance toward solvent change.

(3) The interaction between solute and solvent always occurs at a well-defined site, namely, at the oxygen atom; the remaining coordination sites of the phosphorus atom are blocked off by inert alkyl groups. The incorporation of ethyl groups guarantees efficient electronic shielding without steric hindrance.

(4) Owing to the presence of ethyl groups, triethylphosphine oxide in contrast to other phosphine oxides is sufficiently soluble in all kinds of solvents.

(5) Triethylphosphine oxide is extraordinarily stable. No decomposition was observed after several hours even in solutions of strong protonic acids.

By the interaction of Et_3PO with an acceptor solvent, which attacks at the oxygen atom, the electron density at the phosphorus atom is decreased, which manifests itself in a down-field chemical shift:

$$\begin{array}{c} \mathrm{Et} \\ | \\ \mathrm{Et{-}P{=}O{\rightarrow}Acceptor} \\ | \\ \mathrm{Et} \end{array}$$

The δ values have been extrapolated to infinite dilution, referred to hexane as a reference solvent to which the acceptor number of zero has been assigned, and corrected for the difference in volume susceptibilities between hexane and the respective solvents (δ_{corr}). In order to emphasize the relationship between acceptor properties and their conjugate donor properties, the same substance, namely, antimony(V) chloride, has been used as a standard for both parameters. The δ_{corr} values have been related directly to that for the $Et_3PO{\rightarrow}SbCl_5$ adduct dissolved in 1,2-dichloroethane ($\delta_{corr(Et_3PO\cdot SbCl_5)}$) by arbitrarily assigning this the value of 100.

The acceptor number (AN) (Mayer *et al.*, 1975; Gutmann, 1976*b*) is defined as a dimensionless number related to the relative chemical shift of ^{31}P in Et_3PO in the particular solvent, with hexane as a reference solvent on one hand, and $Et_3PO–SbCl_5$ in 1,2-dichloroethane on the other, to which the acceptor numbers of 0 and 100 have been assigned, respectively:

$$AN \equiv \frac{\delta_{corr} \cdot 100}{\delta_{corr(Et_3PO\cdot SbCl_5)}} = \delta_{corr} \cdot 2.348$$

In contrast to classical solvent parameters such as dielectric constant, dipole moment or polarizability, the acceptor number allows the interpretation of numerous solvent dependent NMR-, IR-, Raman-, UV-spectroscopic, and kinetic data (Mayer *et al.*, 1975, 1977*b*; Gutmann, 1976*b*, 1977*b*).

In a solution of a protonic acid HX one has to assume an equilibrium between the hydrogen-bonded complex[I] and the fully protonated phosphine oxide[II]:

$$\underset{[I]}{Et_3PO{\rightarrow}HX} \rightleftharpoons \underset{[II]}{Et_3POH^+ + X^-}$$

The fact that only one resonance signal is observed is due to rapid exchange between the two species, the observed resonance signal being the weighted average of the signals of the isolated species. Despite differences in acid strengths, nearly the same shift values were measured for the two strongest acids, namely, methanesulfonic and trifluoromethanesulfonic acid, suggesting that Et_3PO is nearly exclusively present in the fully protonated form[II]. In weak protonic acids (e.g., acetic acid), in alcohols, in C—H acidic solvents (CH_2Cl_2, $CHCl_3$) or N—H acidic solvents (formamide, *N*-methyl formamide) the predominant species are most probably the hydrogen-bonded complex[I] and uncoordinated phosphine oxide. This interpretation is in agreement with the observation of considerable line broadening in solutions of trifluoroacetic acid, which has an acid strength intermediate between those of acetic acid and methane- or trifluoromethanesulfonic acid, respectively (Table 2.3). In acetic acid both

Table 2.3. Acceptor Numbers (AN) of Various Solvents

Solvent	AN	Solvent	AN
Hexane (reference solvent)	0.0	PDC	18.3
Diethyl ether	3.9	CH_3CN	19.3
Tetrahydrofuran (THF)	8.0	DMSO	19.3
Benzene	8.2	CH_2Cl_2	20.4
CCl_4	8.6	Nitromethane (NM)	20.5
Diglyme	10.2	$CHCl_3$	23.1
HMPA	10.6	*i*-Propanol	33.5
Dioxane	10.8	Ethyl alcohol	37.1
Acetone	12.5	Formamide (FA)	39.8
N-Methyl pyrrolidionone (NMP)	13.3	Methyl alcohol	41.3
DMA	13.6	Acetic acid	52.9
Pyridine	14.2	Water	54.8
Nitrobenzene (NB)	14.8	$SbCl_5$ in DCE	100.0
Benzonitrile (BN)	15.5	CF_3COOH	105.3
DMF	16.0	CH_3SO_3H	126.1
Dichloroethylene carbonate	16.7	CF_3SO_3H	129.1

hydrogen-bonded adduct and protonated phosphine oxide appear to be present in comparable quantities. The exchange process is sufficiently rapid to give one single-resonance signal, but it is slow enough to cause line broadening. In all other solvents sharp resonance lines are observed.

With the exception of acetic acid there is a correlation between ^{31}P chemical shift and pK_a values. The deviation observed for acetic acid is almost certainly due to the strong association in the liquid state, by which the effective acceptor strength of this solvent is reduced. The chemical-shift values found for the alcohols and CH-acidic solvents agree with the order of increasing acceptor strength in the series $CH_2Cl_2 < CHCl_3 <$ *i*-propanol $<$ EtOH $<$ MeOH $< H_2O$, as obtained by spectroscopic measurements (Balasabramanian and Rao, 1962; Olsen, 1970). within the group of aprotic solvents, hexane and other saturated hydrocarbons, as expected, behave as extremely poor acceptor molecules.

Surprisingly benzene and carbon tetrachloride, which are nonpolar and considered as inert solvents, show definitely stronger acceptor properties than does the polar diethyl ether. This may explain the formation of "pentahalocarbonate" complexes from carbon tetrachloride and tetra-alkylammonium halide (Creighton and Thomas, 1972). The extremely weak acceptor properties of ether solvents, which would not be expected from a consideration of their dipole moments, account for the high reactivity of anions (e.g., carbanions) in these media and hence for the usefulness of these solvents for electrophilic substitution reactions (Cram, 1965; see Chapter 13).

Ketones behave also as comparatively weak acceptors. The electrophilic character of carboxylic acid amides varies over a broad range. Solvents which are capable of hydrogen bonding such as formamide or *N*-methyl formamide show acidities similar to the lower aliphatic alcohols. Substitution of acid hydrogen atoms by alkyl groups drastically decreases the acceptor properties. Well-developed acceptor properties are exhibited within the group of dipolar aprotic solvents, namely, DMSO, acetonitrile, and PDC. The fact that these solvents at the same time possess appreciable donor properties and fairly high dielectric constants accounts for their increasing use in basic research, chemistry, and industry. Hexamethylphosphoric amide is a poor acceptor (Table 2.3) but an extraordinarily strong donor (Table 2.1). These unique properties together with a comparatively high dielectric constant make it particularly suitable as a medium for reactions involving strong and highly reactive bases such as alkoxides, carbanions, etc. The combination of these properties of HMPA also accounts for the observation that (as in liquid ammonia) alkali metals dissolve with formation of comparatively stable blue-colored solutions containing "free" electrons and strongly solvated metal ions (Normant,

1967). The fact that this reaction does not proceed in ethers is due to their lower donor numbers (see Table 2.1). Nitromethane, in contrast to HMPA, shows well-developed acceptor but only poor donor properties.

An approximate estimation of the coordinate bond energies is possible by making use of the formula (Gutmann, 1976*b*)

$$\Delta H = \frac{\mathrm{DN} \cdot \mathrm{AN}}{100}$$

A fair relationship has been found between acceptor numbers and Kosower's Z values (Fig. 2.9). For solvents with dielectric constants below 10 the Z values generally exhibit negative deviations indicating that dielectric polarization effects become important owing to the highly polar character of the ground-state complex (Mayer *et al.*, 1975). In this way the low Z value of acetic acid as compared to water may be explained: $H_2O: \varepsilon = 78.5$, $AN = 54.8$, $CH_3COOH: \varepsilon = 6.2$, $AN = 52.9$. Thus the Z values bear a close relationship to the acceptor properties of a solvent rather than representing a general measure of the polarity or of the ionizing properties of the solvent. It is rather unfortunate that Kosower's otherwise correct ideas about the formation of molecular adducts as a result of solute–solvent interactions could not be applied widely since he failed to recognize the chemical meaning of his polarity scale as an approximate guide for the relative acceptor character of the solvents (Fig. 2.9).

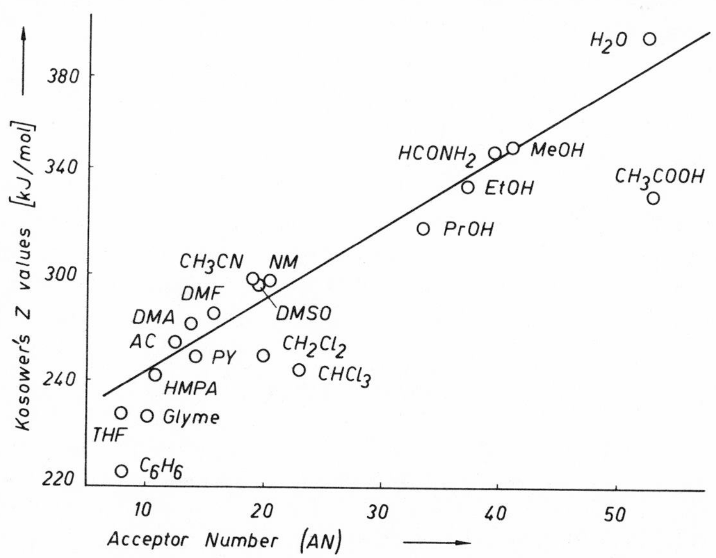

Fig. 2.9. Relationship between the acceptor number of various solvents and Kosower's Z values (from Mayer *et al.*, 1975, courtesy of Springer-Verlag, Vienna, New York).

A relationship also exists between the acceptor numbers and the so-called "*Y*" values of Grunwald and Winstein (Section 2.1), based on the rate of solvolysis of *t*-butyl chloride (Fig. 2.10).

Similar relationships are found between E_T values (Dimroth *et al.*, 1963) and the acceptor numbers (Fig. 2.11). The former represent the energy of the lowest electronic transition of the pyridinium-*N*-phenol betaine. Although this is a neutral compound with a large π-electron system its behavior is expected to be very similar to that of Kosower's pyridinium iodide. Solvent attack occurs electrophilically at the highly basic oxygen atom, which is sterically fairly well accessible, while the nitrogen is again coordinatively saturated. The use of the E_T values as a measure for the acceptor properties of solvents (Krygowski and Fawcett, 1975; see also Chapter 10) is therefore justified.

The free energies of solvation of various halide ions (Alexander *et al.*, 1972; Cox *et al.*, 1974) have been successfully used (together with the solvent donicities) for the precalculation of thermodynamic and kinetic data of various equilibria (Mayer, 1975).

The quantitative characterization of acceptor properties of cations is not possible since these depend to a great extent on their specific solvation. The stronger the cation is coordinated either by solvent molecules or by solutes, the weaker the acceptor property of the cation. The general trend of

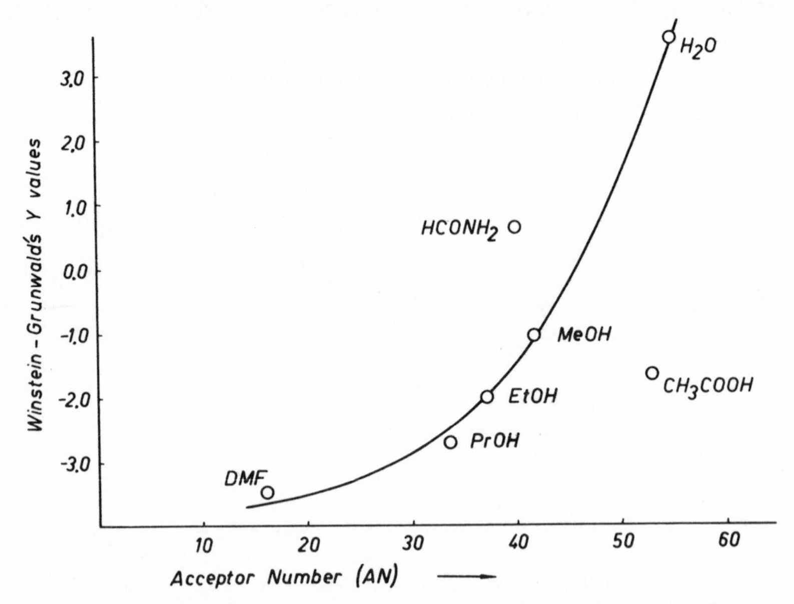

Fig. 2.10. Relationship between acceptor numbers and Winstein and Grunwald's *Y* values (from Mayer *et al.*, 1975, courtesy of Springer-Verlag, Vienna, New York).

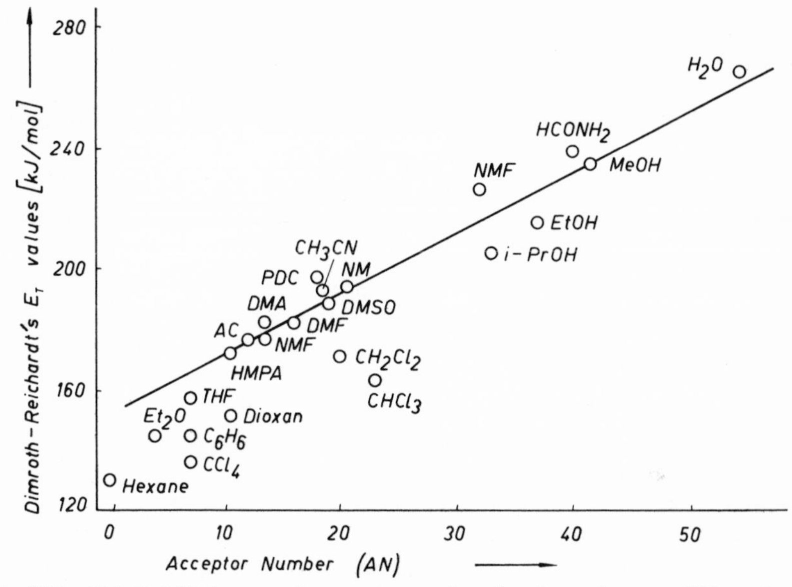

Fig. 2.11. Relationship between the acceptor number of various solvents and Dimroth and Reichardt's E_T values (from Mayer *et al.*, 1975, courtesy of Springer-Verlag, Vienna, New York).

acceptor properties of cations in a given medium may be derived approximately by comparing hydration energies and acidity constants of the hydrated cations in water. In general there is much agreement with the following order of acceptor properties for cations derived from various chemical observations (Schlosser, 1973):

$$Bu_4N^+ < Pr_4N^+ < Et_4N^+ < Me_4N^+ < Cs^+ < Rb^+ < K^+ < Ba^{2+} < Sr^2 < Na^+$$
$$< Ca^{2+} < Li^+ < Mg^{2+} < Be^{2+} < Zn^{2+} < Cu^{2+}$$

Chapter 3
Molecular Adducts

3.1. Theoretical Approaches

The terms "molecular adduct" or "molecular complex" embrace a wide variety of diverse substances ranging from the stable coordination compounds such as the ammonia–boron(III) fluoride complex to the weak complexes known as "electron-donor–acceptor complexes" or "charge-transfer complexes." Different approaches are used for each of these groups of molecular adducts although no sharp border line can be drawn between "strong" and "weak" interactions: The coordination-chemistry approach is applied to the first group and the Mulliken valence-bond method is applied to charge-transfer complexes (Mulliken and Person, 1962), while hydrogen-bonded systems and association phenomena in the liquid state have frequently been interpreted by the electrostatic theory, the inadequacies of which have now been recognized (Schuster, 1973). Molecular complexes are important as intermediates in chemical reactions and weak complex interactions are involved in outer-sphere coordination (Gutmann and Schmid, 1974), ion association, and adsorption phenomena (Gutmann, 1975). Interest in weak complex interactions has developed strongly over the past two decades and their importance in biochemistry and biology is now generally recognized (Rose, 1967; Paetzold, 1975*a*).

Molecular complexation always leads to changes in bond properties. In weak molecular adducts the electronic changes are small and yet the chemical character may be considerably altered in a characteristic and highly specific way.

In the Mulliken treatment (Mulliken, 1950, 1951, 1952*a*,*b*; Mulliken and Person, 1962; Briegleb, 1961; Foster, 1973) the term "charge transfer" is used for an electronic excitation that is characteristic for the complex. This concept has been extremely useful in interpreting the charge-transfer bands in electronic absorption spectra of weak complexes. Unfortunately, the term charge transfer is used alternatively by chemists in describing the net transfer of charge from a donor to an acceptor for the electronic ground state of the complex.

However, the Mulliken treatment, while suitable for the interpretation of the charge-transfer effects within the complex, is not applicable to the ground state (Gutmann and Mayer, 1971; Paetzold, 1975). Experimental and quantum chemical investigations on the properties of the ground state, such as bond energies, dipole moments, infrared, NQR, and Mössbauer spectra confirmed the inapplicability of this concept to various types of complexes, such as $C_6H_6—Br_2$, $C_6H_6—CCl_4$, organic π-complexes and H-bonded complexes (Paetzold, 1975*a*). Some of the simplifications and inadequacies of the resonance and the electrostatic perturbation theories are as follows: Neglect of possible resonance structures, limitation to small values of overlap integrals, problems of localization of dipole and quadrupole moments in reducing the problem to that of nonbonding electron pairs, lack of information of complex geometry in the gas phase and in solution, and uncertainty in the choice of the repulsive potential function.

The quantum mechanical treatment as well as the donor–acceptor approach have in common that they consider the complex as a whole and treat the various types of complexes in the same ways. A donor–acceptor interaction leads to increasing polarities of the bonds originating from the donor and acceptor atoms, respectively. Increasing polarity is related to an increase in fractional positive charge at the acceptor atom and an increase in fractional negative charge at the donor atom.

The original decrease of fractional positive charge at the acceptor atom due to both the charge-transfer and polarization effects is overcompensated by passing over the negative charges, including part of those originally situated at the acceptor atom to other areas of the acceptor molecule. In this way the fractional negative charges of other nuclei in the acceptor component are increased, in particular those terminating the acceptor molecule. This has been described as the *spillover effect* of negative charge from the acceptor atom (Gutmann, 1977*a*).

The original loss of negative charge at the donor atom by charge transfer toward the acceptor molecule is overcompensated by attracting electronic charge from other parts of the donor molecule to the donor atom. In this way the electron density at the donor atom is increased with appropriate changes of fractional nuclear charges in other areas of the donor component. This has been termed the *pileup effect* of negative charge at the donor atom (Gutmann, 1977*a*). Table 3.1 shows the results

Table 3.1. Changes in Fractional Nuclear Charges Due to Complex Formation of Donor Molecules with Chlorine

Complexes	ΔH ($kJ \cdot mol^{-1}$)	Δq_N (donor atom)	$\Delta q_{Cl(1)}$ (acceptor atom)	$\Delta q_{Cl(2)}$	Transfer of charge
$H_3N \rightarrow Cl_{(1)}-Cl_{(2)}$	4.30	−0.003	+0.030	−0.038	0.008
$C_5H_5N \rightarrow Cl_{(1)}-Cl_{(2)}$	3.45	−0.004	+0.026	−0.031	0.005
$HCN \rightarrow Cl_{(1)}-Cl_{(2)}$	1.95	−0.002	+0.018	−0.019	0.001

of calculation by Roothaan's method using the computer program GAUSSIAN-70 with an STO-3G basis set. Although in the systems NH_3-Cl_2, pyridine$-Cl_2$, and $HCN-Cl_2$, charge transfer was found to be small (LaGrange *et al.*, 1977), both the spillover effect at the acceptor atom and the pileup effect at the donor atom are considerable.

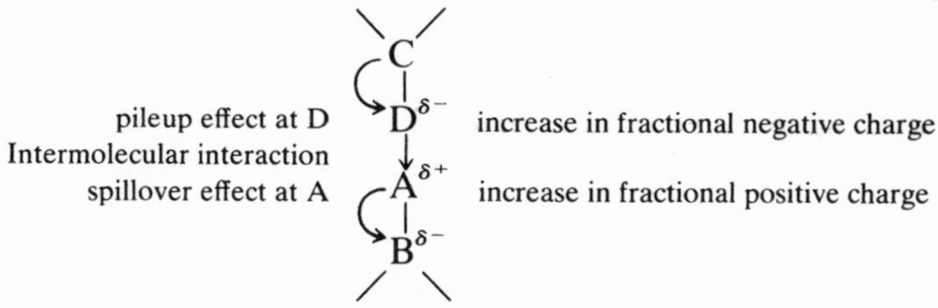

C

pileup effect at D $D^{\delta-}$ increase in fractional negative charge

Intermolecular interaction

spillover effect at A $A^{\delta+}$ increase in fractional positive charge

$B^{\delta-}$

For the pyridine compound, small increases in positive fractional charge were calculated for $C_{(2)}$, $C_{(4)}$, and $C_{(6)}$, and small increases in negative fractional charge were calculated for $C_{(3)}$ and $C_{(5)}$.

For the interaction between water and chlorine, the heat of formation has been calculated at 3.1 kJ/mol with very small charge transfer as compared to the polarization effects (Leroy *et al.*, 1975). The calculated changes in net charges (Leroy and Louterman-Leloup, 1975) show both the spillover effect at the Cl-acceptor atom and the pileup effect at the O-donor atom:

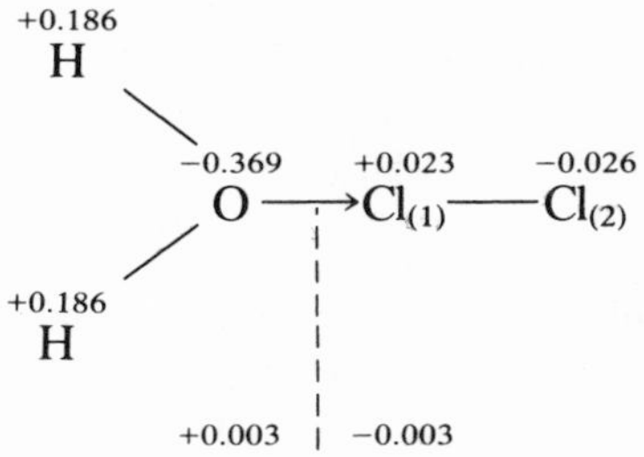

+0.186 H

−0.369 O → +0.023 $Cl_{(1)}$ — −0.026 $Cl_{(2)}$

+0.186 H

+0.003 | −0.003

For the interaction between acetonitrile and phenol at the H-acceptor atom of the phenol molecule, a spillover of 0.145 electron has been calculated, with a pileup of 0.162 electron at the N-donor atom of the cyano group (Baraton *et al.*, 1973).

A theoretical study on the system formaldehyde–chlorine leads to the heat of formation 2.1 kJ/mol of the adduct, with a minimal charge transfer of 0.001 electron (Leroy *et al.*, 1975). Both the spillover and the pileup effects can be seen from the changes in fractional nuclear charges due to complex formation:

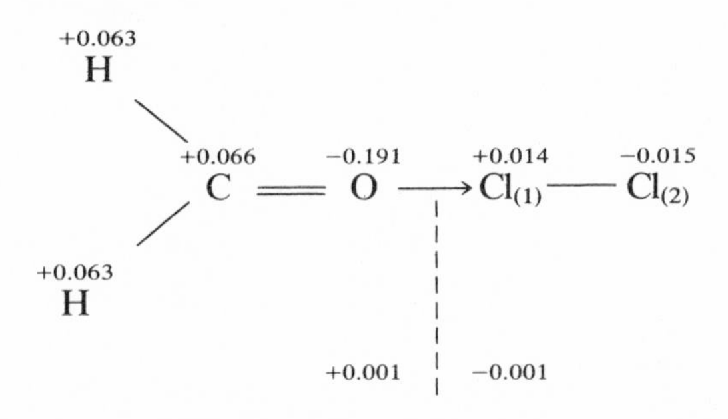

In the course of the formation of the formation of ammonium chloride from gaseous ammonia and hydrogen chloride, a transfer of electronic charge is induced from the hydrogen atoms $H_{(1)}$, $H_{(2)}$, and $H_{(3)}$ of the ammonia molecule to the nitrogen atom, as well as there being a transfer of electrons from the hydrogen atom $H_{(4)}$ of the hydrogen chloride molecule to its chlorine atom. We have therefore the following net charge alternation: more positive ($H_{(1)}$, $H_{(2)}$, $H_{(3)}$), more negative (N), more positive ($H_{(4)}$), more negative (Cl). The calculated increase in the H—Cl bond distance within the complexed acceptor molecule for several ammonia complexes is shown in Table 3.2 (Clementi, 1967).

$H_{(1)}$

$H_{(2)}$—N→$H_{(4)}$—Cl

$H_{(3)}$

A further example for such structural changes is provided by the results of quantum chemical *ab initio* calculations for the possible structures of BH_5 (Hoheisel and Kutzelnigg, 1975). The latter was found stable

Table 3.2. Calculated Increases in A—B Bond Distances of Acceptors AB Due to Coordination by Ammonia

Donor	Acceptor A—B	Increased A—B distance (pm)
NH_3	H—F	40
NH_3	F—F	10
NH_3	H—H	30
NH_3	Cl—Cl	35

in a C_s structure (analogous to that of CH_5^+) by 8 kJ/mol with respect to $BH_3 + H_2$.

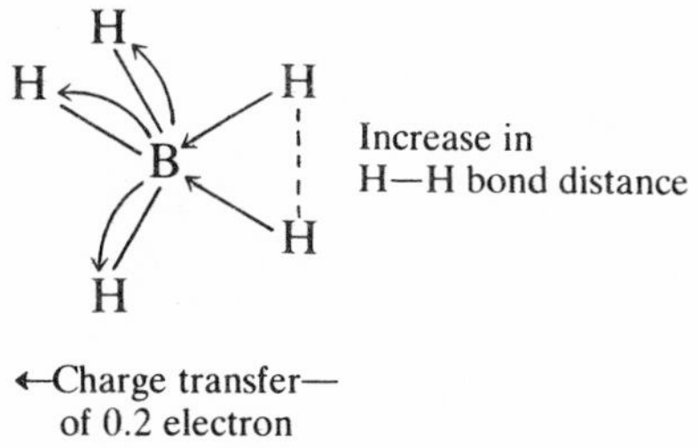

H H H B H H

Increase in H—H bond distance

←Charge transfer— of 0.2 electron

The intermolecular distances are smaller than expected for van der Waals bonds. Both the BH_3 and H_2 subunits in BH_5 are appreciably distorted as compared with the isolated molecules: The out-of-plane angle in the BH_3 subunit is 8° and the H—H bond distance in the subunit H_2 is 148 pm as compared to 140 pm in isolated H_2.

The charge distribution, usually assigned to donor and acceptor atoms after complexation in organic chemistry, e.g.,

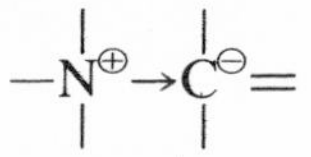

$$-\overset{|}{\underset{|}{N}}{}^{\oplus}\rightarrow\overset{|}{\underset{|}{C}}{}^{\ominus}=$$

is therefore *not in agreement with the facts,* in that for the interaction of a donor with an acceptor, the donor atom is not increased in positive charge and the acceptor atom is not increased in negative charge.

3.2. *Application of the Bond-Length Variation Rules*

For weak molecular complexes, by far the largest contribution to crystal structures has come from Hassel and his co-workers (Hassel and Rømming, 1962). In the majority of the structures, the donor atoms are nitrogen, oxygen, sulfur, or selenium atoms and the acceptor molecules are

Table 3.3. Halogen–Halogen Bond Length in Halogen Molecular Complexes

Compound	Halogen–halogen bond length (pm)	
	Free	In complex
Acetone–Br_2 (chain structure)	228	228
1,4-Dioxane–Br_2 (chain structure)	228	231
1,4-Dioxane–Cl_2 (chain structure)	199	202
1,4-Dithiane–I_2	267	279
1,4-Diselenane–I_2	267	287
Hexamethylenetetramine–2 Br_2	228	243
Trimethylamine–I_2	267	283
Trimethylamine–ICl	232	252
Pyridine–ICl	232	251
γ-Picoline–I_2	267	283

halogen, interhalogen, or halogen-rich haloalkanes. In accordance with the first bond-length variation rule, in the acceptor unit the bond adjacent to the coordinate bond is lengthened (Table 3.3) with lengthening increasing as the intermolecular distance (Table 3.4) is decreased.

In trimethylamine iodine and trimethylamine iodine monochloride (Hassel and Hoppe, 1960) the N→I—I and N→I—Cl systems are linear and the N—I bond distances are 227 and 230 pm, respectively. This is 140 pm less than the sum of the van der Waals radii and only 30 pm greater than the sum of covalent radii. The I—I bond distance in the iodine monochloride complex is 22 pm longer than in the free ICl molecule.

Table 3.4. Intermolecular and Intramolecular Bond Distances (in pm) in Adducts of the Halogens

Bond	Br_2	Acetone–Br_2	Methanol–Br_2	1,4-Dioxane–Br_2
>O→Br	—	282	278	271
Br→Br	228	228	229	231

Bond	I_2	1,4-Dithiane–I_2	Benzyl sulfide–I_2
>S→I	—	287	278
I—I	267	279	282

The increase in halogen–halogen bond distance by complexation is least pronounced when donor and acceptor molecules contain two coordinating sites: Thus the 1,4-dioxane–bromine complex (Hassel and Hroslev, 1954) and the 1,4-dioxane–chlorine complex (Hassel and Strømme, 1959*a*) form two-dimensional structures of endless chains of alternating halogen molecules and 1,4-dioxane molecules. In such compounds the charge-transfer effects are weak or even nonapparent. In the benzene–bromine complex the benzene and halogen molecules form donor–acceptor stacks with the halogen–halogen bond coincident with the perpendicular through the centroid of the benzene-ring plane, and the halogen–halogen bond distance is no different from that found in the free halogen molecules (Hassel and Strømme, 1958).

The same is true for the chain-structured acetone–bromine adduct. Each oxygen atom is coordinated by two bromine molecules and hence the interactions are weaker than one would expect from the donor properties of acetone toward one acceptor molecule. Hence the oxygen–bromine bond distances of 282 pm are considerably longer than those of 271 pm in the dioxane–bromine complex (Hassel and Strømme, 1959*b*; Hassel and Hroslev, 1954). The O—Br bond distance in the acetone–bromine adduct is apparently not short enough to induce a detectable increase in Br—Br bond distance. However, the shorter O—Br bond distance in the dioxane–bromine adduct leads to an (induced) increase in Br—Br bond distance from 228 pm to 231 pm.

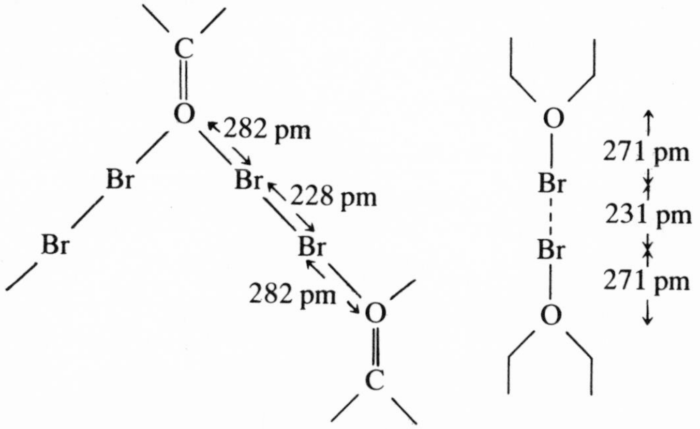

C
O 282 pm
Br Br 228 pm
Br Br
282 pm O
C

O
Br 271 pm
231 pm
Br
O 271 pm

Another example of the lengthening of a homonuclear bond, namely, the O—O bond, by incorporation into a chain structure is provided by the paramagnetic $[(H_3N)_5Co—O_2—Co(NH_3)_5](SO_4)_2(HSO_4)\cdot 3H_2O$. Due to the strong Co—O interactions, the O—O bond is lengthened from 120.7 pm in the gaseous oxygen molecule to 131 pm in the complex ion under consideration (Schaefer and Marsh, 1966). This is longer than the O—O distance of 128 pm in the hydroperoxide ion and shorter than in hydrogen peroxide. It may be mentioned that the complex ion formerly was thought to contain Co(III) and Co(IV), but the structure determination and the ESR spectra have shown that both cobalt ions are equivalent (Ebsworth and Weil, 1959), indicating that the charge of the single unpaired electron is distributed equally over both cobalt nuclei. The assignment of integral oxidation numbers is without meaning in this case.

3.3. Application of the Empirical Parameters

The application of the donor number concept to the charge-density rearrangement in complexes of silanoles and silylamines has been shown in Section 2.2. An example of its application to weak molecular adducts is provided by results from the complexation of trifluoroiodomethane. This molecule contains an iodine–carbon bond with the partial positive charge residing at the iodine atom (due to the strong electron-attracting properties of the CF_3 group). An electron donor attacks at the iodine atom. For the interaction with the strong donor, hexamethylphosphoric triamide, the ΔH value is $\approx$25 KJ·mol^{-1}. The electron densities at the fluorine atoms are increased as is seen from the ^{19}F NMR spectra, which show a shift toward higher field with increased donor number of the solvent.

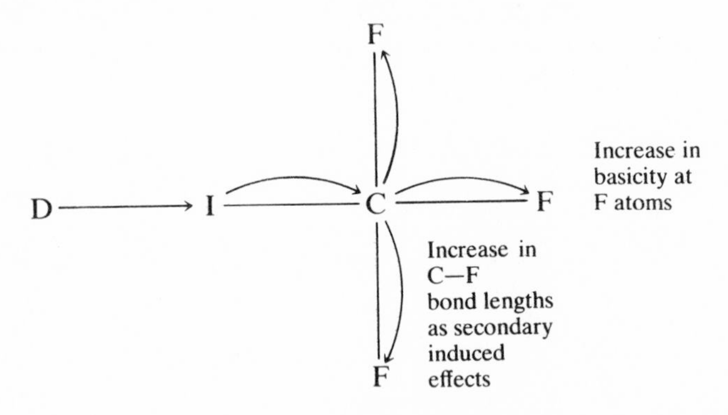

Figure 3.1 shows the linear correlation between the ^{19}F NMR chemical shift and the donor number of various solvents (Spaziante and Gutmann, 1971). No relationship is found between chemical shifts and dipole moments, ionization potentials, polarizabilities, or dielectric constants of

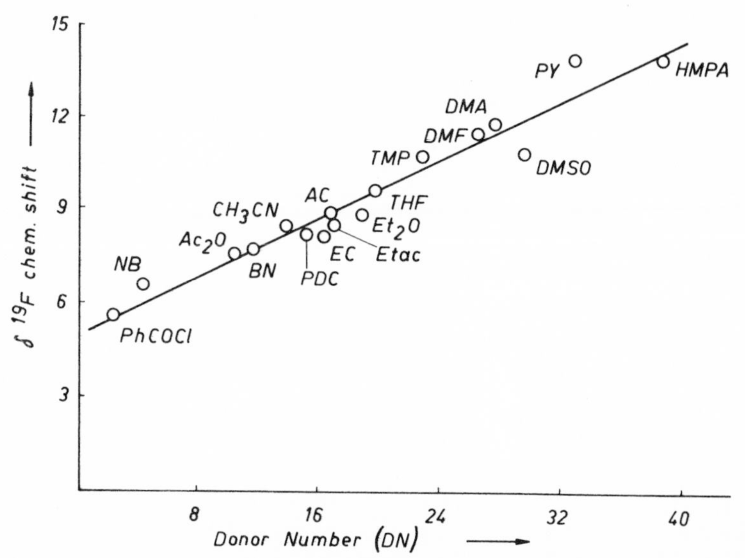

Fig. 3.1. Relationship between the $\delta\,^{19}F$ chemical shift in trifluoroiodomethane complexes and the donor number of the donor component. δ in ppm, related to ICF_3 as external standard (from Gutmann, 1977*a*, courtesy of Springer-Verlag, Vienna, New York).

the donor solvents. It is important to note that unlike other relationships for molecular adducts, the $\delta\,^{19}F$–DN relationship is not confined to complexes with the same functional group, since the relationship holds for different donor groups such as the nitro group, nitrile group, $>C{=}O$, $>S{=}O$, and $\geqslant P{=}O$ groups, as well as for ethers and nitrogen-donating bases.

There is also a relationship between the $\Delta\nu_{OD}$ values of methanol-*d* and the donor number of the solvent (Kagiya *et al.*, 1968).

$$\text{Donor}\longrightarrow {}^{2}_{1}\text{D}\overset{\frown}{\text{———}}\text{O}\text{———}\underset{\text{H}}{\overset{\text{H}}{\text{C}}}\text{———H}$$

Increase in bond length

A less satisfactory relationship is found between pK_b values of the donor and $\log K$ (K is the formation constant of the complex) for iodine

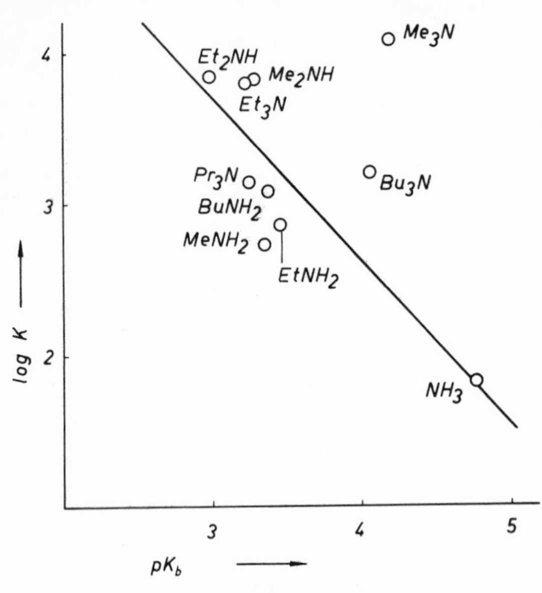

Fig. 3.2. Relationship between log K for complexes of iodine with amines and the pK_b of the amine (from Gutmann, 1977*a*, courtesy of Springer-Verlag, Vienna, New York).

complexes (Herlem and Popov, 1972) (Fig. 3.2), where the halogen–halogen stretching frequencies vary approximately inversely with the strength of the complex: An increase in coordinate bond strength is accompanied by a progressive weakening of the interhalogen bond (Yamada *et al.*, 1972). Thus, the blue shift of the iodine band in iodine complexes is related to donor number of the donor (Fig. 3.3). Coordination by a strong donor to iodine may result in heterolysis of the I—I bond, such as in the presence of excess pyridine, when cations are formed with equivalent N—I bond distances of 216 pm:

$$2PY + 2I_2 \rightleftharpoons PY{-}I{-}PY^+ + I_3^-$$

Thus a molecular adduct is formed by donor–acceptor interactions as long as the adjacent bonds are polarized to such a limited extent that heterolytic bond cleavage does not take place. In other words, a molecular adduct is the result of well-balanced donor–acceptor interactions which are not sufficient for heterolysis of any of the bonds.

Within a donor unit the bond-length variations are related to the acceptor number of the acceptor component. A relationship exists between the ^{13}C chemical shift in the >C=O group of acetone (Maciel and Ruben 1963) and the acceptor number (Mayer *et al.*, 1975). The ^{13}C resonance signals are shifted to lower field with increasing acceptor number (Fig. 3.4) in accordance with the increasing polarity of the C=O double bond.

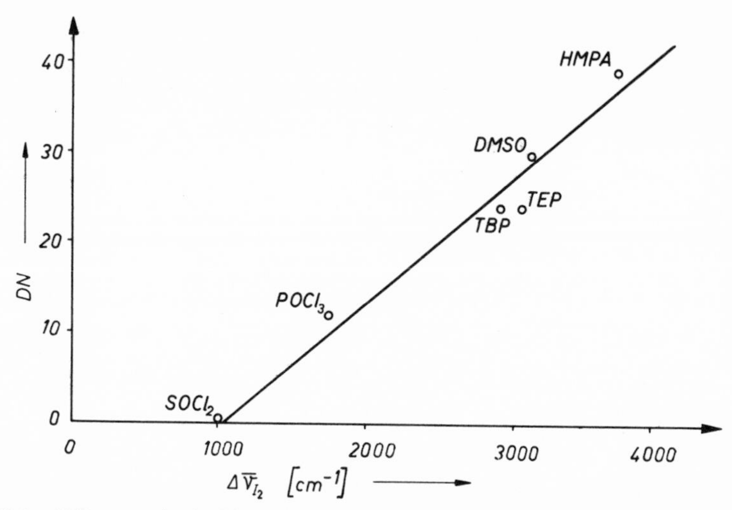

Fig. 3.3. Differences in the blue shift $\Delta\nu_{I_2}$ in iodine in various aprotic solvents as related to the donor numbers of these solvents.

The increase in polarity of the C=O bond of benzophenone and DMF in solvents of increasing acceptor number has been discovered from infrared studies (Allerhand and Schleyer, 1963). The infrared shift has been characterized by the so-called G value,

$$G = \frac{\nu_0 - \nu_S}{\nu_S}$$

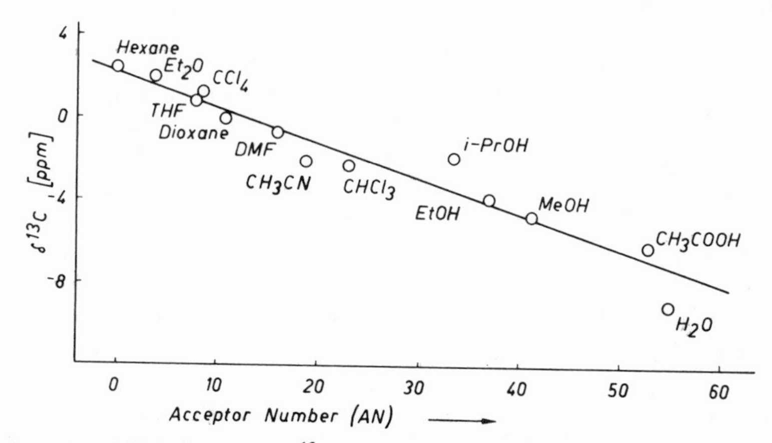

Fig. 3.4. Relationship between the ^{13}C chemical NMR shift of the carbonyl carbon atom in acetone in various solvents and the acceptor numbers of these solvents (from Mayer *et al.*, 1975, courtesy of Springer-Verlag, Vienna, New York).

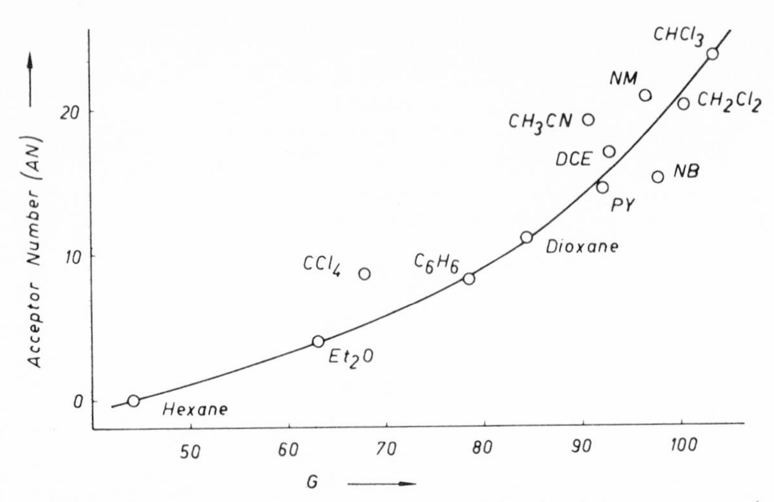

Fig. 3.5. Relationship between G values for the C=O bond and the acceptor number of various solvents.

with ν_0 and ν_S representing the valence frequencies in the gas phase and in the solvent, respectively. The relationship between the G value and acceptor number is shown in Fig. 3.5. Further relationships for carbonyls and related compounds will be discussed in the following sections.

3.4. *Metal Carbonyls and Related Compounds*

The qualitative theory, which states that in metal carbonyls the metal–carbon bonds involve a combination of C→M dative σ-bonding with M→C dative π-bonding back-donation, appears universally accepted. The metal–carbon distances may be considered as due to the intermolecular interactions, by which the C—O bonds have been lengthened: The C—O bond distances are longer in the complexes than in the gaseous CO molecule (Wells, 1962). An inverse relationship exists between the C—O and C—Fe distances for different iron(0)–carbonyl compounds: The shorter the Fe—C bond, the longer the C—O bond (Fig. 3.6).

The actual transfer of negative charge in the course of this type of complex formation takes place from the metal atom toward the ligand. The lower oxidation state of the metal is more strongly stabilized than the higher oxidation state, since complexation leads to a shift to more positive potential values (Gutmann, 1973). This is interpreted as due to an increase in fractional positive charge at the metal, which is greater in the reduced than in the oxidized form. The increase in fractional positive charge may be considered as a "spillover effect" of electronic charge toward the π-bonding ligands, initiated by σ-donation to the metal.

The back-donation theory has been interpreted in a semiquantitative way with regard to the C—O stretching force constants. Rules have been presented for the effects of substituents (Cotton and Kraihanzel, 1962; Cotton, 1964). The carbonyl stretching frequencies are lower in carbonyl complexes than in free carbon monoxide, where the carbonyl stretching frequency is found at 2168 cm^{-1}. This corresponds to a lower bond order of complexed carbon monoxide, in particular when strong electron donor ligands such as phosphines or amines are coordinated at the metallic coordination center. Such ligands increase the availability of *d* electrons of the metal for back-donation and hence the metal–carbon bond distances are shortened, with subsequent induced lowering of the carbon–oxygen bond order, which may even approach a value nearly corresponding to a C=O bond present in ketonic carbonyls.

Back-donation is enhanced by lowering the oxidation number of the metal, and this is reflected in progressive lowering of the C—O stretching

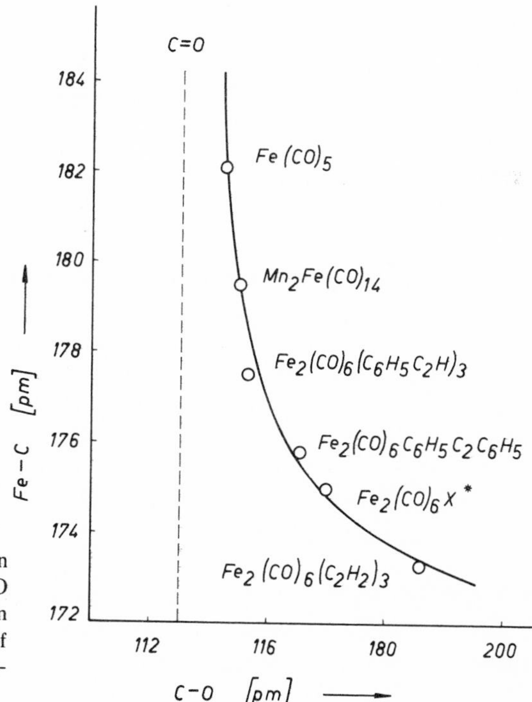

Fig. 3.6. Relationship between the Fe—C distance and the C—O distance in iron carbonyls (from Gutmann, 1975, courtesy of Elsevier Publishing Co., Amsterdam).

frequencies in isoelectronic and isostructural species (Cotton and Wilkinson, 1967):

$\overset{0}{Ni}(CO)_4$	$\overset{-I}{Co}(CO)_4^-$	$\overset{-II}{Fe}(CO)_4^{2-}$
$\approx 2060\ cm^{-1}$	$\approx 1890\ cm^{-1}$	$\approx 1790\ cm^{-1}$
$\overset{+I}{Mn}(CO)_6^+$	$\overset{0}{Cr}(CO)_6$	$\overset{-I}{V}(CO)_6^-$
$\approx 2090\ cm^{-1}$	$\approx 2000\ cm^{-1}$	$\approx 1860\ cm^{-1}$

The specific intensities of the species $V(CO)_6^-$, $Cr(CO)_6$, and $Mn(CO)_6^+$ in THF (Beck and Nitzschmann, 1962) show a linear dependence upon the formal oxidation state of the central metal atom (Kettle and Paul, 1972). The observed shifts, produced by the solvent, of the C—O stretching frequencies are much larger than theories based on the variation of the dielectric constant of the solvent would suggest (Barraclough *et al.*, 1961; Beck and Lottes, 1964). No relationship has been found between the shifts and the solvent polarities. For example, nonpolar carbon tetrachloride produces a larger shift than polar diethyl ether! However, since the oxygen atom of the carbonyl group acts as a donor toward a solvent, a relationship does exist between the shifts and the acceptor number of the solvent. In Fig. 3.7 the carbonyl vibrations of molybdenum hexacarbonyl have been plotted vs. the acceptor number of various solvents: The C—O bond is lengthened as the acceptor number of the solvent increases (Gutmann, 1977*b*).

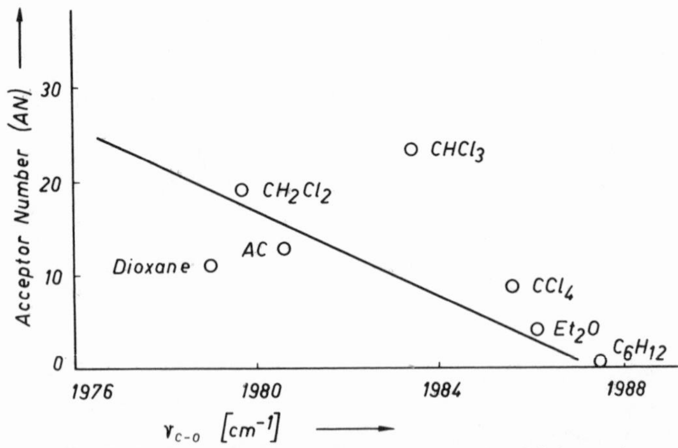

Fig. 3.7. Relationship between the IR carbonyl vibrations of molybdenum hexacarbonyl in various solvents and the acceptor numbers of these solvents (from Gutmann, 1977*b*, courtesy of Springer-Verlag, Vienna, New York).

In carbonyl halides, however, the shifts observed are in the opposite direction (Barraclough *et al.*, 1961). Although the shifts are small, some form of the solute–solvent interaction has been postulated. A tentative explanation was based on Bellamy's suggestion, for a similar phenomenon with NOCl, that there is preferential solvation of the halogen atom by the solvent (Bellamy, 1958).

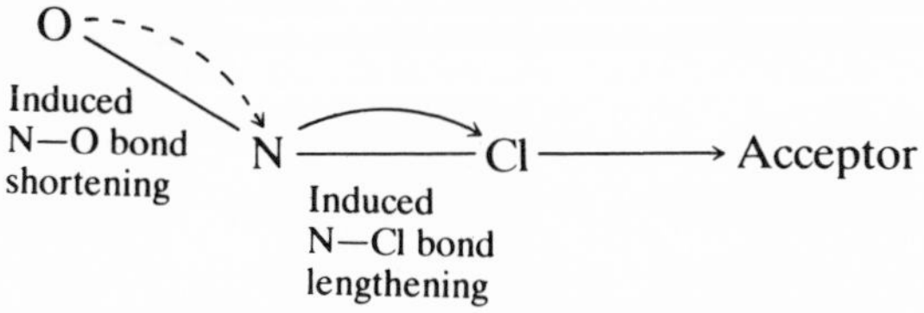

O

Induced N—O bond shortening

N —— Cl ——→ Acceptor

Induced N—Cl bond lengthening

For a number of ruthenium nitrosyls of type $Ru(NO)L_2Cl_3$ ($L = PR_3$, AsR_3, SbR_3, $R = Et$, Bu) (Beck and Lottes, 1964) in aprotic solvents, differences in frequencies are linearly related to the differences in the acceptor number of the solvent (Gutmann, 1977*b*). The interaction of the coordinated chlorine atoms with acceptor solvents decreases the electron density at the metal atom to the extent that its ability for back-donation is drastically decreased. This leads to lengthening of the nitrogen–metal bond with subsequent decrease of the NO bond distance:

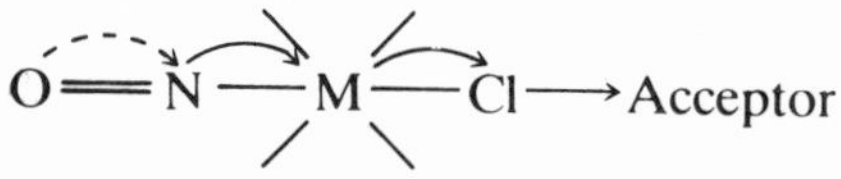

$$O{=}N{-}M{-}Cl \longrightarrow \text{Acceptor}$$

The changes in bond differences are considerably less in alcoholic solvents (Fig. 3.8). This is explained by additional hydrogen-bonding interactions with the oxygen atoms of the NO groups, which leads to increasing ON distances. In that way the decrease in NO bond distance, due to the acceptor attack of the solvent at the halogen atoms, is partly compensated (Gutmann, 1977*b*).

The electronic effect induced within a bridging CO group is particularly great. In the dimeric carbonyl cobalt(0),

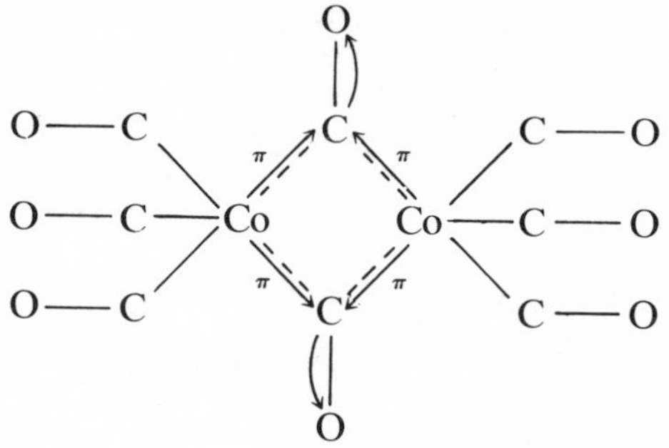

O

O—C C C—O

O—C—Co π π Co—C—O

O—C C C—O

O

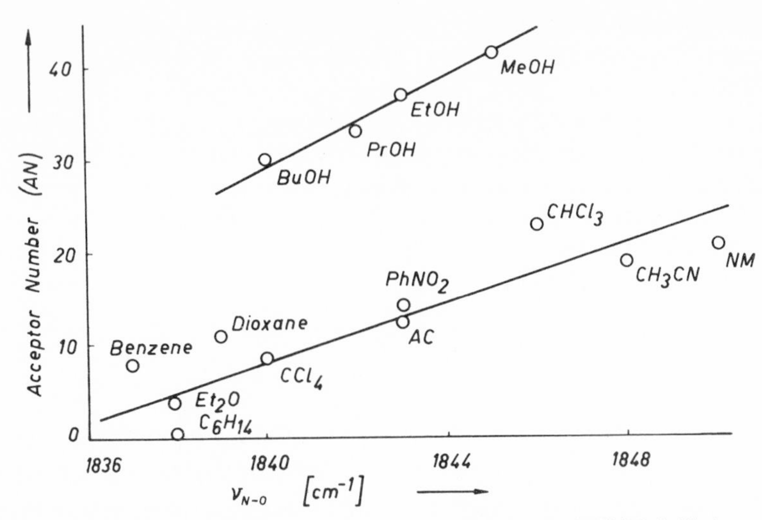

Fig. 3.8. Relationship between the IR nitrosyl vibrations in $Ru(NO)(PR_3)_2Cl_3$ in various solvents and the acceptor numbers of these solvents (from Gutmann, 1977*b*, courtesy of Springer-Verlag, Vienna, New York).

the C—O bond lengths are 116 pm for the terminal and 120 pm for the bridging ligands. The same trend is found for the variation of C=C bond lengths by coordination. The C—C bond length of 119 pm in free diphenylacetylene is increased to 146 pm in the compound $Co_2(CO)_6Ph_2C_2$ (Sumner *et al.*, 1964). As a secondary electronic effect the C—C_{Ph} bond distance is increased from 140 to 142 pm. The C—O bonds are considerably longer than in $Co_2(CO)_8$, namely, 123 pm. The Co—C bond distance of 175 pm is shorter than that in $Co_2(CO)_8$(180 pm).

Ph
C
π π
$(CO)_3Co$ $Co(CO)_3$
π π
C
Ph

The polarizations of the coordinated C—O groups are reflected in their ability to coordinate aprotic acceptor molecules, such as trimethyl aluminum (Kotz and Turnipseed, 1970). The aluminum salt Al[π-cp)$W(CO)_3]_3$, 3THF, was shown by an X-ray single-crystal structure determination to contain three $[(\pi\text{-cp})W(CO)_3]^-$ anions, each coordinated through one carbonyl oxygen to the aluminum (Burtlich and Petersen, 1970; Petersen *et al.*, 1971). By this type of coordination a further increase in C—O bond distance is induced, as evidenced by a shift in the infrared ν_{CO} band from 1729 cm^{-1} in the parent compound $Mo(phen)(CO)_2$ to 1633 cm^{-1} in the Et_3Al adduct; similar features are found when coordination of acceptor molecules takes place at a bridging C—O group (Nelson *et al.*, 1969).

3.5. *Molecular Adducts of Organic Carbonyl Compounds and of Phenols*

The intermediate formation of molecular adducts has been recognized as decisive for the course of numerous reactions (Pfeiffer, 1922). Pfeiffer stated:

> Da die chemischen Kräfte nur in Entfernung von atomaren und molekularen Dimensionen wirken, so folgt ohne weiteres, daß 2 Moleküle, die miteinander in Reaktion treten, sich zunächst einander anlagern, worauf dann in der so entstandenen Molekülverbindung intramolekular die eigentliche Reaktion einsetzt.
>
> [Since chemical forces act only at distances comparable to atomic or molecular dimensions, it follows immediately that two molecules that react with one another first take up an appropriate relative orientation, and then the actual reaction takes place intramolecularly in the molecular adduct so created.]

The consideration of the intermediate compounds from the point of view of donor–acceptor interactions has, however, been retarded, since the suggestion by Bennett and Willis (1929) has not been accepted. They considered the union within a molecular compound by means of covalent linkages and showed this to be consistent with the chemical observations made for various types of organic reactions. Even 35 years later the following statement was made (Andrews and Keefer, 1964):

> For a variety of reasons including the fact that X-ray diffraction studies have been clearly established that the separation distances between the solid components by far exceed normal covalent bond lengths, this proposal need no longer be considered seriously.

It is obvious, however, that there are no criteria for what constitutes the limit of length of a "normal" covalent bond.

Nucleophilic addition is one of the basic reactions in organic chemistry. For example, it is involved in the formation and hydrolysis of peptides, carboxylic acid esters, carboxylic acid anhydrides, and acetals. The general course of the reaction may be represented as follows: The nucleophilic lone pair of a donor attacks the acceptor carbon of the carbonyl compound (a). The donor–acceptor interaction leads to lengthening of the adjacent bonds; e.g., when an amine attacks, the C—O and the N—H bonds are lengthened. The increase in basicity of the oxygen atom of the CO group and the increase in acidity of the N—H hydrogen atom leads to proton transfer. By proton abstraction from the N atom its donor properties are increased, while the C—O bond is weakened by having accepted the proton. Hence the elimination of a molecule of water is possible from the tetrahedral intermediate molecular adduct (b) to yield the amide (c).

$$\underset{\text{(a)}}{R'{-}\overset{H^{\delta+}\leftarrow}{\underset{H}{N^{\delta-}}}\!\longrightarrow\!\overset{O^{\delta-}}{\underset{R}{C^{\delta+}}}{-}OH} \rightleftarrows \underset{\text{(b)}}{R'{-}\underset{H}{N}{-}\overset{HO}{\underset{R}{C}}{-}OH} \longrightarrow \underset{\text{(c)}}{R'{-}\underset{H}{N}{-}C\genfrac{}{}{0pt}{}{\diagup O}{\diagdown R}} + H_2O$$

The analysis of structural data pertaining to intermolecular N→C=O interactions have shown that for decreasing N→C bond distance the intramolecular C—O bond distance increases (first bond-length variation rule), and the carbonyl carbon is increasingly displaced from the plane of its three ligands and pulled toward the donor atom (Bürgi *et al.*, 1973). A scan of the crystallographic literature revealed many examples of intermolecular and intramolecular O→C=O interactions (Bürgi *et al.*, 1974): Structural data on the donor keto oxygen O→C=O contacts in crystals reveal that for O→C bond distances shorter than 300 pm the carbon atom is displaced from the plane of the carboxyl group (with substituents) toward the attacking donor oxygen atom (Bürgi *et al.*, 1974). This intramolecular displacement from planarity (the pyramidalization or rehybridization to tetrahedral carbon) tends to increase as the intermolecular O→C bond distance decreases. The plot of the intermolecular O→C bond distance vs. the deviation from planarity Δ shows—despite considerable scatter—a definite tendency for the displacement Δ to increase with decreasing intermolecular distance. The curve obtained (Fig. 3.9) lies well below the N→C=C curve in agreement with the relative donor properties of the alcoholic or keto oxygen on one side and the amine nitrogens on the other side.

The structure of the tetrahedral subunit with two oxygen atoms attached to the same carbon atom represents an extreme case of O→C=O interaction where the addition reaction has proceeded to completion. We have a large and heterogeneous collection of examples of several types (diols, ketals, hemiketals, acetals, hemiacetals) which introduce unpredictable sources of scatter into any possible correlation of the data. The differences observed in bond lengths and bond angles, taken individually, may appear as insignificant in a statistical sense and may be due to the geometrical environment in each of the crystals. The results are,

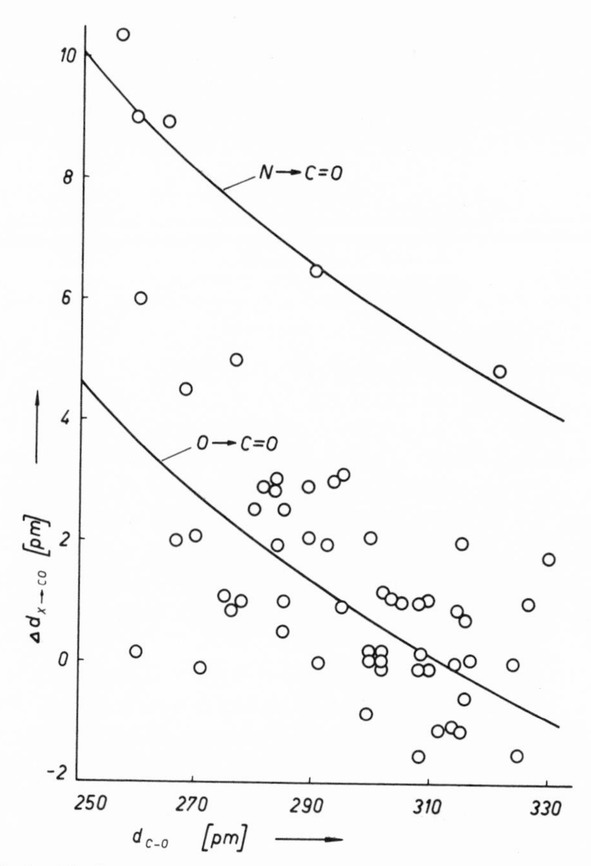

Fig. 3.9. Relationship between intermolecular X→CO and intramolecular C—O distances of acetone in the presence of various donors.

however, extremely useful in showing trends that may furnish a structural basis for the discussion of chemical reactivity (Bürgi *et al.*, 1974). For pyrrole as the hydrogen donor and a variety of carbonyl compounds, a regular correlation exists between the shift $\Delta\nu_{NH}$ and the stretching frequency of the carbonyl group ν_{CO}, with 30 compounds of widely different structure (Bellamy and Pace, 1963) showing increasing intramolecular C—O and H—N bond distances due to increasing intermolecular interaction (first bond-length variation rule):

$$\rangle C{=}O \rightarrow H{-}N \langle$$

Analogous plots have been otained for phenol (Widom *et al.*, 1957; Gramstad and Fuglerik, 1962; Middaugh *et al.*, 1964) or phenylacetylene

(Cook, 1958; Baker and Harris, 1960) as hydrogen donors. Likewise, in mixtures of tetramethylurea and phenols the O—H and C—O bond distances are longer than in the pure components, and the frequency shifts of the ν_{OH} and ν_{CO} vibrations vary proportionally to each other (Muller *et al.*, 1972):

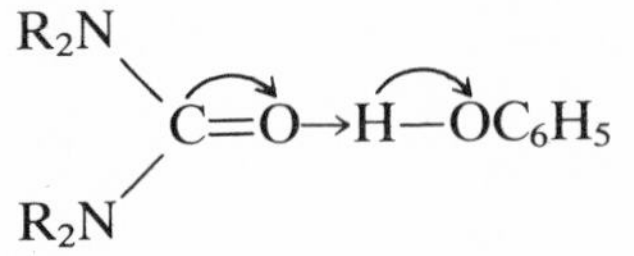

$$(R_2N)_2C{=}O \rightarrow H{-}OC_6H_5$$

Donor–acceptor interactions between Et_2NH and phenols give rise to increasing H—O bond distances and hence to increasing donor properties at the oxygen atoms as the acidity of the phenol increases (Duterme *et al.*, 1968). The situation is analogous for the triethylamine–phenol systems (Clotman *et al.*, 1970). By strong interaction with the amine, ionization may take place to give triethylammonium cations and phenoxide anions, the latter being capable of binding phenol molecules to give new "pseudomacromolecules" of definite structure (Duterme *et al.*, 1968), which may be represented as

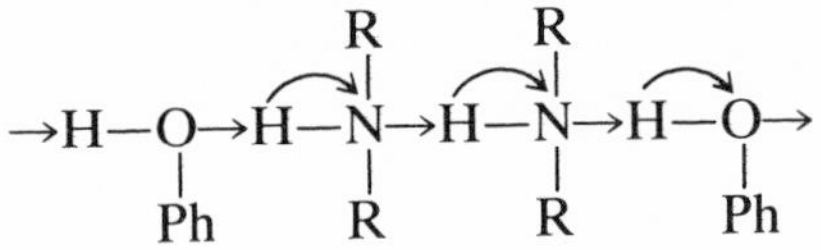

$$\rightarrow H{-}O(Ph)\rightarrow H{-}NR_2\rightarrow H{-}NR_2\rightarrow H{-}O(Ph)\rightarrow$$

The physical properties of such systems differ remarkably from those which might be expected from an ideally statistical distribution of the component molecules. For example, in mixtures of amines and phenols the viscosities and the conductivities are considerably greater than expected (Felix and Huyskens, 1975; Felix *et al.*, 1975).

For the reaction of substituted anilines with different thenoyl and furoyl chlorides in benzene a linear correlation exists between ν_{CO} and the log K value for adduct formation (Alberghina *et al.*, 1973). The most likely structure for the transition state involves the attack of the lone pair of the amino group at the carbonyl carbon atom, similar to that proposed for the benzoylation of aniline (Venkataraman and Hinshelwood, 1960):

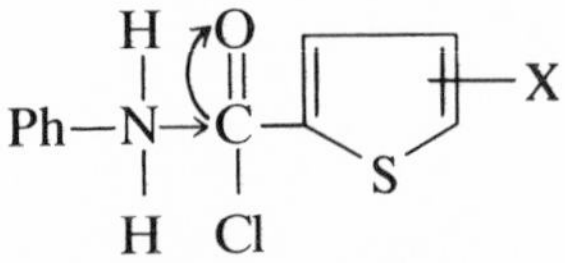

$$Ph{-}NH_2\rightarrow C(=O)(Cl){-}C_4H_2S{-}X$$

A theoretical study on the substituent effect on the thiophene ring leads to a linear relationship between the net charge at the carbon atom of the carbonyl group and ν_{CO}. The slope of this plot is the same whatever approximation is adopted and independent of the substituent position in the ring (Paoloni *et al.*, 1975). According to this theoretical correlation, a linear relationship is expected for the rate constant, and this is actually the case (see Chapter 12).

The differences in spectral parameters of acetone in different environments are related to its differences in reactivity (Bukowska and Kecki, 1975*a*): Addition of certain salts causes the stretching frequency of the C—O band to decrease, corresponding to lengthening of the C—O bond distance. The C—C—C bands increase, corresponding to shortening of the C—C bonds (Borucka *et al.*, 1973). In solutions of perchlorates the changes follow the order of increasing acceptor strength of the hard cations: The lengthening of the C—O bonds and the shortening of the C—C bonds follow the order $Na^+ < Ba^{2+} < Sr^{2+} < Li^+ < Mg^{2+}$. The integral intensity of the C—O band is increased and that of the symmetric C—C—C band decreased, the changes again following the above order. The splitting of both bands indicates that in the solutions two kinds of acetone molecules exist (Borucka and Kecki, 1971). The relative frequency changes of the ν_{1CO} in the acetone band vs. the relative frequency changes of the ν_{4CCC} band in acetone electrolyte solutions are shown in Fig. 3.10.

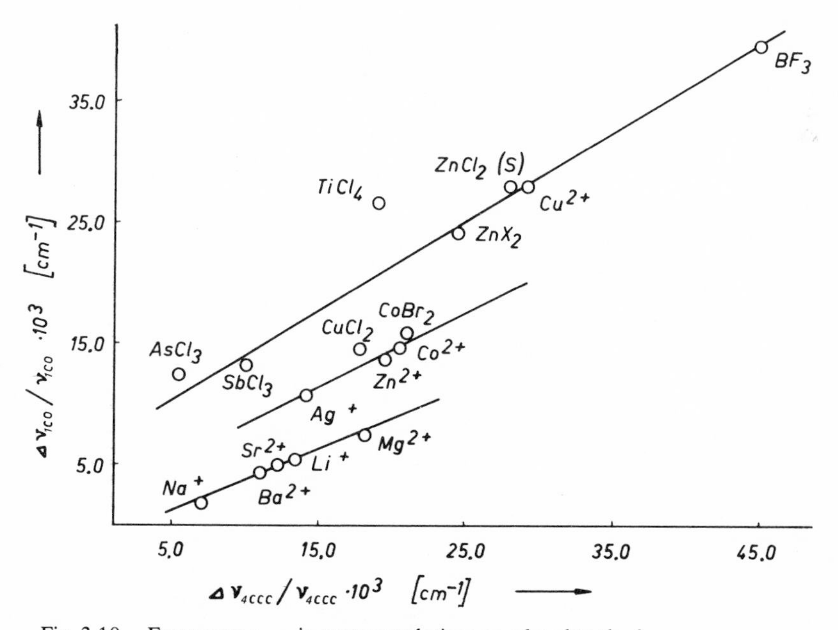

Fig. 3.10. Frequency ν_{1CO} in acetone solutions as related to the frequency ν_{4CCC}.

The solutions may be classified into three groups: (a) alkaline and alkaline-earth-metal perchlorates, (b) silver, cobalt, and zinc perchlorates, and (c) copper perchlorate, zinc halides, and several covalent halides. Condensation reactions take place only in solutions of the third group. The pronounced structural changes provoked by the solutes seem to be an essential condition for the aldol condensation to occur. Zinc halides, which are not fully ionized in acetone (Dellwaulle, 1955), are particularly suitable for the condensation reaction, which may be represented by simultaneous attack of zinc at the oxygen atom and of chloride at the carbon atom of the CO group. Copper perchlorate, which promotes the condensation reaction, causes the largest splitting of the C—O band, copper chloride, which does not, causes the smallest splitting of the C—O band (Bukowska and Kecki, 1975*b*; Kecki and Gulik-Krzywicki, 1964).

Chapter 4
Bond-Length Considerations in the Crystalline State

4.1. Molecular Lattices

The formation of a molecular lattice from the gaseous molecules may be considered due to intermolecular donor–acceptor interactions, which induce characteristic changes in intramolecular bond lengths. The structural features in molecular lattices are phenomenologically analogous to those found in molecular adducts and they can be interpreted by the first bond-length variation rule (Gutmann, 1975). While the iodine–iodine distance in gaseous iodine is 267.6 pm (Ukaji and Kuchitsa, 1966), the intramolecular iodine–iodine bond distance in the crystalline state is found to be slightly greater, namely, 272 pm (Bolhuis *et al.*, 1967). From a study of quadrupole coupling, it has been concluded that in crystalline iodine the 354 pm intermolecular contacts contain about 9% of covalent bonding (Townes and Dailey, 1952). The intermolecular contacts may be considered as inducing an increase in intramolecular iodine–iodine bond lengths as may be schematically represented for one dimension as follows:

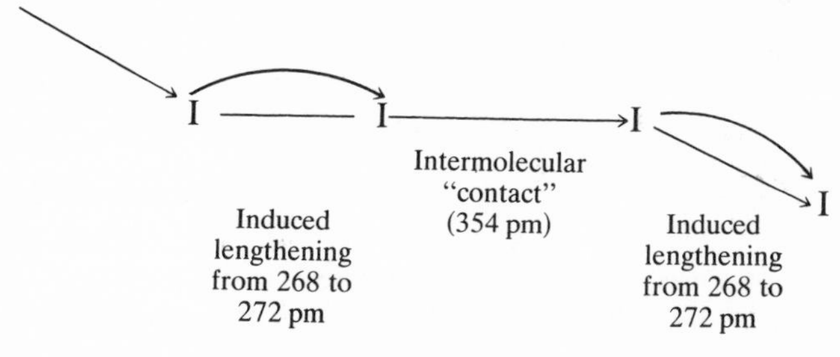

This representation describing the formation of the molecular lattice from the gaseous molecules expresses the different functionalities of iodine atoms connected by the same type of bond; it is analogous to that for the triiodide ion complex formation from the iodide ion and molecular iodine: $I^- \rightarrow \overset{\frown}{I—I}$. Since the iodide ion acts as a considerably stronger donor than molecular iodine, the induced polarization of the I—I bond is so great as to lead to equally long I—I bond distances of 293 pm in the triiodide ion (Migchelsen and Vos, 1967). The differences in donor properties of iodine and the iodide ion are also evident from their different solubilities and enthalpies of solvation in water.

The induced charge-transfer effects are, as may be expected, slightly more pronounced in iodine bromide. The I—Br bond distance is 247 pm in gaseous IBr and 252 pm in the solid state (Knobler *et al.*, 1971), in accordance with the first bond-length variation rule.

In analogy to the features in molecular adducts, we may expect an inverse relationship between the intermolecular and the intramolecular bond distances within a crystal lattice. The limiting factors are: (a) The intermolecular interactions are too weak to give measurable intramolecular effects, (b) the intermolecular interactions are so strong that inter- and intramolecular bond distances and properties become equal and hence indistinguishable.

4.2. *Ionic Crystals*

Although it may appear strange to treat an ionic crystal like a molecular adduct, this approach was used to describe the formation of an ionic crystal from the gaseous molecules (Gutmann, 1975). The changes in bond lengths incurred in the formation of an alkali halide crystal may be considered as the result of intermolecular interactions between gaseous alkali metal halide molecules and the growing crystal. Alkali metal ions are known to function as stronger acceptors and halide ions as stronger donors than molecular iodine. Hence the induced bond-length variation effects should be greater in ionic crystals than in the molecular lattice of iodine. Table 4.1 shows the metal–halogen bond distances in the crystalline lattice and in the gaseous molecules. According to the electrostatic approach the increase in internuclear distances is explained by the increasing polarizing power of the alkali metal ions from Rb^+ to Li^+ and the increase in polarizabilities from Cl^- to I^-. Increasing covalent character of the bond is accompanied by decreasing bond distance.

The donor–acceptor interpretation considers the induced cooperative effects involved in changing from the gaseous to the crystalline state as

Table 4.1. Alkali Metal–Halogen Bond Distances in Alkali Halide Diatomic Molecules and within the Crystal Lattice

Halide	M—X bond distance[a] (pm)		Δd (relative increase in M—X bond distance, %)[b]
	vapor	crystal	
LiBr	217.04	274.7	26.5
NaBr	250.20	298.1	19.1
KBr	282.07	329.3	16.8
RbBr	294.48	343.4	16.8
LiI	239.19	302.5	26.6
NaI	271.15	323.1	19.1
KI	304.78	352.6	15.8
RbI	317.69	366.3	15.1
KCl	266.66	314.7	18.0
KBr	282.07	329.3	16.8
KI	304.78	352.6	15.8
RbCl	278.68	328.5	17.9
RbBr	294.48	343.4	16.8
RbI	317.69	366.3	15.1

[a] From Wells (1962). [b] From Gutmann (1975).

based on the relative donor and acceptor properties of the ions. According to the third bond-length variation rule, the Na—Cl bonds are increased in polarity as the coordination number is increased, with the Na^+ ions acting as acceptors and the Cl^- ions as donors, as shown for one dimension only (see also Section 5.8):

$\longrightarrow Na^{\delta+} \text{———} Cl^{\delta-} \longrightarrow Na^{\delta+} \text{———} Cl^{\delta-} \longrightarrow$

Induced effect | Interaction | Induced effect

The relative increase in M—X bond distance within the bromides of different alkali metal ions is related to the differences in acceptor properties of the alkali metal cation, which are increasing in the same order as the relative M—X bond distances are increased, e.g.: $Rb^+ < K^+ < Na^+ \ll Li^+$. Likewise, for a given cation the effect of the relative donor properties of the halide ions increases from iodide to chloride in the same way as the M—X bond distances increase within the group of halides with a certain cation (Table 4.1).

The interatomic distances of alkali halide vapors obtained from early electron diffraction studies were consistently higher than those now derived from microwave spectroscopy, and this is interpreted as due to the presence of polymeric molecules in the vapor in the earlier work. Mass spectrometric analysis of the ions shows the presence of dimers, trimers, and in some cases tetramers, particularly for lithium halides (Berkowitz and Chupka, 1958).

The quantitative aspects of the electrostatic treatment for hard–hard ionic interactions may be superior to those of the present approach. On the other hand, the donor–acceptor concept describes at least qualitatively the structural changes irrespective of the interpretation of the bonding forces. Indeed, this concept can be applied to the formation of a metallic lattice. For example, the Ag—Ag bond distances in the Ag_2 molecule in the gas phase are 260 pm (Shin-Piaw *et al.*, 1966), whereas in the crystalline state the Ag–Ag distances are 289 pm.

4.3. Silicates

The variations in Si—O bond lengths in different silicates are correlated with the Pauling bond strength $\bar{s}$ (Pauling, 1929), which is defined as the valence Z of a cation divided by its coordination number v (Baur, 1971; Brown and Shannon, 1973). According to the Pauling electrostatic valence principle, the sums $P = \sum \bar{s}$ of mean bond strengths around the cations and anions are approximately equal to their valence. This concept is derived from an ionic model of chemical bonding, but it can be applied to situations where the bonding is primarily covalent. The Pauling bond strength is used as an approximate measure of the potential of the cation on the anion in an undistorted coordination site. The Pauling theory expresses the differences in covalency of Si—O bonds within the anionic groups due to varying the cationic species.

Noll (1963) has demonstrated the usefulness of the application of the electronic theory to silicates and organopolysiloxanes. His starting point is the three-dimensional net work of SiO_2. Substitution of Si^{4+} ions by cations of lower charge is reflected in variations of bond lengths and physical properties. This was explained in terms of variations of inductive effects due to differences in coordinating interactions between the metal cations and the silicate anions. The Si—O bond distance is shortest in SiO_2, namely, 160 pm because $\geqslant$Si—O$\leftarrow$ groups are strongly coordinated by Si^{4+} ions. Substitution of the latter by ions of lower charge leads to a decrease in the coordinating interaction, since the acceptor properties of the metal ions decrease with decrease in charge, e.g., the —O—Si bond is stronger than the —O—Al bond. With reference to the acidity of Si^{4+} the substituted metal ions are considered as electron donors toward the O—Si bond, and this is related to the decrease in Si—O bond distance. As a secondary effect, an increase in the Si—O—Si bond distances is observed, both effects increasing with increasing Noll's "electron donor" properties of the metal ions, e.g., with decreasing acceptor properties.

Since from the point of view of coordination chemistry a metal ion is considered an electron acceptor rather than an electron donor, Noll's correct ideas may be reformulated in the following way: The Si—O bond length in the hypothetical SiO_4 group should be relatively small and it is expected to be increased by increasing acceptor strength of the cation. In other words the "intermolecular" $—O{\rightarrow}M^{n+}$ bond strength should be related to the acceptor property of the cation and according to the first bond-length variation rule inversely related to the "intramolecular" adjacent Si—O→ bond length.

In disilicates this effect induces further a decrease in Si—O—Si bridging distances as the result of secondary inductive effects:

M^{n+} M^{n+} M^{n+}

O—Si—O —Si—O—Si—

Increase in Si—O bond distances by increasing acceptor properties of M^{n+} (primary inductive effect)

Decrease in Si—O—Si bond distances by increasing acceptor properties of M^{n+} (secondary inductive effect)

The differences in bond length are reflected in differences in chemical reactivities. Small differences in bond length may provide remarkable differences in chemical behavior. Acetic acid heterolyzes the Si—O—Si bonds to a greater extent in wollastonite than it does in hypersthene (Weitz *et al.*, 1950). Both compounds are chain silicates that differ, apart from the

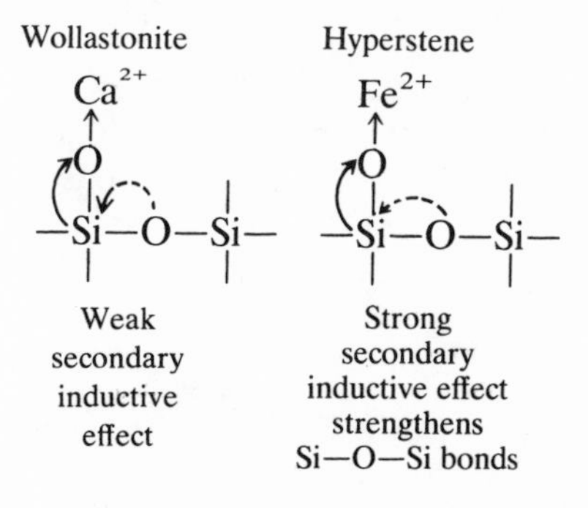

periods of the $[SiO_3]_n^{2n-}$ chains, by the fact that wollastonite contains calcium ions which are weaker electron-pair acceptors than the Mg^{2+}—Fe^{2+} ions in hypersthene. Accordingly, in the lattice of wollastonite, the Si—O—Si bonds are more strongly polar, the Si—O—Si bond distances are greater, and the terminal O—Si bond distances are smaller than in hypersthene. The higher electron density at the bridging oxygen atoms in wollastonite provides a higher basicity than in hypersthene. This allows stronger acceptor attack by hydrogen ions in wollastonite, by which the Si—O bonds within the chain are lengthened and finally heterolyzed:

$$-\mathrm{Si}-\underset{\downarrow\atop H^+}{\mathrm{O}}-\mathrm{Si}-$$

Likewise, the old mineral chemical fact that decomposition of plagioclase by acids is increased by increasing aluminum content and is greater than that of potash feldspar, shows the strong influence exerted by substitution of Si^{4+} by Al^{3+} in the anionic group (Noll, 1963): The basicities of the oxygen atoms are increased. A further example is the fact, which has also been known for a long time, that common augites, in which $[Si, Al]_2O_6$ chains are present, are considerably more readily cleaved by acids than enstatite $Mg_2[Si_2O_6]$. Investigations on the decomposition by acid of potash feldspars have shown that contrary to earlier views, acids not only dissolve the potassium ions out of the lattice, but, in addition, decompose the $[Si, Al]O_2$ network (Noll, 1963). Both the case of heterolytic cleavage of the silicate anion framework and the high degree of dispersion of the dissolved silicic acid are related to the relatively strong polarity of the (Si, Al)—O—Si bond in the feldspar lattice (Noll, 1963).

The bridging oxygen atom in silicates may be compared to that in siloxanes. Hexamethyldisiloxane is attacked by acceptors such as BF_3, while hexachlorodisiloxane remains unaffected (Noll, 1963): In the former the Si—O—Si bonds are more strongly polar and hence the bridging oxygen atoms are stronger donors than in the latter compound:

$$(CH_3)_3Si-\underset{\downarrow\atop BF_3}{O}-Si(CH_3)_3$$

Noll (1963) considers organopolysiloxanes as "weakened silicate melts," which are subject to attack by hydroxyl ions:

$$\geqslant\mathrm{Si}-\mathrm{O}-\underset{\uparrow\atop OH^-}{\mathrm{Si}}\leqslant \rightleftharpoons -\mathrm{Si}-\mathrm{O}^- + \mathrm{HO}-\mathrm{Si}\leqslant$$

$$\geqslant\mathrm{Si}-\mathrm{O}^- + \mathrm{K}^+- \rightleftharpoons \geqslant\mathrm{Si}-\mathrm{OK} \xrightarrow{\mathrm{HOSi}\equiv} \geqslant\mathrm{Si}-\mathrm{O}-\mathrm{Si}\leqslant + \mathrm{KOH}$$

Noll (1963) has also given an electronic interpretation of the magmatic crystallization. The Si—O bond is most stable in SiO_2 and is inductively labilized by substitution of Si^{4+} by cations of lower acidity. In the course of the crystallization of naturally occurring silicatic magmas the formation of crystalline phases appears to have taken place in a sequence which is characterized by the smallest possible inductive weakening of the Si—O—Si bonds by the metal ions taken up.

In silica-rich melts polyanions are present and this is in agreement with the mineralogical phenomenon that silicate crystals with anions of high dimension number* are more abundant than isolated groups.†

4.4. *Bond-Length Variability in Complex Compounds*

The remarkable distortability of coordination polyhedra of copper(II) complexes has been reviewed recently (Gažo *et al.*, 1976). Correlations exist between axial and equatorial bond lengths d_a and d_e respectively, in CuO_6, CuN_6, and CuN_4O_2 chromophores, in that decreasing d_a is reflected in increasing d_e and vice versa (Fig. 4.1). It has been concluded that transition between the various types of coordination in the chromophores CuO_6 is practically continuous in the range $d_a = 190$ to 350 pm and $d_e =$ 190 to 215 pm. This is a manifestation of the distortability of the coordination polyhedra of Cu(II) (Gažo *et al.*, 1976) following the first bond-length variation rule.

This phenomenon is also found for Cu(II) complexes with different ligands. As may be seen from Fig. 4.2 an analogous relationship exists for square planar complexes in the crystalline state. Here the axial interactions correspond to intermolecular interactions and those within the planar units correspond to intramolecular interactions (Gažo *et al.*, 1976):

```
   Br   ¦   NH3
     \  ¦  /
       Cu
  de  / ¦ \
 H3N    ¦   Br
    da  ¦    ¦
        ¦  da¦
        Br   ¦   NH3
          \  ¦  /
            Cu
           / ¦ \
       H3N   ¦   Br
```

*The dimension number denotes how many dimensions of the molecule are infinite.

†This is analogous to the higher stability of water aggregates than that of monomeric molecules in the condensed states (see Section 6.1).

The variations in bond lengths in peroxotitanium(IV) dipicolinates, with colors ranging from deep red to yellow-orange, have been related to their colors (Schwarzenbach and Girgis, 1975). The relationship between the bond-length variations within the structural units is indeed in consonance with the first bond-length variation rule: The greater the donor property of the axial ligand, the shorter the axial bond and the longer the titanium–peroxide bond.

For octahedral complexes of the type SnX_4L_2 a correlation exists between the Sn—X stretching frequency and the strength of the Sn—L bond: The stronger the Sn—L bond, the lower the Sn—X frequency (Ohkaku and Nakamoto, 1973). The observed frequency sequence for

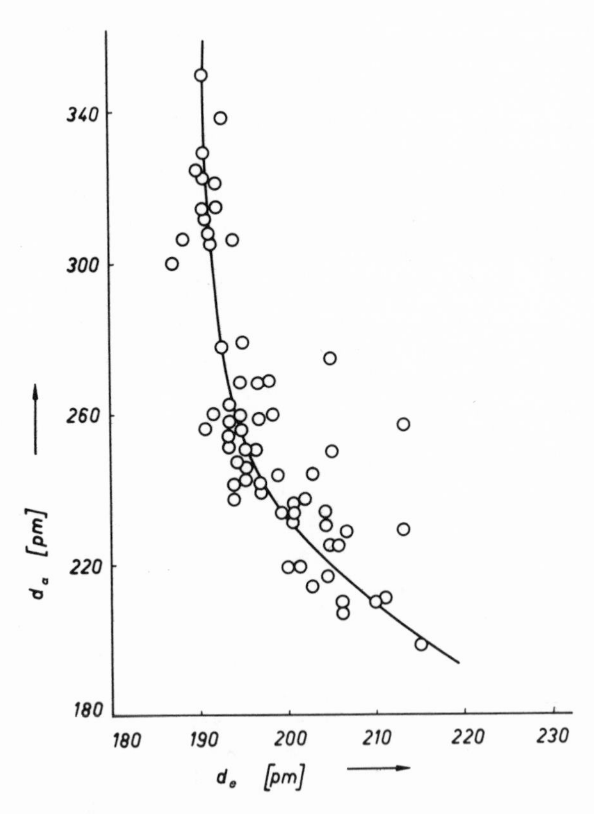

Fig. 4.1. Relationship between axial distances d_a and equatorial distances d_e in CuO_6, CuN_6, and CuN_4O_2 chromophores (from Gažo *et al.*, 1976, courtesy of Elsevier Publishing Company, Amsterdam).

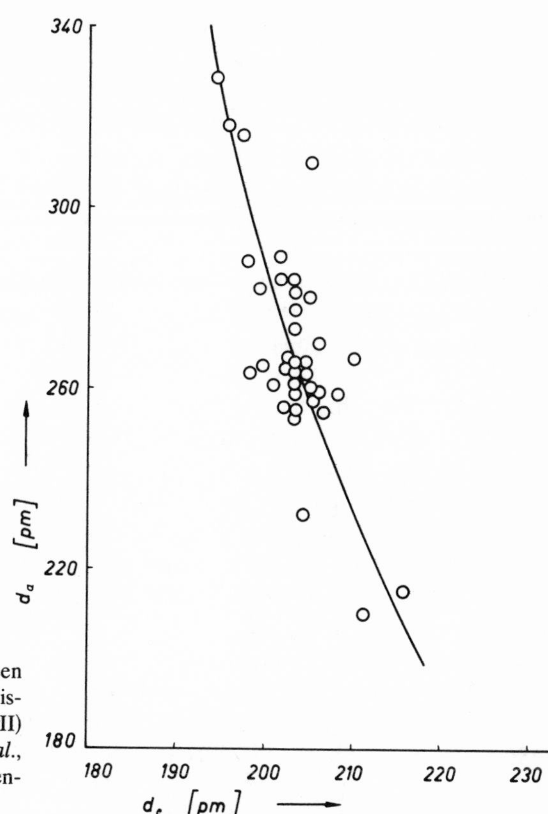

Fig. 4.2. Relationship between distances d_a and equatorial distances d_e in crystalline copper(II) complexes (from Gažo *et al.*, 1976, courtesy of Elsevier Scientific Company, Amsterdam).

L = Et_2O ≈ THF > MeS, Et_2S > Me_2S suggests that the Sn—L strength increases in going from oxygen to selenium donors (Ruzicka and Merbach, 1976).

4.5. *Bond-Length Variations under Pressure*

Many materials, whether they have originally ionic, nonionic, or molecular lattice structures, are transformed into the metallic state by the application of a sufficiently high pressure, and indeed this can be expected to be true of all materials. Even quite modest increases in pressure can affect interatomic distances, spectral transition, and formal oxidation states, and many other phenomenological parameters, e.g., can increase the coordination number. Various attempts have been made in an effort to establish relationships between pressure and these phenomenological parameters, but none of them accounts satisfactorily for all of the observed features. This is almost certainly because of the absence up to now of a model which is capable of interpreting the facts without concerning itself with too detailed an interpretation of the binding forces. However, it will be shown here, after a brief survey of the present situation, that the extended donor–acceptor concept seems to successfully provide such a model (Gutmann and Mayer, 1976).

The most useful rule in describing the effect of pressure on solids is the so-called "pressure-coordination rule" (Neuhaus, 1964), according to which the coordination number is increased with pressure. Various crystal-structure transformations follow this qualitative rule at different pressures and temperatures. An exception to this rule is known for ytterbium (Hall *et al.*, 1963): The cubic face-centered modification (coordination number = 12) of the metal is transformed at 40 kbar into a cubic space-centered structure (coordination number = 8).

The *pressure–homology rule* states that the crystal structure of the heaviest homolog is obtained for the lower homologs by an increase in pressure (Neuhaus, 1964). For example, the crystal structure of cesium chloride (coordination number = 8), stable at room temperature, is obtained for NaCl, KCl, and RbCl under pressure. The required pressure is greater the lighter the constituents. The pressures required increase in the series RbCl < KCl < NaCl as the cation radius is decreased (Klement and Jayaraman, 1967). The same trends are seen in the more covalent series, InSb < GaSb < AlSb < and ZnTe < ZnSe < ZnS (Klement and Jayaraman, 1967).

In molecular lattices, chain lattices, and layer structures, the coordination numbers are found to be increased by applying pressure, and at very high pressure a highly coordinated metallic packing is obtained. For example, phosphorus is converted into a structure in which its homologous element, antimony, is stable at ordinary pressure (Jamieson, 1963*b*).

Within the homologous series, C—Si—Ge—Sn, all of the elements are stable (or metastable) in the diamond structure (coordination number = 4). Pressure converts these nonconductors or semiconductors into metallic modifications with a coordination number of 6 [which is the stable modification of tin at room temperature and atmospheric pressure (Alder and Christian, 1961; Jamieson, 1963*a*; Barnett *et al.*, 1966)].

The basic effect of pressure is of course to increase the density of the material. Hence a decrease in bond lengths should be expected. The *pressure–distance paradox* expresses the fact that this may not be true of *all* interatomic distances, since the increase in pressure may in fact cause a *lengthening* of the shortest internuclear distances in the crystal (Kleber, 1967). For example, the transformation of coesite (SiO_2), with a coordination number of 4, into stishovite (SiO_2), with a coordination number of 6, is accomplished at $\gg$100 kbar and at $\gg$1200°C (Stishov and Popova, 1961; Akimoto and Synono, 1969). As expected the density increases (from 2.93 to 4.28 g/cm^3, i.e., by 46%) but the Si—O bond lengths are *increased* (from 161.3 pm in coesite to 177.8 pm in stishovite, an increase of 10.2%). Another well-known example of this paradox is the transition of graphite into diamond (Bundy, 1962): The C—C distances in diamond are 8.7% *greater* than those within the graphite layers, although the density is of course considerably increased in going from graphite to diamond due to the far greater *shortening* of the interlayer distances initially present in graphite.

In order to account for this effect the electronic transitions provoked by increase in pressure must be considered, and it is interesting to note that changes in electronic energy levels are found to occur in charge-transfer complexes as well as in transition metal complexes. Jørgensen (1959) has analyzed a number of spectra of hexacoordinated transition metal ions in relation to pressure. In K_2OsBr_6 the peaks at 22,200 and 16,750 cm^{-1} are shifted distinctly to lower energy with increasing pressure; at the same time the splitting increases drastically. Spectroscopic changes occur as a result of the pressure-induced transformation of tetrahedral cobalt(II) complexes in solution into octahedral arrangements (Lüdemann and Franck, 1967, 1968), and indeed are used as evidence for this transformation. In many compounds of iron(III), the application of pressure causes a change in the spectroscopic oxidation state of the iron(II); i.e., an apparent reduction is observed. The mechanism of this reduction is thought to involve the transfer of an electron thermally from a ligand nonbonding level to the metal d_π orbitals (Drickamer and Frank, 1973).

We may consider the action of pressure as paralleling that of an electron donor, in that the electron density is increased in internuclear areas of initially low-electron density. This formal analogy allows the application of the bond-length variation rules (Gutmann and Mayer, 1976).

We may start from the third bond-length variation rule, which relates the increase in coordination number to the increase in adjacent bond lengths. In a molecular lattice a distinction between intermolecular and intramolecular bonds is made. In a layer lattice, such as graphite, an analogous distinction may be made between interlayer bonds and intralayer bonds (Gutmann and Mayer, 1976).

In both of these cases the concept of induced bond polarization may be applied as follows: Pressure increases the electron densities in the first place along the intermolecular and interlayer bonds, respectively, which are considerably shortened and hence a geometrical rearrangement with increase in coordination number may take place. Either the shortening in intermolecular bond length or the increase in coordination number induces lengthening of the intramolecular or intralayer bond lengths, respectively. The observed increase in density is due to the fact that the decrease in intermolecular bond lengths is greater than the induced increase in intramolecular bond lengths. From this point of view the pressure–distance paradox is a logical consequence of the donor–acceptor approach!

In crystalline iodine, pressure leads to changes in macroscopic properties characterized by an increase in metallic properties. This suggests an increase in coordination number with simultaneous decrease in intermolecular and increase in intramolecular bond length, to the extent that both become equally long. The decrease in intermolecular bond length is expected to be greater than the increase in intramolecular bond length, as is observed for the formation of the triiodide ion from iodine and the iodide ion. In crystalline iodine the intermolecular I—I bond distance is 354 pm and the intramolecular I—I bond distance is 272 pm (Bolhuis *et al.*, 1967). The formation of the I_3^- ion may be considered as due to the replacement of an I_2 molecule by an iodide ion: $I^- \rightarrow I—I$. Although the ionic radius of the iodide ion is considerably greater than the covalent radius of the iodine atom in the molecule, the "intermolecular I—I bond distance" is decreased to a greater extent, namely, from 354 to 293 pm, than the intramolecular I—I bond distance is increased, namely, from 272 to 293 pm.

Structural data are available for the high-pressure transformation of graphite into diamond (Kleber, 1967). In the graphite structure the intralayer C—C bond distances may be treated as intramolecular bonds and those between the layers as intermolecular "bonds." The intermolecular distances of 335 pm in graphite are drastically decreased to 154.4 pm when transformation into the diamond structure is taking place, while the intramolecular bond distances are lengthened from 142 to 154.4 pm, and at the same time the coordination number is increased from 3 to 4. The intramolecular gain in bond length is by far less pronounced from 3 to 4. The intramolecular gain in bond length is by far less pronounced than the decrease in intermolecular bond length and hence the density is increased from 2.22 to 3.51 g/cm^3.

The situation is analogous for the transformation of the hexagonal B-12 type structure of boron nitride into the cubic B-3 (zincblende) structure (borazon) at ≧60 kbar and 1300°C. The interlayer distances are drastically decreased from 334 pm to 157 pm, while the intralayer B—N distances are less dramatically increased from 145 pm to 157 pm (Kleber, 1967). The net effect accounts for the increase in density from 2.30 to 3.45 g/cm^3.

In white phosphorus the intramolecular P—P bond distances are 221 pm and the intermolecular distances considerably longer. At 12 kbar black phosphorus is produced, with the intramolecular distances being increased to 222, 224, and 231 pm, although no change in coordination number is involved. The interlayer distances in black phosphorus are 359 pm. When this is transformed into the As-type structure the interlayer distance is reduced to 327 pm with the coordination number unchanged. More remarkable changes occur when the coordination number is increased from 3 to 6 under a pressure of 125 kbar: All of the P—P internuclear distances are found to be 238 pm, and this smaller than the interlayer bond lengths and greater than the intralayer bond lengths in black phosphorus (Jamieson, 1963*b*; Krebs, 1968).

In a coordination lattice, the effect of pressure will lead to increasing bond distances when the coordination number is increased. This may be considered as the result of the shortening of the homonuclear distances within the crystal lattice, which are usually not considered as chemical bonds. If these homonuclear distances are treated like intermolecular bonds, the same approach as presented for a molecular lattice may be applied.

Both zinc sulfide and zinc selenide are converted from the zincblende type (coordination number = 4) into the sodium chloride structure (coordination number = 6) at 117 kbar and 100 kbar, respectively. In zinc sulfide the "intramolecular" Zn—S bond distance is increased from 233.9 pm to 250 pm when the "intermolecular" Zn—Zn bond distance is decreased to a greater extent, namely, from 382 to 353 pm. In zinc selenide the Zn—Se bond distance is increased from 246 pm to 254 pm and the Zn—Zn bond distance is considerably decreased, namely, from 401 to 359 pm (Smith and Martin, 1965).

In the wurtzite and zincblende structures (coordination number = 4) of silver iodide, stable at atmospheric pressure, the Ag—I bond distances are 278 pm and 280 pm, respectively. At 4 kbar the NaCl-type structure (coordination number = 6) becomes stable, in which the Ag—I bond distances are remarkably greater, namely, 303 pm. Increase in pressure to 100 kbar contracts all of these distances to 283 pm with a decrease in molar volume from 33.8 to 27.6 $cm^3 \cdot mol^{-1}$. Thus the transformation with an increase in coordination number from the low-pressure modification into the high-pressure modification involves an increase in Ag—I bond distances by 8.4%, but further increase in pressure does not produce another geometrical rearrangement and hence all of the equidistant bonds within the crystal lattice are shortened (Basett and Takahashi, 1965).

At ordinary pressure, silicon displays the coordination number of 4 toward oxygen in most of the silicate structures. The transformation of

SiO_2 into an octahedral arrangement is another example for the interpretation of the pressure–distance paradox by the bond-length variation rules. Coesite (coordination number = 4) is converted into stishovite (rutile type, coordination number = 6) at ≧100 kbar and 1200°C (Stishov and Popova, 1961; Akimoto and Syono, 1969). By this process both of the homonuclear distances are decreased, namely, the Si—Si bond distances (intermolecular) from 301 to 267 pm, which is close to that in oxygen-free K_4Si_4 (Si—Si bond distance = 246 pm) (Busmann, 1961), as well as the O—O bond distances (intermolecular) from 263 to 251 pm. On the other hand the "intramolecular" Si—O bond lengths are increased by 10.2% from 161.3 to 177.8 pm (Zoltai and Buerger, 1959). The latter is overcompensated by the shortening of the Si—Si and O—O bond distances and hence the density is increased by 46.1% from 2.93 to 4.28 g/cm^3. On the other hand, the Si—O bond distance in the quartz–coesite transformation (Zoltai and Buerger, 1959) is only slightly increased from 160.7 to 161.5 pm when the coordination number of 4 remains unchanged; the density is increased from 2.65 to 2.93 g/cm^3.

In magnesium silicates under normal pressure, the coordination number is 6 and the Mg—O bond distance is 210 pm. Pyrope, $\overset{[8]}{Mg_3}(Al\overset{[6]}{Mg_{0.5}}Si_{0.5})(SiO_4)_3$, is synthesized at 30 kbar. In this structure Mg has the coordination number of 8 toward oxygen and the Mg—O bond distance is 225 pm (Ringwood, 1967).

It has been mentioned that NaCl is converted from the six-coordinate structure into the eight-coordinate cesium chloride type by pressure. Indeed the Na—Cl bond distances are greater in the latter than in the former. Precise structure data are available for the NaCl-type and CsCl-type structures of rubidium chloride, the latter being formed by a pressure of 11 kbar at 25°C. In the six-coordinate structures the (intramolecular) Rb—Cl bond distance is 329 pm and the (intermolecular) Rb—Rb bond distance is 465 pm while in the eight-coordinate structure the former is greater, namely, 371 pm, and the latter is smaller, namely, 429 pm (Wyckoff, 1965).

Application of pressure to metallic structures of coordination number = 12 usually results in lattice contraction with simultaneous delocalization of electrons. For example, in the cubic face-centered cerium metal the internuclear distances of 362 pm are contracted to 340 pm under pressure at 20 kbar.

The situation is paralleled in asymmetrically hydrogen-bonded solids, such as ice. IR measurements show that an increase in pressure causes shortening of the O···O bond distance, and a decrease in O—H stretching frequencies (Hamann and Linton, 1975), corresponding to an increase in O—H internuclear distances (see also Section 6.2).

Chapter 5
Interface Phenomena

5.1. Lattice Contraction at Clean Crystal Surfaces

Many natural and technological processes take place within or on the surfaces of ordered aggregates or crystals. A crystal surface is characterized by the discontinuation in structural features. It may be considered as an area of lattice distortion capable of appreciable lattice deformations (Boehm, 1966). The coordination number of a surface atom is smaller than that of an atom within the crystal lattice. Application of the third bond-length variation rule implies a shortening of the bonds originating from the surface atoms. This conclusion should also be applicable to the surface area of a liquid and hence this model may be useful in obtaining a structural interpretation for the phenomenon of surface tension.

Lennard-Jones (1928) predicted that lattice parameters near and perpendicular to the surface should be smaller as compared to those within the crystal lattice, but unfortunately this was not confirmed by early experiments (Finch and Wilman, 1937). Indeed truly clean surfaces are hard to obtain, since the highly reactive surface area tends to react with constituents of the atmosphere such as oxygen or water. Under normal conditions many surfaces are covered by oxides or hydroxides, by which the effect of bond contraction at and beneath the surface area is greatly distorted. Strong adsorption (strong intermolecular interaction) may even lead to greater bond length than within the bulk of the crystal lattice (first bond-length variation rule).

When the first experimental indication for lattice contraction in alkali metal halide microcrystals—although not entirely convincing—was obtained by means of electron-diffraction experiments (Boswell, 1951), no reference was made to the postulate of Lennard-Jones.

Further experimental evidence has been given by the results of structure determinations of surfaces cleaned by ion bombardment and the annealing method (Lander and Morrison, 1963). They showed that the surface atoms are rearranged from bulk positions especially for the diamond structures of the semiconductors germanium and silicon. At a clean cleavage plane of silicon the Si tetrahedra are distorted to that extent that the interatomic distances between surface Si atoms are considerably smaller than within the crystal lattice. Low-energy electron-diffraction patterns obtained from Ge(100) and Ge(111) clean surfaces suggest for the latter an atom defect of about 25% in the top layer.

More detailed information has been provided by the LEED technique within the last few years. For aluminum the data for the contraction of the (110) plane vary between 5 and 15% (Laramore and Duke, 1972; Martin and Somorjai, 1973; Aberdam *et al.*, 1975, 1976). In Jagodzinski's laboratory, for the upper layer of the silver (110) plane a real contraction of 6 to 7% has been found (Wolf, 1972; Alff, 1976); namely, 134 pm as compared to 144.6 pm in the bulk. On gold a hexagonal superstructure is formed with a slight contraction normal to the surface from 249 to 239 pm (Moritz, 1976). The surface contraction by 4% implies that the upper hexagonal layer no longer remains plain. For molybdenum the upper (001) plane distance is contracted by about 11.5% (Ignatiev *et al.*, 1975). The interplane distance, $d_z^s = 139$ pm, corresponds to a Mo—Mo distance of 262 pm along the (111) direction, as compared to 272 pm in the normal lattice, so that the internuclear distance in the (001) surface plane is 3.7% smaller than within the bulk crystal. Van Hove and Tong (1976) reported for wolframium (100) a contraction of the upper-layer distance of 10 pm. On both the titanium and zinc (0001) surfaces the distances between the two outermost layers are contracted by 2% (Shih and Jona, 1976; Unertl and Thapliyal, 1976).

The results for copper indicate a small contraction in the (111) planes (Laramore, 1974; Kleiman and Burkstrand, 1975). Marcus *et al.* (1975) reported for the (110) surface plane a contraction of 5%, which has been corrected by the same group of authors to even a slight expansion (Demuth *et al.*, 1975).

On a fresh (111) cleavage plane of silicon the tetrahedra are distorted to the extent that the Si—Si bond distance is contracted from 384 to 260 pm. The Si atoms in the neighboring layers appear also to be affected, although to a considerably smaller extent. Heating to 700°C leads to the development of a surface structure with "warped" six-membered rings (Lander and Morrison, 1962). Within the rings the Si—Si bond distances are 220 pm as compared to a normal value of 235 pm (Lander and Morrison, 1963; Lander, 1964).

For lithium fluoride some indications show a contraction of 10 pm between the two outermost fluoride layers and for a contraction of 35 pm of the upper lithium layer; hence the uppermost lithium and fluoride layers are no longer in plane, since they have been separated by 25 pm (Laramore and Switendick, 1973). The investigation of the ZnO (wurtzite-type) surface planes revealed a contraction of 20 pm for the outermost layer distance on the (0001) plane (Lubinsky and Duke, 1976).

The surface structural features are related to their reactivities, since decreasing internuclear distances at the surface correspond to increasing surface energies and hence to increasing reactivities.

5.2. Crystal Size and Lattice Deformation

The lattice parameters at the surface do not change abruptly with the bulk dimensions. Instead there is a smooth transition zone consisting of layers which differ but slightly from each other as the electronic effects are transmitted from the surface through an appreciable number of layers. The transition zone will depend on the polarizabilities of the bonds as well as on the crystal size. As the particle size is decreased, the surface area is increased in relation to its volume: The lattice contraction will be greater and the area of induced lattice deformation will become greater.

It has also been suggested to call such lattice changes internal pressures (Schroer *et al.*, 1970): The crystal can be considered under increasing stress as its surface area is expanded relative to its volume. The lattice distortions can therefore be expected to increase with decreasing particle size. The number of layers which can be distorted, e.g., the size of the transition zone, has not been measured directly.

Even in an inert gas crystal the second and deeper layers contribute a certain amount to the surface energy (Alder *et al.*, 1959). In an alkali halide crystal the ions are significantly displaced from their normal position down to the fifth layer (Benson *et al.*, 1961), and even in the fifth layer the ions are in a different energy state from that in the body, which implies that they do make a contribution to the surface energy (Brunauer, 1965). This sort of distortion was suggested long ago (Weyl, 1951). It is also known that any two preparations of the same crystalline material never possess exactly the same total energy. For instance, there is a relationship between the heat of dissolution of tobermovite gel (a calcium silicate hydrate) and the specific surface area.

Experimental information about the size of the distorted transition zone is provided by the work of Weiss (1975), who has shown that the information contained at the surface of an aluminosilicate structure, as determined by appropriate LEED experiments, is transmitted to further layers of SiO_2 when they are allowed to grow on the surface. The remarkable result of this study is that the original information was not completely lost before the formation of 20 molecular layers of SiO_2.

The size of the transition zone will also depend on the extent of distortion in the upper layer: The smaller the crystal, e.g., the greater the surface relative to the volume, the greater the distortion. Hence a microcrystal consisting of a small number of layers should never reach the lattice dimensions of the bulk crystal!

Boswell (1951) measured the lattice parameters of very small crystals of alkali metal halides by means of electron diffraction and compared them to those found in crystals of "normal" size. Below 15,000 pm of particle size the lattice parameters were found to decrease as the particle size became smaller (Fig. 5.1). Lattice contractions in microcrystals have been observed by several other authors (Berry, 1952; Garvie, 1966, 1967; Karioris *et al.*, 1967; Schroer and Nininger, 1967; Mays *et al.*, 1968) and a theoretical foundation has been provided by Koutecký (1964).

Thus a microcrystal can no longer be considered as the regular repetition of elementary cells of absolutely identical dimensions, since the lattice parameters vary slightly in distance from surface to center. For very small crystals the dimensions of the elementary cells may even be in the center somewhat smaller than within the macrocrystal of the same chemical composition.

I must therefore state that microcrystals of identical analytical composition differ in structural parameters and hence in chemical properties as long as they differ in size. The chemical potential of a microcrystal is therefore not only a function of its chemical composition and structure type but also of its particle size. The smaller the microcrystal the more its structural individuality is developed. Grinding of crystals below a certain particle size leads to a variety of individually different crystallites. The increased reactivity by decrease in particle size is applied in powder metallurgy, where the compressed particles are made to fuse together well below the melting point. Use is also made in mineral processing. For example, $CuFeS_2$ is leached more readily after attrition grinding. The increased reactivity with decreasing particle size is also responsible for the fact that a certain crystalline material cannot be ground below a certain particle size (the so-called *milling equilibrium*).

A surprising result obtained from the impact milling of crystalline mineral substances derives from the fact that the impact milling not only comminutes the substances but also mechanically disturbs or affects the lattice structure. There appears to be a strong stressing, distortion, or defect formation in the crystalline particles resulting from impact milling

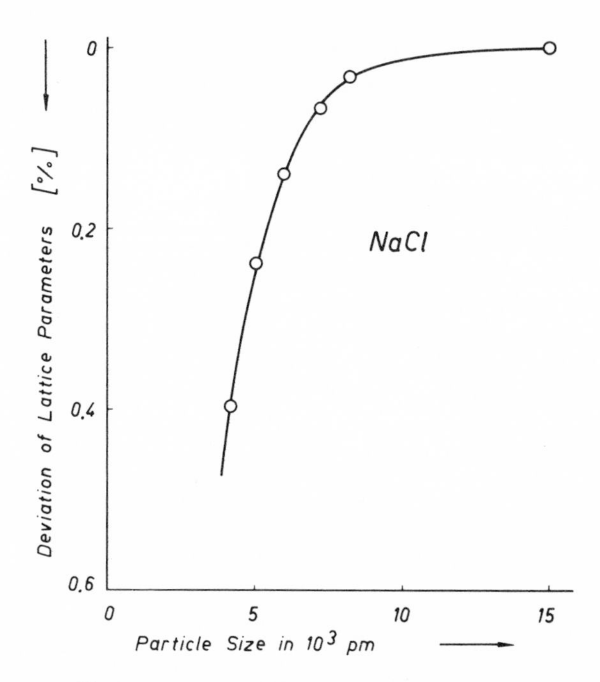

Fig. 5.1. Decrease of lattice parameters of NaCl crystals with decrease in particle size (from Boswell, 1951, courtesy of The Institute of Physics, London).

(Gerlach *et al.*, 1973). As the grinding time in the oscillating mill increases, a strong line broadening of the X-ray pattern is observed: The longer the treatment in the mill, the greater the distortions of the crystal lattice.

Chalcopyrite after treatment in the oscillating mill can be extensively dissolved at redox potentials of about 320 mV, while dissolutions of samples pretreated in other ways require much stronger oxidizing conditions.

The mechanical activation generates not only distortions, but also defects in the crystal lattice, and the chalcopyrite takes up so much energy that fast dissolution in dilute H_2SO_4 is possible. The crystal-lattice defects will also depend on particle size and on particle shape as the position of the disturbance relative to the phase boundary will influence the structural surface pattern.

The so-called "mechanochemical" reactions (Hedvall, 1952; Peters, 1962) are another manifestation of increased reactivity by decrease in particle size. For example, grinding of a mixture of dry dimethylglyoxime and nickel carbonate gives red-colored nickel dimethylglyoxime (Peters and Pajakoff, 1962).

It is therefore apparent that the structural and chemical properties do not only depend on the state of aggregation (Chapter 4): The particle size must be considered both in the solid and in the liquid state.

5.3. Adsorption Phenomena

The formation of an ad-layer is the first step in adsorption and crystal-growth processes. Two distinct types of adsorption are usually differentiated, namely, physical and chemical adsorption (Hair, 1967; Little, 1966). Physical adsorption or physisorption is regarded as due to forces of the van der Waals type. The bonds between adsorbent and adsorbate are weak, usually of the order below 20 $kJ \cdot mol^{-1}$. This is of the same order of magnitude as the heat of vaporization of the adsorbate, lending credence to the concept of weak physical bonding (Hair, 1967). Chemical adsorption or chemisorption is characterized by high heats of adsorption and energies of activation usually of the order of 60 to 80 $KJ \cdot mol^{-1}$.

Infrared investigations (see below) revealed the same characteristic structural changes in the adsorbate for both types of adsorption differing only in magnitude. For this reason any choice of a border line between these two kinds of interactions is rather arbitrary! An exact differentiation between physisorption and chemisorption is therefore unprofitable and will not be pursued in this presentation.

Our present knowledge on adsorption is still primitive since it is a much more complex process than imagined in the past (May, 1970). The present situation has been characterized by Yates (1974):

> The outcome of the joint efforts between chemists and physicists, experimentalists and theorists will be a new level of understanding of the electronic nature on solid surfaces and adsorbed layers, a level comparable to that of understanding in structural inorganic and organic chemistry.

A large variety of problems related to the nature of adsorption processes have been studied by infrared spectroscopy and in principle, at least, it is possible to determine from spectral data in detail the chemical functionality of a surface (Leftin and Hobson, 1963). Adsorption on a surface—an intermolecular interaction—leads to increasing intranuclear distances both in the adsorbate and in adsorbent. We shall first discuss the effects on the surface of the adsorbent.

The surfaces that have been most studied by the infrared technique are those of silica materials. Even after careful dehydration hydrogen atoms remain on the surface and these are referred to as "free hydroxyl groups" or "isolated silanol groups" (Benesi and Jones, 1959). The terminal hydrogen atoms are acidic and readily adsorb basic materials. Even benzene is adsorbed due to the interaction between a surface hydroxyl group and the donor π-electron system of the aromatic molecule. The spectrum of the adsorbed benzene is little, but distinctively different from that of the liquid (Galin *et al.*, 1962). Nitrobenzene behaves in an analogous manner, while aromatic amides and phenols appear to interact by means of their donor nitrogen and oxygen atoms, respectively, toward the acceptor hydrogen atoms of the silica surface (Hair, 1967). Horill and Noller (1976) followed the changes in OH valence frequency by adsorption of a variety of organic donor molecules on aerosile. The $\Delta\nu_{OH}$ values were found to have a linear relationship to the donor number of the adsorbate (Fig. 5.2):

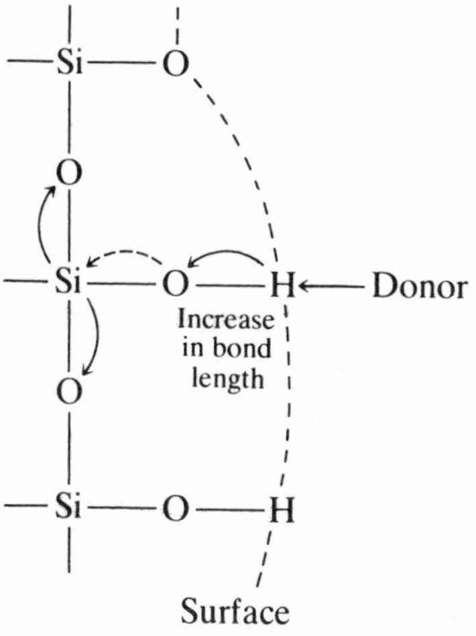

—Si—O
O
—Si—O—H ← Donor
Increase in bond length
O
—Si—O—H
Surface

No relationship was found for the ionization potential of the donor. The fact that no correlation was found with the heats of adsorption suggests that adsorption does not occur exclusively at the OH groups.

The adsorption of oxygen or of iodine on (111) silicon and germanium surface planes at low temperatures leads to disappearance of "superstructures" (Moesta, 1968). The contraction found at a clean Mo (001) plane no longer remains after adsorption has taken place (Honigmann, 1958: Jette and Foote, 1935). A quantum chemical model calculation for the adsorption of benzene on platinum leads to an analogous conclusion (Lennard-Jones, 1928). The increasing internuclear distances on a metal surface lead to a weakening of the metal–metal bonds, which has been described as "demetallization" (Miessner *et al.*, 1976). The effects will be greater the stronger the interaction, the thicker the surface layer, and the greater the bond polarizability within the adsorbate.

Since adsorption leads to increasing bond distances within the surface area of the adsorbent and hence to decreasing electron density, a corresponding decrease must be involved in desorption. When a gas is adsorbed on a solid surface, a change in pressure leads to changes in adsorption

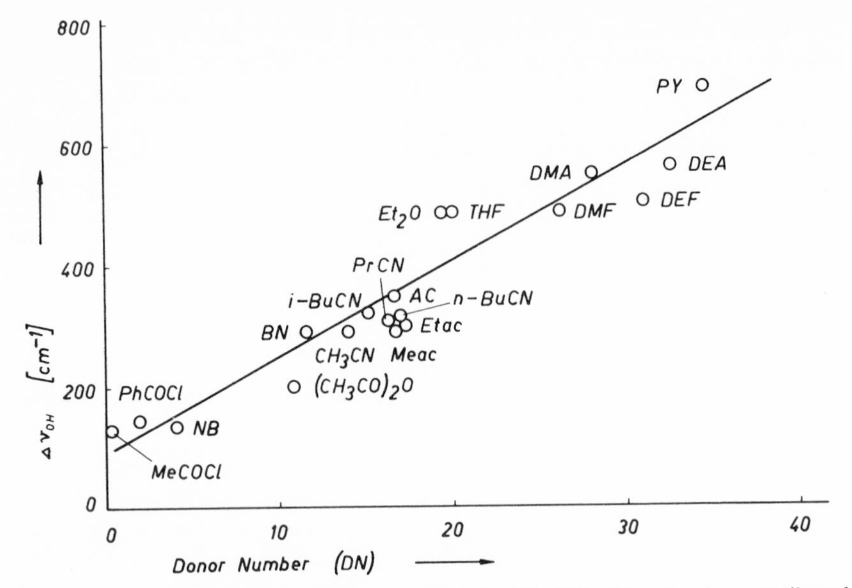

Fig. 5.2. Correlation between shift in wave number of the OH valence band on aerosile and the donor number of the adsorbate (from Horill and Noller, 1976, courtesy of Akademische Verlagsgesellschaft, Frankfurt).

and hence in surface bond distances! This means that a crystal surface is not rigid, but rather distortable.

The adsorption of acetylenes and derivatives on alumina and silica is readily explained in terms of the donor–acceptor concept: IR spectra suggest that chemisorbed acetylene and methylacetylene are held normal to the surface since the acidic protons of the adsorbate interact with the oxygen donor sites at the surface. On the other hand, the nonacidic dimethylacetylene is held parallel to the surface, as the π electrons of the C≡C bond behave as donor and interact with the acceptor sites of the alumina surface (Yates and Lucchesi, 1961).

We shall now proceed to examine the bond-length variations induced within the absorbed species: Adsorption of triphenylchloromethane on barium sulfate provokes heterolysis of the C—Cl bond as the spectrum of the adsorbate becomes identical to that of the triphenylcarbonium ion in sulfuric acid solution (Kortüm *et al.*, 1958):

$$—Ba^{2+} \leftarrow Cl — CPh_3$$

Adsorption   Induced heterolysis

The reverse situation was found when symmetrical trinitrobenzene was adsorbed on magnesium oxide, where it was red in contrast to adsorption on CaF_2. The spectrum of the adsorbed species was very similar to that obtained in alkaline ethanol solution suggesting that the oxide ions of the magnesium oxide surface functioned as donor groups for the chemisorption of trinitrobenzene.

5.4. *Surface Layers*

The thermodynamic theory (Bauer, 1958; Hirth and Pound, 1963) of nucleation has been very useful, but it is limited in that it does not describe the electronic changes involved.

Extremely thin layers *in vacuo* may be considered as microcrystals with a relatively large surface area. The lattice dimensions are expected to be smaller than those of the corresponding bulk crystalline material, the differences being greater the greater the surface area as compared to its volume, e.g., the smaller the thickness of the layer. Experimental evidence for the dependence of the lattice contraction from the thickness has been given by Suhrmann *et al.* (1960) by X-ray measurements on nickel and copper films (Table 5.1).

Schroer *et al.* (1970) measured the Mössbauer isomer shift of ^{197}Au in samples of gold microcrystals supported in gelatin. The experimental isomer shift correlates well with the lattice contraction observed in these microcrystals, with decreasing thickness of the gold layer (Fig. 5.3). A lattice contraction of about 1% has also been observed on thin gold (001) layers by means of high-resolution transmission electron microscopy (Krakow and Ast, 1976).

Table 5.1. Lattice Constants, a, of Copper and Nickel in Layers of Different Thickness d

Copper		Nickel	
d (pm)	*a* (pm)	*d* (pm)	*a* (pm)
Bulky crystal	361.49	Bulky crystal	353.90
60.000	361.29	216.000	352.11
50.000	361.18	80.000	351.81
10.000	360.74	70.000	351.57
		25.000	351.35

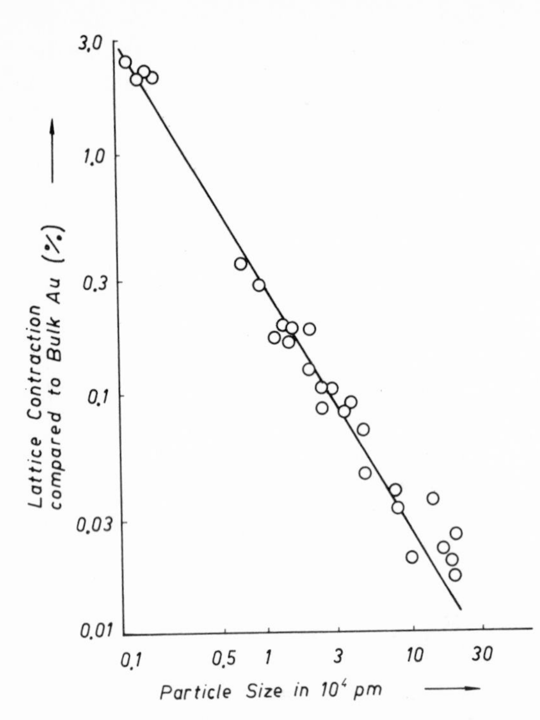

Fig. 5.3. Lattice spacing in gold microcrystals as a function of the particle size (from Schroer *et al.*, 1967, courtesy of The American Physical Society, New York).

The second important factor in determining the lattice constant in a thin layer is the extent of the (intermolecular) interaction with the constituents of the surface area to which it is attached. The consideration of the lattice-geometrical aspects is important, but even more important are the electronic effects. According to the first bond-length variation rule the lattice parameters within the ad-layer will increase by increasing binding to the original surface layer. The lattice dimensions observed in the ad-layer will be mainly the result of two effects: The distances are decreased, the thinner the layer and the weaker the bonding of the adsorbate layer on the original surface. In a film of given composition and of given thickness a lattice contraction occurs for weak interactions with the substrate, whereas a lattice expansion is expected for strongly bonded ad-layers.

Thus, a thin deposit of silver on carbon is bonded so weakly that the Ag—Ag distances in the surface layer are shorter than in the bulk crystal (de Planta *et al.*, 1964), a situation comparable to the gold crystals supported in gelatin, as mentioned above. On the other hand the Ag—Ag distances are found to be appreciably longer than in the silver bulk crystal after silver has been deposited on silicon (Niedermayer, 1970), where it is bonded strongly.

The relationship between the strength of interaction between the different layers and surface energy and hence of reactivity appears to be of great importance for the understanding of the supporter effect in heterogeneous catalysis (Chapter 15) as well as of adhesion and epitaxy phenomena.

5.5. *Epitaxy Phenomena*

For epitaxy phenomena—the oriented growth of crystals on a surface—no general theory is available. The experimental difficulties in reaching conclusions are due to uncontrolled impurity effects for which the following sources are possible: (a) impurities contained in the matrix enriched in the course of vapor deposition, (b) impurities contained in the substance, and (c) impurities from the "inert gas" atmosphere.

For epitactic growth the properties of the surface are important, as an "ad-layer" is primarily formed. The final condensation occurs on the ad-layer, and hence its properties will be decisive. It has been mentioned that the consideration of the lattice-geometrical aspects is not sufficient in accounting for the observed surface features. Various facts cannot be explained from the geometrical point of view, for example, the influence of the hardness of a crystal, the orientation effects of ice on lead iodide, cadmium iodide, calcium sulfate, and graphite (Kleber and Weis, 1958), or the different epitactic behavior of amino acids on quartz (good epitactic growth) and on the isotypical aluminum phosphate with nearly the same lattice parameters (no epitactic growth) (Seifert, 1956).

When silver vapor is deposited on a clean Ge(100) surface plane, the first step is the formation of a loose and unordered layer on which oriented insulas can grow. The Ag—Ag bond distances within the layer of two-dimensional periodicity are longer in the direction of the diagonal of the (100) plane than in the perpendicular direction (Niedermayer, 1970).

Spiegel (1967) investigated the growth of silver layers on (111) planes of silicon by the low-energy electron diffraction method. At 200°C an adsorption layer of low order is formed, which up to 0.75 atomic layer does not show any adaptation between layer and original surface. At 250°C an Si(111)–$\sqrt{3}$Ag structure is produced, in which each silver atom is very strongly bonded to three germanium atoms, which withdraw greatly the electron density from the silver atoms. The Ag layer of two-dimensional periodicity is desorbed only above 900°C. The Ag—Ag bond distances in the (111) plane are extremely long, namely, 660 pm, as compared to 290 pm in a silver bulk crystal. Further condensation of silver on this strongly adsorbed layer occurs only in the presence of a large excess of silver, and a three-dimensional layer is formed only after 18 layers of silver has been deposited.

A similar change of reactivity in the adsorbate layer by increasing coverage has been found for the interaction of carbon monoxide on palladium. At one-half monolayer coverage on Pd(100), CO switches from a structure with bonding to definite lattice sites to a structure that is out of registry with the substrate lattice (Tracy and Palmberg, 1969). The adsorption energy decreases sharply at this point (Yates, 1974).

Five phases have been observed with aluminum on Si(111) surfaces, ranging from one-third to one "monolayer" (Lander and Morrison, 1964). The α-Si(111)–$\sqrt{3}$Al phase is formed with a one-third monolayer coverage when the sample is maintained above 500°C. Decomposition with evaporation of aluminum is slow below 900°C. An almost perfectly oriented large-crystal (epitaxial) layer having the aluminum structural parameter (smaller than that of silicon by 25%) is obtained after at least five monolayers of Al have been deposited. Thinner films were more disordered.

All of these findings are in agreement with the bond-length variation rules (Chapter 1): Strong interactions between Si surface atoms and Ag atoms lead to a corresponding lengthening of the Ag—Ag bond distances within the silver monolayer (first bond-length variation rule); the greater the Ag—Ag bond distances, the lower the surface energy is and hence the slower the growth of further layers. The following layers of silver are also different from the bulk features, although to smaller degrees. This continues until a point in the layer is reached that is equal to the parameters of the bulk crystal.

The situation is analogous for the adsorption of oxygen on (110), (111), or (100) copper surface planes, where various superstructures are produced on each of the planes (Ertl, 1967; Siddiqui and Tompkins, 1962; Delchar and Tompkins, 1966). The adsorption of oxygen starts quickly and is soon retarded (Rhodin, 1953). Within the first layer the O—O bond distances are nearly twice as long as compared to those in Cu_2O, the structure of which is not produced (Pritchard, 1963).

Long bond distances in the first layer are related to strong bonding on the original surface, and they also provide its high resistivity. In this way a structural and electronic interpretation is provided for the passivation phenomena on metal surfaces!

An intermediate layer is also formed when sodium is evaporated on a tungsten crystal (Chen and Papageorgeopopoulos, 1970). After the deposition of 0.5 atomic layer of sodium atoms the Na—Na internuclear distances are found to be twice as long as the W—W internuclear distances in the (111) direction. Each of the sodium atoms appears to be bonded to

Table 5.2. Type of Structure of the Initial Deposit for Various Deposit–Substrate Combinations

	Substrate Material		
Deposit	LiF	NaCl	KBr
CsCl	NaCl	NaCl	NaCl
CsBr	NaCl	NaCl	NaCl
CsI	NaCl	NaCl	NaCl
TlCl		CsCl	NaCl
TlBr	NaCl	NaCl	NaCl
TlI	NaCl	NaCl	NaCl

four tungsten atoms. Up to 0.8 atomic layer, additional sodium atoms are incorporated in this layer up to a distance of 0.8 atomic layer, before a second layer is produced. The second layer seems to form a two-dimensional periodic region, with the Na—Na internuclear distances of 391 pm being greater than the double atomic radius of Na (384 pm). This again suggests that the W—Na bonds are so strong that the Na—Na bonds in the layer are lengthened correspondingly. After further deposition of sodium the Na—Na bond distances, as expected, are shortened, the shortest distance of 354 pm being smaller than in metallic sodium (371 pm).

When layers having up to a 1000-pm thickness of cesium chloride are deposited from the vapor phase on a lithium fluoride crystal (NaCl type), the cesium chloride is found to crystallize in the NaCl type as well (Schulz, 1951) (Table 5.2). The internuclear distances in the deposits of cesium chloride in the NaCl structure are shorter than in the cesium-chloride-type structure (Table 5.3), although the lattice energies for both types of configurations appear to be nearly the same.

Table 5.3. Interatomic Distances (in pm) in Cesium and Thallium Halides

Salt	NaCl type	CsCl type
CsCl	347	356
CsBr	362	372
CsI	383	395
TlCl	315	332
TlBr	329	344
TlI	347	364

5.6. Solid-Water Interfaces

Drost-Hansen (1969) has emphasized the formation of specific water structures at interfaces. Extremely thin liquid-water layers are considerably more reactive than bulk water. Below a certain thickness of the liquid layer the reactivity is progressively enhanced by a decrease in thickness of the film, since the electronic effects of the donor–acceptor interactions between solid surface and film cannot be distributed over a sufficiently great number of water molecules.

The high reactivity of a thin water layer seems to be responsible for the formation of the concentrated solution, which was originally thought to be "poly water." This is formed when water is condensed in capillary tubes of silica 2 to 4 μm in diameter (Deryagin, 1970). The resulting solution contains constituents of the material of the reaction tube, such as silicic acid, sodium, potassium, calcium, and magnesium ions (Schuller, 1973).

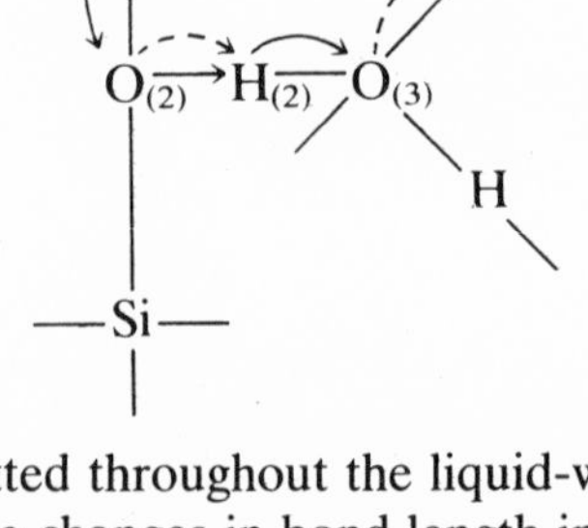

The donor–acceptor interpretation is as follows: The interactions between a silicon surface atom and an oxygen atom $O_{(1)}$ of the water surface lead to lengthening of both the Si—$O_{(2)}$ and the $O_{(1)}$—$H_{(1)}$ bonds. The increased electron density at $O_{(2)}$ allows a strong interaction with the $H_{(2)}$ of another water molecule, with subsequent further lengthening of Si—$O_{(2)}$ and lengthening of $H_{(2)}$—$O_{(3)}$. In the simplified model $O_{(3)}$ interacts with $H_{(1)}$, initiating further cooperative effects, e.g., shortening of $O_{(1)}$—Si, further lengthening of Si—$O_{(2)}$, shortening of $O_{(2)}$—$H_{(2)}$, lenthening of $H_{(2)}$—$O_{(3)}$, shortening of $O_{(3)}$—$H_{(1)}$, and lengthening of $H_{(1)}$—$O_{(1)}$. In this way the electronic cycle may be repeated with final heterolysis of Si—$O_{(2)}$, $H_{(2)}$—$O_{(3)}$, and $H_{(1)}$—$O_{(1)}$. The electronic effects are, however, transmitted throughout the liquid-water structure (see Section 6.2) and hence the changes in bond length in the water surface layer remain small. In a capillary tube of 2 μm radius, the number of water layers from surface to surface is approximately 14,800.

As soon as the process of dissolution has set in, the solute particles will contribute to further appreciable changes in the water structure with enhancement of the donor–acceptor properties within the surface region.

5.7. *Other Surface Phenomena*

The formation of a strongly bonded layer appears also to be decisive for adhesion phenomena. A polymer molecule absorbed at a surface is different in configuration and free enthalpy from a molecule in a bulk polymer matrix. A gradient in polymer configuration and energy exists from the interface up into the bulk polymer. The first polymer molecule absorbed does not conform to the surface topography since it absorbs at sites along the chain, but during adsorption both its configuration and its free enthalpy change. The next molecule that is absorbed also changes but to a smaller degree. The stronger the deformation in the inner layer and the greater the induced structural effects in the subsequent layers, the greater the adhesion between the two phases.

Many liquids are nonspreading on high-energy surfaces because the molecules adsorbed on the solid form a film whose critical surface tension of wetting is less than the surface tension of the liquid itself (Hare and Zisman, 1960). A liquid which is unable to spread upon its own adsorbed and oriented monolayer has been named "autophobic" liquid. Polar liquids which are not autophobic have surface tensions which are less than the critical surface tension of wetting of their adsorbed monolayers. Examples for nonautophobics are aliphatic hydrocarbons, aliphatic ethers, polyethers, polymethylsiloxanes, and perfluoroalkanes. Esters, which have only one polar group, spread on all metals completely, while there are no such effects on glass, silica, and α-Al_2O_3.

The structural rearrangement of solution structures at liquid–liquid interfaces is another problem, which should, at least in principle, be interpretable by the donor–acceptor concept, and liquid–liquid junction potentials may be considered as resulting from specific electronic interactions.

5.8. *Crystal Growth*

The above considerations lead to a reinterpretation of crystal growth, which occurs at the phase-boundary crystal-gas or crystal-solution. Crystal-growth phenomena have been described from the morphologic, thermodynamic, and kinetic point of view. A donor–acceptor interpretation, and hence a general formulation from the point of view of a chemical reaction, has not been provided,* although the inverse reaction, the dissolution of a crystal, is frequently given in this way.

We shall very briefly consider the crystal growth, on a clean NaBr surface, from gaseous NaBr molecules. Alkali-metal halides in the gaseous state have considerably shorter internuclear distances than in the crystalline lattice; in the dimeric gaseous species metal–halide distances are greater than in the monomeric molecules and smaller than in the crystalline state. The Na—Br bond distances are 250 pm in the monomeric, 257 pm in the dimeric gaseous form (Akišin and Rambidi, 1960), and 298 pm in the crystal; the distances at the surface area are expected to be shorter than within the bulk crystal.

When a monomeric sodium bromide molecule is condensed to the surface of crystalline sodium bromide, the Na—Br bond distances are increased at both the crystal surface and, to a greater extent, within the condensing molecules. The electronic changes lead to increase of all Na—Br bond distances to various degrees:

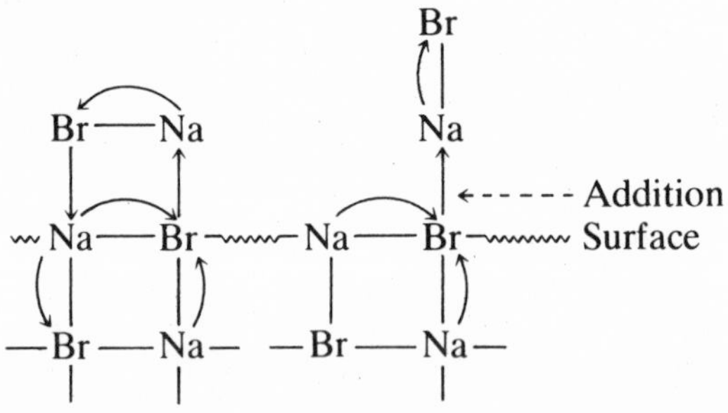

Br
Br—Na Na
Addition
~Na—Br~Na—Br~ Surface
—Br—Na— —Br—Na—

After the growth of a few further layers the internuclear distances at the original surface will have been increased to the normal Na—Br bond distances in the crystal. The internuclear distances of the molecules attached to the surface area are increased to a much greater extent depending on the coordination numbers developed at the surface area. They definitely will be smaller than within the crystal.

The description of crystal growth from solution is more complicated, since desolvation of both the solute particles and the crystal surface area must be considered. In principle the reaction may be regarded as a substitution reaction, in that solvent molecules at the crystal surface are replaced by solute particles.

*Only a crystallization process associated with change in molecular conformation has been suggested to be pictured as a "rudimentary form" of chemical reaction, in particular where different isomers are involved (Curtin and Engelman, 1972).

The solvation of a crystal depends on its size: The smaller the crystal the larger its surface area as compared to the crystal volume. It has been shown that the activation energy E_a is a function of the relative strengths of the bonds to be cleaved and of those to be formed (Szabó, 1962; Szabó and Bérces, 1968):

$$E_a = \sum_i D_{i(\text{cleaving})} - \alpha \sum_j D_{j(\text{forming})}$$

For a small crystal, less energy is required for cleaving the surface bonds since they have been lengthened by strong hydration; e.g., the first term is small for a small crystal. A small crystal is more strongly hydrated than a large crystal and hence the second term is greater for the large crystal. Less energy is needed for dissolution of a small crystal than for a large crystal, and more energy is provided by growth of the big crystal. This provides an alternative explanation for the fact that in a suspension of large and small crystals in a saturated solution, the smaller crystals are dissolved while the larger crystals are growing.

Chapter 6
Molecular Association in the Liquid State

6.1. Association by Hydrogen Bonds

"Structural order" within a liquid is related to the intermolecular interactions between the molecules constituting the system and hence to their respective donor and acceptor properties. In a "pure" solvent, such as liquid hydrogen fluoride, the structural arrangement of the solvent molecules may be regarded as due to the amphoteric solvent properties and hence due to interactions between the solvent molecules. The HF molecule can act as an acceptor through the hydrogen atom as well as a donor through the fluorine atom. Thus hydrogen bonds are established between the constituent molecules.

Hydrogen bonding between HF molecules is also known in the vapor phase. In dimeric hydrogen fluoride, H—F→H—F, the H—F bonds are longer than in the monomeric molecules (in accordance with the first bond-length variation rule), and hence the acceptor property of the terminal hydrogen atom, as well as the donor property of the terminal fluorine atom, is greater than in the monomeric species. Hence the addition of further HF molecules takes place to give a chainlike structure. According to CNDO/2 calculations the zig-zag structure offers the most stable arrangement (Schuster, 1973). According to an *ab initio* treatment of a linear crystal the structural and the spectral changes involved in going from the isolated vapor phase dimer to the crystal are substantially larger than those involved in the formation of the dimeric from the monomeric species (Karpfen and Schuster, 1976).

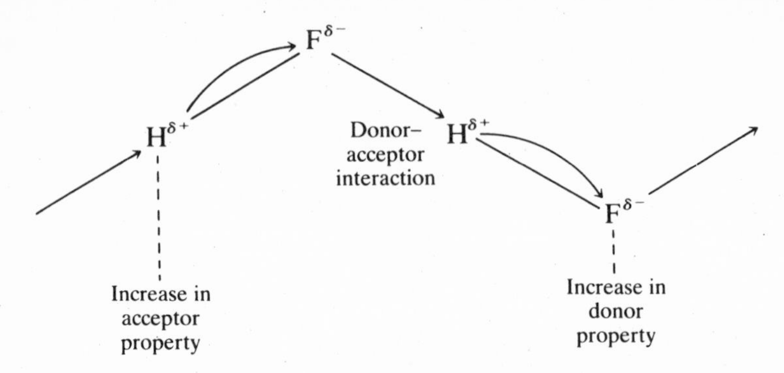

In liquid water the structural arrangement is also characterized by hydrogen bonding. The following changes in charge densities have been calculated for the dimerization of water molecules (Hankins *et al.*, 1970):

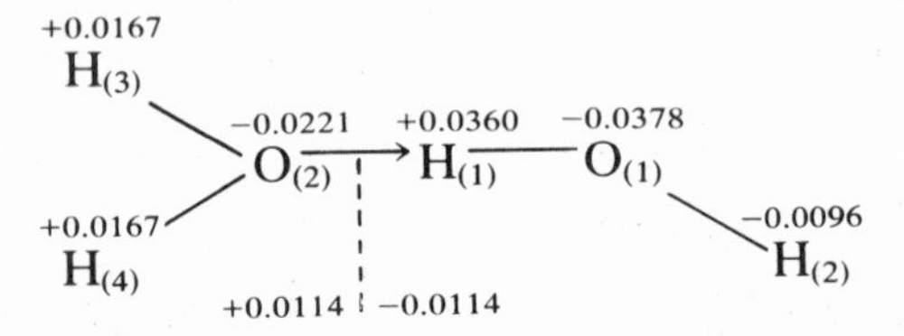

The mean energy ΔE of the hydrogen bonds is increased with increasing number of associated molecules (*cooperative effect*) (Schuster, 1973). Even more significant is the energy $\Delta E_{(n-1)\to n}$ for addition of the "last" solvent molecule, which is always higher than ΔE and reflects the increasing coordinating properties of both the terminal hydrogen and the oxygen atoms, respectively (Table 6.1).

The O—H bond distance in a water molecule in the gas phase is 95.8 pm (Herzberg, 1945) and in ice 99.0 pm (Cross *et al.*, 1937). Although direct experimental access to the individual bond distances in liquid water is not available, there is a relationship between the O···O and the O—H

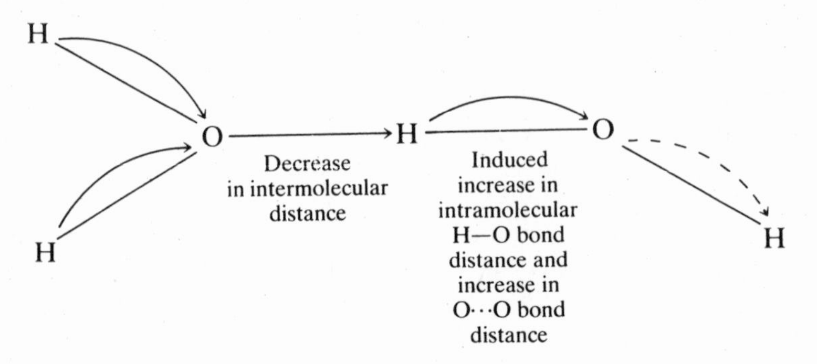

Table 6.1. Mean Energy of the Hydrogen Bond ΔE and Mean Energy for Addition of the Last Solvent Molecule $\Delta E_{(n-1)\to n}$ for Linear Chains of H_2O and HF Molecules as a Function of Chain Length

Number of molecules in chain	H_2O		HF	
	$-\Delta E$ (kJ/bond)	$-\Delta E_{(n-1)\to n}$ (kJ/bond)	$-\Delta E$ (kJ/bond)	$-\Delta E_{(n-1)\to n}$ (kJ/bond)
2	36.1	36.1	39.5	39.5
3	40.1	44.4	45.2	51.1
4	42.8	46.1	48.5	54.8
5	45.4	47.1	50.3	56.1

bond distances in crystalline hydrates (Lundgren and Olavsson, 1976). In agreement with the first bond-length variation rule the formation of a coordinate O→H link (usually denoted O···H) induces an increase in adjacent H—O bond length: The shorter the O—H bond distance the shorter the O···O and the greater the H—O bond distance; e.g., the stronger the intermolecular interaction, the more symmetrical becomes the H bond, which is formed between the oxygen atoms (Fig. 6.1).

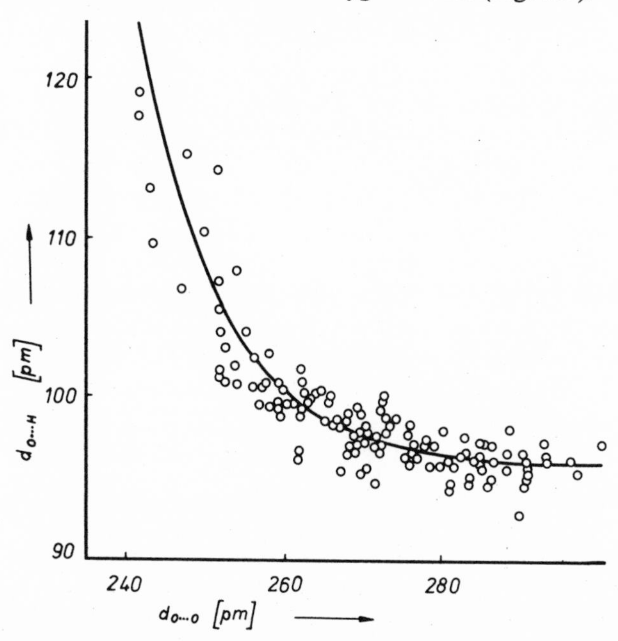

Fig. 6.1. Relationship between O—H and O···O bond distances in crystalline hydrates (from Lundgren and Olavsson, 1976, courtesy of North-Holland Publishing Company, Amsterdam).

Water and aqueous solutions are characterized by a continuous absorption in the IR spectra (Zundel, 1976), which show the simultaneous presence of differently polarized hydrogen bonds in the liquid, e.g., a continuous array of different structural units of slightly different energies. Associated with the different internuclear distances are different orientation effects, as seen from the wide distribution in bond angles. The most probable value for the deviation of H bonds from linearity is not in the region between 0 and 5°, but rather between 5 and 10°, with deviations up to 70° (Pedersen, 1974; Hasegawa and Noda, 1975; Zundel, 1976). The deviation from linearity is greater, the longer the O···O bond distance (Ferraris and Franchini-Angela, 1972) (Fig. 6.2). Linearity of hydrogen bonds (at least in crystals) appears rather the exception than the rule! The hydrogen bonds are extremely easily polarizable (Weidemann and Zundel, 1970, 1972, 1973), in that their polarizabilities are about two orders of magnitude larger than usual bond polarities.

6.2. *Liquid Water and Its Solutions*

In the absence of direct experimental information about the structure of liquid water, various models have been developed. Efforts to verify or to invalidate any of these hypotheses are hampered by lack of a general

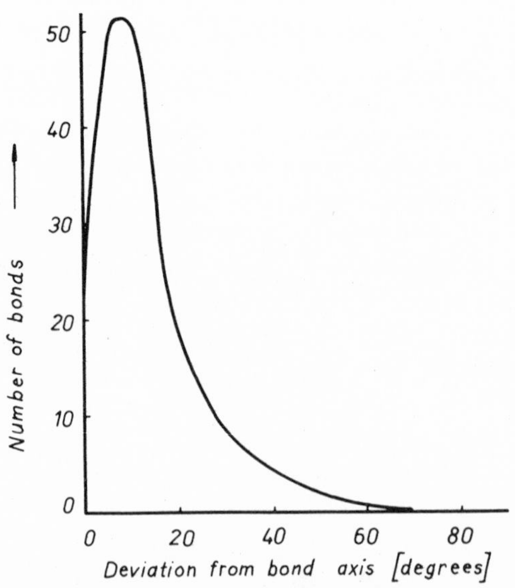

Fig. 6.2. Deviation of hydrogen bonds from linearity.

theory of liquid state (Eisenberg and Kauzmann, 1969). Considerations of the structure of liquid water have been based mainly on two approaches, neither of them rigorous. One of them consists in formulating a model by the methods of statistical mechanics, which usually involve massive approximations (Franks, 1967). The second is to deduce aspects of the liquid structure from the macroscopic properties of water.

"Structural order" in liquid water is regarded as due to the presence of hydrogen-bonded clusters, which have limited half-life periods (Frank and Wen, 1957) in the order of magnitude of 10^{-11} s. A distinction has been made according to the differences in relaxation times. The so-called "*I* structure" (instantaneous structure) refers to a time interval that is short compared to the period of oscillation τ_v, which is roughly 10^{-13} s in ice. No experimental technique to obtain information on this structure is available. The so-called "*V* structure" (vibrationally averaged structure) is ascribed to the features observed in a time interval longer than the period of oscillation, but shorter than the half-life time (displacement time) of the clusters (Eisenberg and Kauzmann, 1969). The "*D* structure" (diffusionally averaged structure) is applied to a time interval that is long compared to the displacement time of approximately 10^{-11} s. Most of the experimental information, for example, the thermodynamic properties of water, its volume, heat capacity, compressibility, X-ray pattern, or NMR data, are characteristic of the *D* structure, to which we shall refer in this section whenever the term "structure" is used.

A distinction is frequently made between the discontinuum or mixture models, which are also called "two-state models" and "continuum models." All of them are intuitive and heuristic and none of them has a firm and physical foundation (Krindel and Elizier, 1971). They differ from each other by virtue of the assertion that the separate contributions to any property from the several species are, in principle, measurable. If this assertion is abandoned, the differences between the models boil down to semantic niceties (Rice, 1975). Both features must be present in the true liquid, which is more complex than any of the simple models proposed (Nemethy, 1974). In this sense, the controversy reflects the limitation of the ways in which we describe water. It should disappear as better models are developed (Nemethy, 1974).

A common feature of the "two-state models" is that they require at least two distinctively different species of water, namely, a bulky species, the so-called clusters made up of a framework of hydrogen-bonded units, as well as a dense species, such as monomeric (vaporlike) water molecules.

The "gas hydrate" models have been derived from the hydrate structures of xenon and other nonpolar substances of sufficiently small molecular size (Stackelberg, 1949). In these hydrates the water molecules form polyhedra, which constitute cages containing the nonpolar molecule (Stackelberg and Müller, 1954). Pauling (1959, 1960) assumed liquid water to consist of a clathrate network, the cages occupied by monomeric water molecules that are non-hydrogen-bonded and free to rotate. This proposal has been superceded, since X-ray data are inconsistent with this assumption (Danford and Levy, 1962) and there are no gas hydrates known with H-bonding solutes.

Frank and Wen (1957) proposed a model that involves an equilibrium between hydrogen-bond forming and breaking. They postulate that the formation of the bonded structure in liquid water is a cooperative process in which "flickering clusters" of varying extent form, relax, and reform in a temporal sequence and in a spatial pattern determined by the energy fluctuations, which are constantly taking place (Frank and Quist, 1961).

The so-called "continuum models," namely, the distorted hydrogen-bond model of Pople (1951) and the closely related random network of Bernal (1964), avoid the difficulty of insisting on the presence of appreciable amounts of monomers in equilibrium. The hydrogen bonds are regarded as distorted to varying degrees rather than as either intact or broken as in mixture models (Pople, 1951). The hydrogen bonds in liqud water are considered as flexible in that they can bend continuously. The free energy of the hydrogen bond is assumed to be a continuous and smooth function of the bonding angle, so that every bonding angle corresponds to a possible temporary equilibrium state of a molecule. This model is characterized by a long-range order in liquid water, which is highly influenced even by small amounts of solute.

Solutes are usually classified with respect to their charges, namely, as cations, anions, or neutral species. Another classification has been given on thermodynamic grounds, namely, solvation of the first kind and solvation of the second kind (Hertz, 1970). Since we are concerned in this section with the liquid structures, we shall make use of the classifications "*hydrophilic*" and "*hydrophobic*" solutes (Bene and Pople, 1970), although their usefulness has been questioned (Jackson and Symons, 1976).

Hydrophilic solutes or "structure breakers" interact more strongly with water than do the water molecules between each other. They are readily soluble and give rise to hydration layers that show structural features different from those of the bulk water. IR and NMR spectra, although relatively insensitive to the presence of salts, do indicate changes which simulate the effect of a decrease in temperature of the pure water spectrum (Bernal and Fowler, 1933). This suggests a higher order in the areas surrounding the solute hydrophilic solutes. The "structure temperature" of a solution is defined as the temperature at which the spectrum of pure water equals that of the solution at room temperature. Although most cations belong to this group, tetraalkylammonium or trialkylsulfonium ions are not included. Many anions are structure breakers, for example, the fluoride ion, but there are others which do not behave in analogous ways, such as the perchlorate or the tetraphenylborate ion. On the other hand, there is a number of neutral molecules, which are structure breakers, for example, amines or phenols. All of the structure breakers have in common the ability to be hydrated by water, e.g., to interact with water molecules by exercising either a donor or an acceptor function. Representative structural data will be discussed in Section 7.3.

The role of donor–acceptor interactions for neutral solutes in water can be seen from the relationship between H-bond energy and donor strength of the solvate molecules. The frequency shift of the OH stretching vibration of phenol increases in the series, $NM < NB < CH_3CN < Etac < AC < dioxane < Et_2O < DMF < PY$ (Luck, 1974), thus following increasing donor numbers of the solvents.

Hydrophobic or "structure-making" solutes are sparingly soluble in water, since the intermolecular forces between water and solute molecules are weaker than the H-bond interactions between the water molecules (Luck, 1976*a*, *b*). The IR spectra of their solutions equal those of water at higher temperatures and hence indicate a corresponding structural rearrangement as "structure makers." The dissolution of such molecules causes a positive change (loss) in entropy. Examples of this type of solute in water are xenon, oxygen, and various ions which resist hydration, such as NBu_4^+ or BPh_4^- ions. The solute molecules or ions are placed in holes of the water structure, which are adaptable to the requirement of the solute molecule for free rotation.

From the solutions, crystalline hydrates may be obtained, which have been structurally characterized. Two types of solid-gas hydrate structures may be distinguished (Stackelberg and Müller, 1954). Structure I contains 46 water molecules in the unit cell, which has a lattice constant of 1200 pm; this unit cell contains 6 big and 2 small holes. Structure II contains 136 water molecules in the unit cell, which has a lattice constant of 1730 pm; this unit cell contains 20 big and 8 small holes. In both cases the small holes usually remain empty, although they may be occupied by small molecules, such as oxygen, nitrogen, or carbon dioxide. The presence of such gas molecules in the small holes increases the melting points of the crystalline hydrates up to 15°C (Stackelberg and Müller, 1954) and also increases the surface tension (Sobol *et al.*, 1976).

It is usually anticipated that the rearrangement of water molecules around either a structure-breaking or structure-making solute is restricted to a limited number of water molecules (Schuster *et al.*, 1975). There is, however, good reason for assuming that a greater number of water molecules are involved in the hydration structures than is usually thought (Gutmann *et al.*, 1977):

(1) A recent experimental study on hydration of the lithium ion reveals that after incorporation of 90 water molecules the properties of the water molecules are still slightly different from those in pure water (Narten *et al.*, 1973). The results of micelle formation with polyethylene oxide derivatives indicate hydration numbers up to 200 and hence a long-range distribution of the hydration structure (Bene and Pople, 1970). From the partition coefficient of *p*-cresol between cyclohexane/water, a hydration number of 90 has been found in a 0.1 M Na_2SO_4 solution (Bene and Pople, 1970). In a 1 M $Al_2(SO_4)_3$ solution nearly all of the water molecules available are engaged in the hydration structure (Luck, 1976*b*).

(2) Calculations on a simplified model for the structures of successive water molecules, arranged in a hypothetical chain commencing from an alkali-metal ion, showed that with increasing number of water molecules, which should simulate the number of hydration layers surrounding the cation, the inner-sphere metal–oxygen bond distance is decreased. A limiting bond energy appears to be attained only after addition of six layers of water molecules (Burton and Daly, 1971).

(3) Drost-Hansen (1969) emphasizes that structural elements may be induced in water by the presence of certain solutes. Evidence for particularly long-range structural stabilization has been inferred from the rheological properties of an aqueous solution of the copper salt of cetylphenyl ether sulfonic acid, which retains elastic properties in a 0.002% solution (Booij, 1949). This solution has a solute concentration of approximately $2 \cdot 10^{-6}$ mol/liter and thus it contains about 20 million water molecules for 1 solute molecule. Hence the approximate mean separation between solute molecules is on the order of more than 100 water layers!

(4) Hammes and Schimmel (1967) have shown from an ultrasonic study of polyethylene glycol in aqueous solution that the relaxation process may be attributed to a cooperative process in the local hydration structures, in which considerable amounts of water appear to be involved.

(5) A formal analogy exists for the structural features involved in growing successive SiO_2 layers on an aluminosilicate surface (Weiss, 1975; see also Section 5.2): The structural effects due to the Al—O bonds can be seen up to within 20 SiO_2 layers! Since the polarizabilities of hydrogen bonds in water are considerably greater than that of a normal bond (Zundel, 1976), the number of water layers transmitting information from a solute should be substantially greater.

(6) From the minimum radius of the silica capillary tube required for the production of the concentrated solution, which was originally thought

to be "poly water," the number of water layers that are structurally and chemically affected by the silica surface is estimated to be 7400! (See Section 5.6.)

Unfortunately minor structural differences within the hydration structure are beyond the accuracy of measurement. The possible drastic effects of extremely small differences in structures can be seen by comparing the effects of H_2O and D_2O on most organisms (Hübner *et al.*, 1970).

The driving forces for the reorganization of the structural pattern due to addition of a solute are both the cooperative and the polarization effects between solute and surrounding water, as well as mutual polarization of the individual solvation spheres. The high polarizabilities of the hydrogen bonds are enhanced by decreasing field strength, and hence polarization effects increase with increasing distance from a charged or from a polar solute. In principle there are no clear-cut border lines between the solvation-sphere structure and the completely undisturbed solvent structure (Gutmann *et al.*, 1977).

At this point it is important to note that a water structure consisting exclusively of water molecules is purely hypothetical. All of the water models which have been proposed ignore the fact that "pure liquid water" is incapable of existence: Even in highly purified water both hydrated hydrogen ions and hydrated hydroxyl ions are present in equilibrium. One of such ions is expected to be present within a cube formed from edges, each edge being occupied by 820 water molecules. In addition to these ions normal water contains CO_2, which with water partly forms hydrogen ions and hydrogen carbonate ions. A second important effect is due to the fact that water, saturated by air, constitutes at 0°C a 1.25×10^{-3} M solution of oxygen, nitrogen, and carbon dioxide. This corresponds to approximately 44.000 water molecules for 1 solute molecule, and for a spherical hydration shell there are approximately 18 layers of water molecules around each gas molecule.

I prefer therefore to consider liquid water as a highly differentiated, highly organized, and rather flexible "pseudomacromolecule" containing both structure-breaking ions and mobile holes (Gutmann *et al.*, 1977). The holes are partly occupied by the constituent molecules of air and are partly vacant. To some extent this description corresponds to that provided by the flickering cluster model, since the boundaries connecting the contacts between the clusters surround unoccupied areas, which are mobile as the clusters change their relative positions. The pictorial representations for an arbitrary section of the water structure, according to the flickering cluster model and the hole model, are given in Fig. 6.3. The hole model for solutions of hydrophobic solutes requests the existence of interfacial regions within

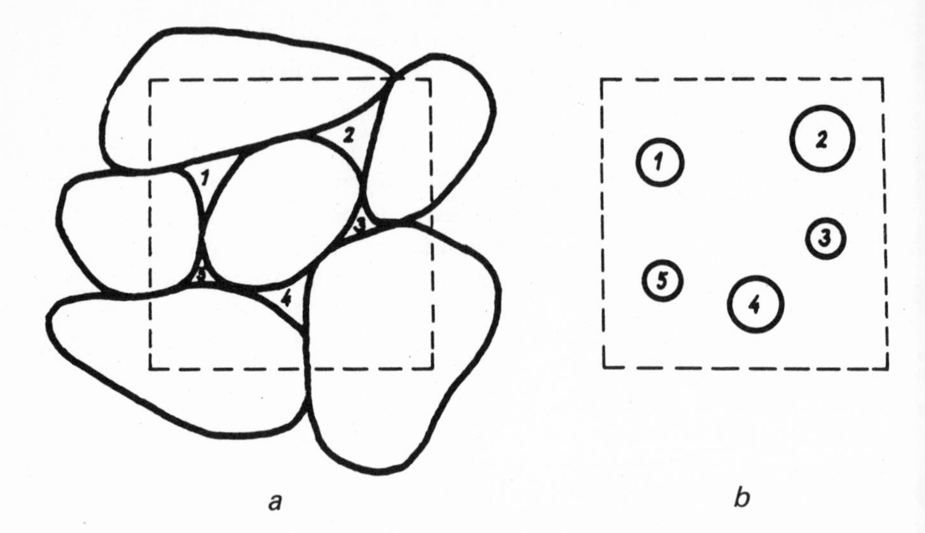

Fig. 6.3. Schematic representation of the (a) cluster model (with contact-free zones numbered) as compared to that of the (b) hole model (with holes numbered) (Gutmann *et al.*, 1977).

the bulk, e.g., internal surface areas, which surround the holes. Each inner-surface area is characterized by hydrogen bonds, which are shorter than those of the successive layers. This is supported by the results of NMR chemical shift measurements of aqueous solutions of hydrocarbons which show an upfield shift, denoting a shortening of hydrogen bonds (Englefeld and Nemethy, 1973), and supported as well by thermodynamic data (Cox *et al.*, 1974).

The three-dimensional pattern of collective structures is composed of structural elements which differ slightly from spatial layer to spatial layer. The greatest O···O bond distances will appear in the regions which are most remote from the surface areas. The situation is analogous for structure-breaking solutes: In their immediate environment the hydrogen-bond distances are shortest and will increase as their distance from the solvate's solute is increased.

We shall therefore consider each structure maker and each structure breaker, including each vacant hole, as a *structure-regulating center.* Their specific nature, their number, and their relative positions to each other will be decisive for the bond-length variation spectrum within the liquid. At very short distances between A and B the number of water layers will be small and hence the bond lengths relatively short (Fig. 6.4). By moving B to B′ with a greater intercenter distance the number of intercenter water molecules is increased. The maximum value of the O···O bond distances is also increased. By further increasing the intercenter distance, e.g., by moving B to position B″ (Fig. 6.4), the number of hydrogen bonds, which are less differentiated, is increased, whereby the maximum values for the O···O bond distances are but slightly increased. In other words the region of "nearly undisturbed water structure," which is usually neglected in current considerations of the water structure, has been widened.

Figure 6.5 shows the hydrogen bond-length spectrum for the three different cases considered in Fig. 6.4. For different mean distances of the structure-regulating centers, A—B, A—B′, and A—B″, different curves will be found. The structural pattern becomes more specific as the number of structure-regulating centers is decreased, e.g., as their mean distances increase. At the same time the maximum O···O bond distances are also increased. The peak areas in this diagram characteristic for the structure of the solution represent the regions, which are completely neglected in any short-range order treatment!

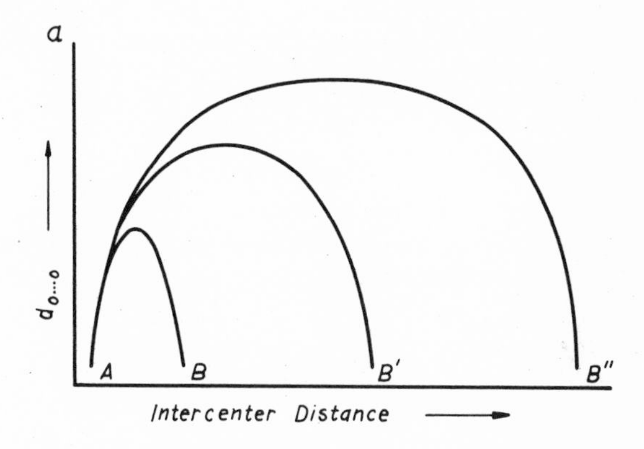

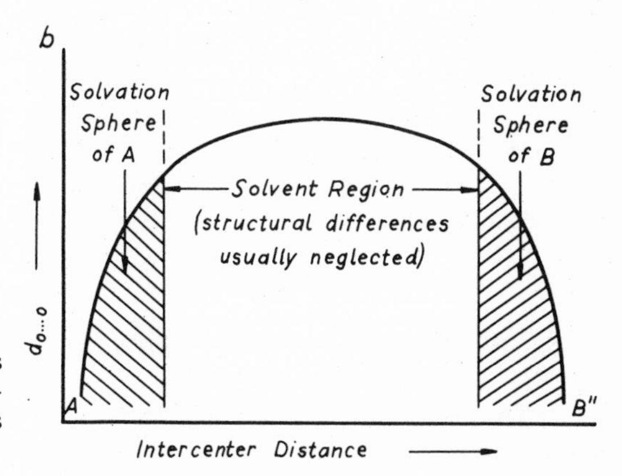

Fig. 6.4. O···O bond distances as a function of the distance between structure-regulating centers (Gutmann *et al.*, 1977).

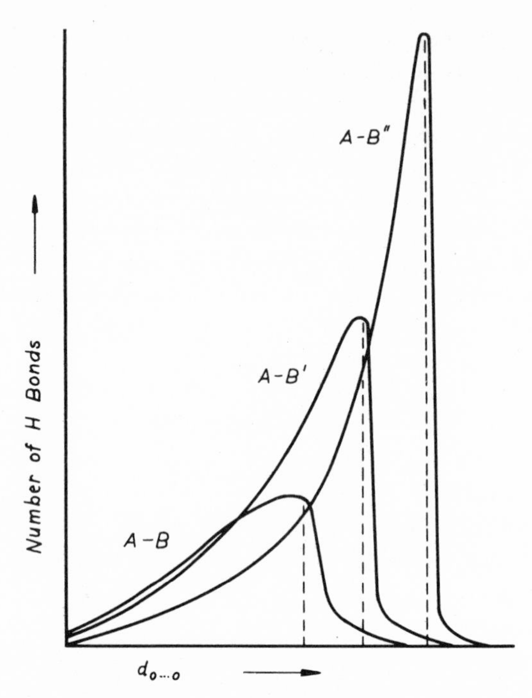

Fig. 6.5. Distribution of O···O bond distances in aqueous solution for different mean distances of structure-regulating centers A—B, A—B′, and A—B″ (see Fig. 6.4) (Gutmann *et al.*, 1977).

This model is also in agreement with the dynamic features of water and its solutions. For a dissociative mechanism the rate of exchange is a function of the bond strength. We should therefore expect the signal for the relaxation times to be sharper the more dilute the solution and to broaden as the concentration is increased. The spectrum of expected lifetimes (Barthel *et al.*, 1970) corresponds in a qualitative way to the hydrogen bond-length spectrum: The greater the number of structure-regulating centers, e.g., the more concentrated the solution, the broader the bond-length spectrum and the smaller the maximum bond length. This is paralleled by a broadening of the relaxation times spectrum and by a decrease in relaxation times.

6.3. *Association in Aprotic Liquids*

The existence of hydrogen bonds is not a necessary condition for association in the liquid state. The association in liquid hydrogen fluoride may be described either by hydrogen bridges between fluorine atoms or by fluorine bridges between hydrogen atoms (Gutmann, 1956). In aprotic solvents, such as in bromine(III) fluoride, fluorine bridging is involved in the association of solvent molecules. By the charge-density redistribution due to the association process, the properties of the terminal atoms are changed: The fluorine atoms are enhanced in donor properties and the bromine atoms in acceptor properties as a consequence of the increased polarities of intramolecular bromine–fluorine bonds, leading eventually to self-ionization with formation of BrF_2^+ and BrF_4^- ions (Woolf and Emeléus, 1949):

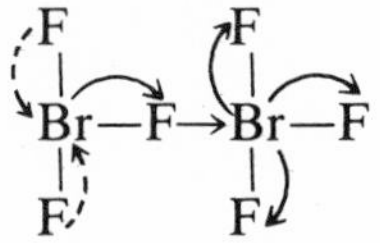

F F
| |
Br—F→Br—F
| |
F F

Heterolysis of a Br—F bond is favored by strong donor–acceptor interactions, for example, by adding a strong acceptor, such as antimony(V) fluoride:

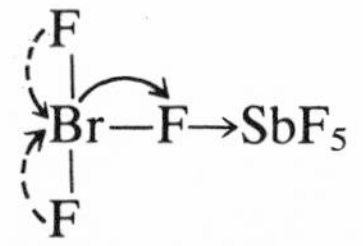

F
|
Br—F→SbF_5
|
F

In crystalline $ICl_3 \cdot SbCl_5$ the I—Cl bond distances adjacent to Cl→Sb bonds are longer than the terminal ones giving discrete ICl_2 and $SbCl_6$ groups (Vonk and Wiebenga, 1957).

Liquid $SeOCl_2$ is more strongly associated than liquid $SOCl_2$ (Gutmann, 1968). Accordingly the increase in the Se—O bond length is greater than that of the S—O bond length in going from the gas phase into the liquid state, as can be seen from the differences in the respective force constants (Hopf and Paetzold, 1972) (Table 6.2).

Likewise, $(SeO_3)_4$ reacts with arsenic(III) chloride to give an insoluble polymer (Dostal and Toučin, 1974), while no reaction occurs with SO_2, which is not amphoteric. Raman-spectrographic results indicate the

Table 6.2. Relative Change of the Force Constants of SeO and SO Bonds Going from the Gas Phase to the Liquid Phase

Species	$f_{(gas)}$	$f_{(liquid)}$	$-\Delta f/f_{(liquid)}(\%)$
$SeOCl_2$	7.67	7.0–7.1	≈9
$SOCl_2$	9.89	9.6–9.8	≈2

presence of As—O—Se bonds. According to the donor–acceptor presentation the formation of the coordinate bond between an oxygen atom of a selenium trioxide molecule and an arsenic atom induces a corresponding increase in donor properties at the chlorine atoms bonded to arsenic, as well as an increase in acceptor properties at the selenium atoms:

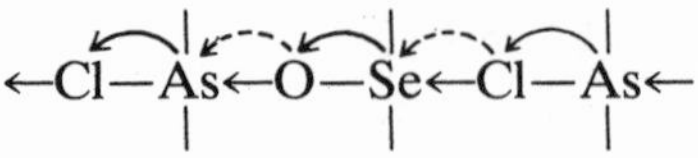

$$\leftarrow Cl-As\leftarrow O-Se\leftarrow Cl-As\leftarrow$$

These changes favor the formation of coordinate bonds —Se←Cl—, yielding a further increase in bond strength of both As—O and Se—Cl bonds. This type of interaction is not restricted to one dimension, as indicated in the formulation. The bonds between $AsCl_3$ and SeO_3 would be weaker if the amphoteric properties of the molecules were developed to a lesser extent. Self-association is also known to be present in liquid dimethyl sulfoxide, as the oxygen atom acts as donor toward sulfur atoms of other DMSO molecules, which act as acceptors (Schläfer and Schaffericht, 1960; Parker, 1962).

Indeed association is related to the amphoteric properties of the molecules, and hence the structural differences within the constituent molecules between the gas phase and liquid phase are expected to be greater the stronger the intermolecular association in the liquid state.

A striking example for new structural units in aprotic solvent mixtures is provided by the system DMSO—SO_2 (Harrison *et al.*, 1976). In this system many metals are readily oxidized and dissolved with the formation of metal disulfates and of dimethyl sulfide. Neither DMSO nor SO_2 alone can oxidize metals, while in the mixture entirely different pseudo-macromolecules are formed due to solvent–solvent interactions such as

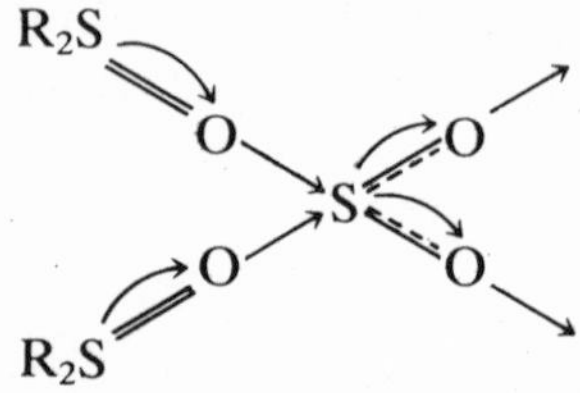

R_2S O O S O O R_2S

Heterolysis with changes in oxidation numbers

Raman-spectrographic studies revealed that the S—O bonds of SO_2 are lengthened to a greater extent than the S—O bonds of the DMSO molecules. Both the coordinating and the redox properties of the system are drastically altered due to the interactions between different solvent molecules. Interaction with a metal gives rise to electron transfer, with simultaneous heterolysis of the S—O bonds originating from DMSO molecules to give metal sulfate and dimethyl sulfide:

$$Me + 2R_2SO + SO_2 \longrightarrow MeSO_4 + 2R_2S$$

Chapter 7

Ion Solvation

7.1. General Considerations

In the absence of a detailed knowledge of the molecular structures of ionic solutions, various models have been proposed with the aim of calculating important macroscopic properties of solutions. An enormous amount of experimental data has been provided, a survey of which can be found in monographs devoted to the subject. References to them as well as to recent review articles have been given by Schuster *et al.* (1975).

Born (1920) and later Bjerrum and Larsson (1927) developed a theoretical approach to ion–solvent interactions based on a rather simple electrostatic model. The ions are considered as rigid spheres of radius r and charge z immersed in a continuous medium of a dielectric constant ε. According to this "sphere-in-continuum" model both cations and anions of equal charge and radius should be equally solvated in a medium of given dielectric constant and of given dipole moment. The application of this model is limited to large spherical ions which remain practically unsolvated in the particular solvent, such as $Bu_4N^+ \cdot BPh_4^-$ in propylene carbonate or in mixtures of nitrobenzene and carbon tetrachloride (Hirsch and Fuoss, 1960). The continuum model is not applicable to most electrolyte systems (Mayer and Gutmann, 1972, 1975; Justice and Justice, 1976). The macroscopic Born model fails also in the prediction of electrode potentials and of most hydration enthalpies (Rosseinski, 1965); the apparent success of Morf and Simon's (1971) calculations for 27 cations was only possible by using values which were not experimental but indirectly inferred or guessed (Rosseinski, 1971), and it is unlikely that this model can be applied to nonaqueous solutions.

The main limitation of the simple hard-sphere-in-continuum model is that it completely ignores the structural aspects as well as the charge redistribution within solute and solvent. Barthel (1976) divided ΔG^0_{solv} into an electrostatic and a nonelectrostatic (structural) component and found the latter to be considerable and in most cases greater than the electrostatic term.

In the discontinuous or chemical treatment (Bernal and Fowler, 1933) the central ion with its nearest neighbors constituting the so-called first solvation shell is regarded as a rigid entity moving in the bulk liquid of the solvent. This approach proved successful in determining thermodynamic properties of single ions in different solvents. These quantities cannot be obtained by purely thermodynamic methods, and approximation procedures must be used. One of the approaches is based on the Born equation, or modifications thereof, and requires an exact knowledge of ionic radii (Desnoyers and Jolicoeur, 1969), which is lacking.

While the free enthalpies and entropies of solvation are accessible for salts, such data cannot be obtained by experiment for single ions. The estimation of single-ion free enthalpies requires extrathermodynamic assumptions.

One approach is based on the assumption that changes in enthalpy or free enthalpy of transfer from one solvent to another will be equal for certain pairs of bulky ions, such as $AsPh_4^+ \cdot BPh_4^-$ or $NR_4^+ \cdot BBu_4^-$ (Grunwald *et al.*, 1960; Parker, 1967; Haberfield *et al.*, 1969; Parker and Alexander, 1968).

A similar approach is based on the assumption that the redox potentials of certain redox systems, such as ferrocene–ferrocinium ion are essentially independent of the nature of the solvent (Koepp *et al.*, 1960). Recent investigations have shown that this assumption gives consistent data in aprotic solvents (Gritzner, 1977*a*), although this data cannot be applied to aqueous solutions (Duschek and Gutmann, 1973*a*).

A third approach assumes that the liquid junction potential between two solvents can be rendered negligible if a salt bridge of a 0.1 M solution of Et_4N picrate is employed between them (Owensby *et al.*, 1974).

7.2. *Relationships between Thermodynamic Solvent Transfer Quantities and Empirical Solvent Parameters*

Free enthalpies of transfer from a reference solvent for cations are related to the donor number of the solvent, whereas free enthalpies of transfer for anions in different solvents are related to the acceptor number of the solvent. Thus a linear relationship exists between the donor number

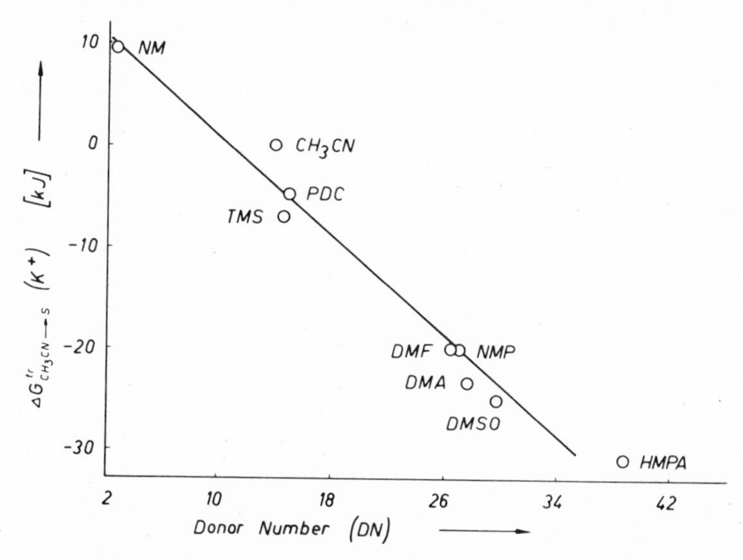

Fig. 7.1. Correlation between the free enthalpies of transfer of the potassium ion in various solvents obtained from EMF measurements using a salt bridge of 0.1 M Bu_4NClO_4 (Owensby *et al.*, 1974) and the donor number of these solvents (Mayer, 1977).

of the solvent and the free enthalpies of transfer for the potassium ion (Mayer, 1977). The data in Fig. 7.1 are based on the assumption of negligible liquid junction potential of the cell $K(Hg)/KClO_4$ in solvent $S//Et_4NPic$ in $CH_3CN//KClO_4$ in $CH_3CN/K(Hg)$ (Owensby *et al.*, 1974).

A relationship exists also between the standard potential of the hydrogen electrode in various solvents and the donor number of the solvent (Bauer and Foucault, 1976) (Fig. 7.2).

Figure 7.3 shows the relationship between the acceptor number of the solvent and the free enthalpies of solvation for the chloride ion in the respective solvent (Mayer *et al.*, 1975, 1977).

Such relationships offer the opportunity for the estimation of free enthalpies of solvation for cations and anions in a solvent of given donor number and acceptor number, respectively.

7.3. *Charge-Density Rearrangement by Ion Solvation*

Charge-density rearrangements due to ion solvation have been calculated by the MO method (Schuster *et al.*, 1975). Whereas the interaction between a neutral donor and a neutral acceptor molecule is characterized by a gain in negative fractional charge at the donor atom, which is due to

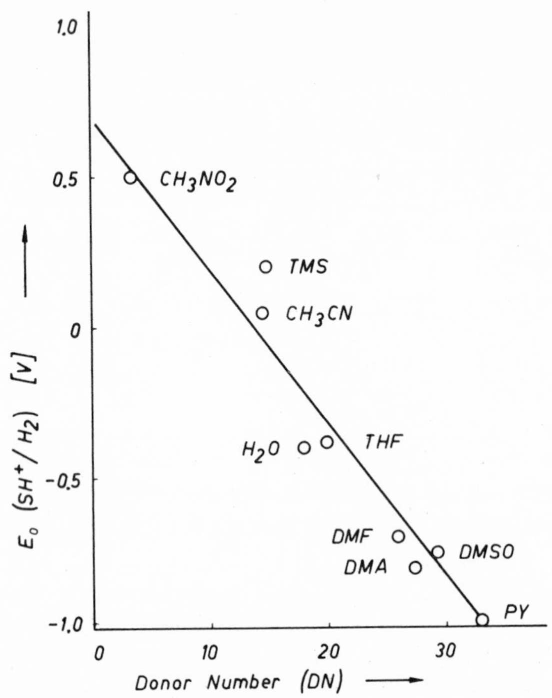

Fig. 7.2. Relationship between the standard redox potential of the hydrogen electrode in various solvents and the donor number of these solvents.

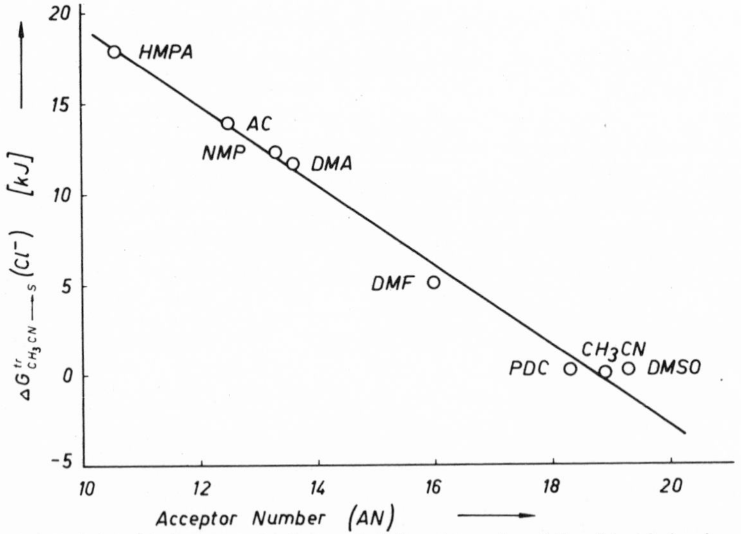

Fig. 7.3. Relationship between the free enthalpies of transfer of the chloride ion in various solvents (Alexander *et al.*, 1972) and the acceptor number of these solvents.

electron withdrawal from other parts of the donor unit ("pileup effect," Section 3.2), the latter possibility is no longer given for an atomic anion as donor (Gutmann, 1977*a*). For example, the fractional negative charge at the fluoride ion is decreased by hydration (no pileup effect). However, within the coordinated water molecules a "spillover" of fractional negative charge takes place from the acceptor hydrogen to the other parts of the hydration sphere.

The two hydrogen atoms of a water molecule coordinated to the fluoride ion are no longer equivalent as has been shown by the nuclear magnetic relaxation rates of ^{19}F in aqueous KF solutions (Hertz and Raedle, 1973): Hydration involves one hydrogen bond for each water molecule (Schuster and Preuss, 1971; Diercksen and Kraemer, 1970).

The extent of charge transfer for the formation of a hydrated fluoride ion is considerably greater than for association of solvent molecules in pure liquid water (Schuster, 1969, 1970; Russegger *et al.*, 1972). The negative unit charge of the gaseous fluoride ion is reduced from −1.0 to −0.887 by withdrawal of 0.113 electronic charge; the negative fractional charge at the acceptor H atom is reduced by 0.122 electronic charge, while a gain in fractional negative charge of 0.149 electron is found for the oxygen atom of the coordinated water molecule. In agreement with the bond-length variation rules the $H_{(1)}$—O bond is lengthened and the O—$H_{(2)}$ bond is shortened.

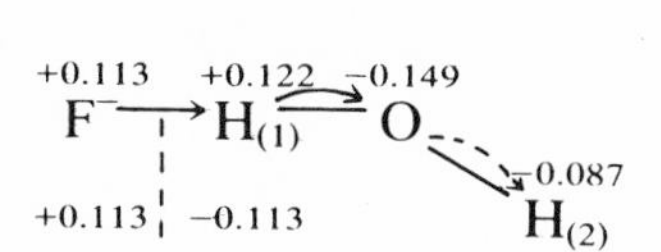

In the course of solvation of an atomic cation by donor molecules, it is not possible for the cation to pass the gain in electron density on to other areas of the acceptor unit. Hence there is no spillover effect and there is instead a decrease in fractional positive charge at the cation, whereas a pileup effect at the donor atoms of the solvent molecules takes place.

Solvation of the lithium ion by four molecules of formic acid, as derived from Raman-spectrographic results (Rode, 1975), leads to lengthening of the C—O and the C—H bonds, whereas the C—O bond is shortened. *Ab initio* calculations indicate the following changes in atomic gross charges (Rode *et al.*, 1975): The positive unit charge of the gaseous lithium ion is reduced by gaining 0.007 electron to become +0.993 charge units. Despite this small charge transfer between ion and solvent molecules, there is a considerable pileup of negative fractional charge at the oxygen

Table 7.1. Changes of Fractional Nuclear Charges, Δq_O at the O Atom and Δq_C at the C atom, of the C=O Group in the Presence of Cations

Cation	Ligand	Δq_O (electron)	Δq_C (electron)
Li^+	HCHO	−0.293	+0.117
Li^+	HCOOH	−0.239	+0.137
Li^+	DMF	−0.284	+0.113
Na^+	HCHO	−0.213	+0.095
Be^{2+}	HCHO	−0.707	+0.272
Be^{2+}	DMF	−0.700	+0.243

donor atom with appropriate increases in positive fractional charges at the C atom, as well as at the oxygen and hydrogen atoms (Table 7.1):

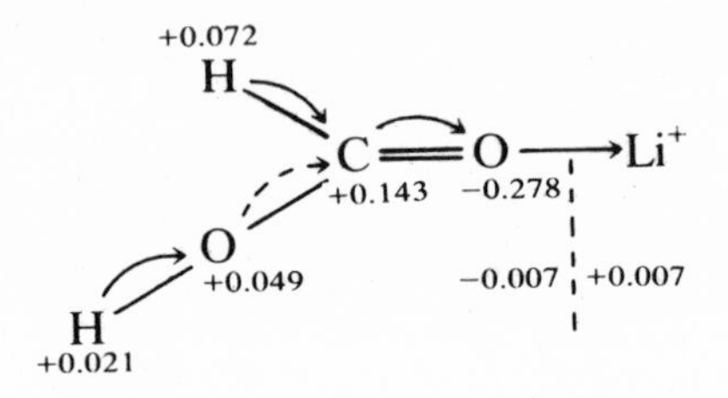

7.4. *Outer-Sphere Effects in Cation Solvation*

Both the geometry and the energy of the inner-core–solvate complex is changed by outer-sphere solvation (Verwey, 1941). The changes are due to the electron redistribution: Between the first and second coordination sphere hydrogen bonding is stronger than between uncoordinated water molecules. The electron drift from the oxygen atoms of the second-sphere water molecules toward the hydrogen atoms of the first-sphere water molecules is continued through the oxygen atoms toward the cation. This leads to the following changes: decrease in metal ion–inner-sphere bond distance, decrease in positive fractional charge at the coordination center, increase in H—O bond distance in the inner-sphere bonded water molecules, a smaller increase in O—H bond length within the water molecules coordinated in the second hydration sphere, and increase in acidity at the terminal hydrogen atoms. This has also been called the "outer-sphere effect for donor ligands" (Gutmann, 1973; Gutmann and Schmid, 1974).

The bond distances within the hydrated cation,

as compared to those in uncoordinated water units

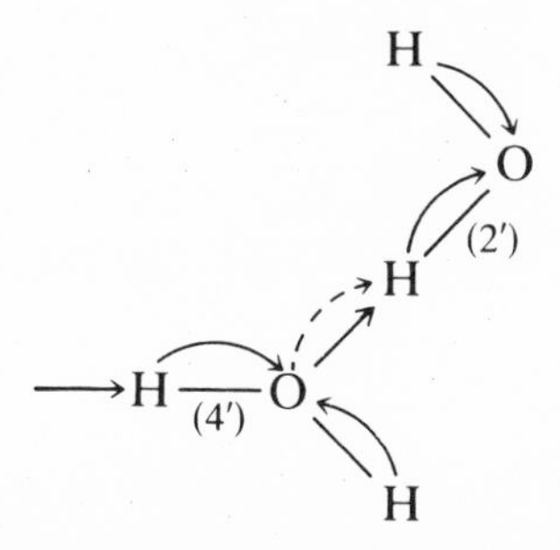

are expected to be as follows: Bond (2) is longer than bond (2′) and bond (4) is longer than bond (4′); bond (4) is not as long as bond (2). All these effects are cooperative in working toward the same direction: Bond (1) is further shortened by the electron flow from the outer-spheres toward the inner core. The extent of these cooperative effects is decisively influenced by the strength of the original coordinate bond (1), e.g., by the electron-pair acceptor properties of the coordination center. The stability of the hydration structure requires further suitable steric arrangements for the molecules within each hydration sphere.

Both the individual effects of the coordination center and the cooperative effects between the subsequent hydration spheres are responsible for the final structural arrangement of the hydrated complex ion, which is characteristic and highly specific for each kind of ion (Fitzgerald *et al.*, 1968).

The strength of the inner-sphere (metal ion–solvent) coordinate bond is determined by the acceptor properties of the metal ion, the donor properties of the solvent, and the outer-sphere effects, and hence by the solute concentration and the presence of other solutes.

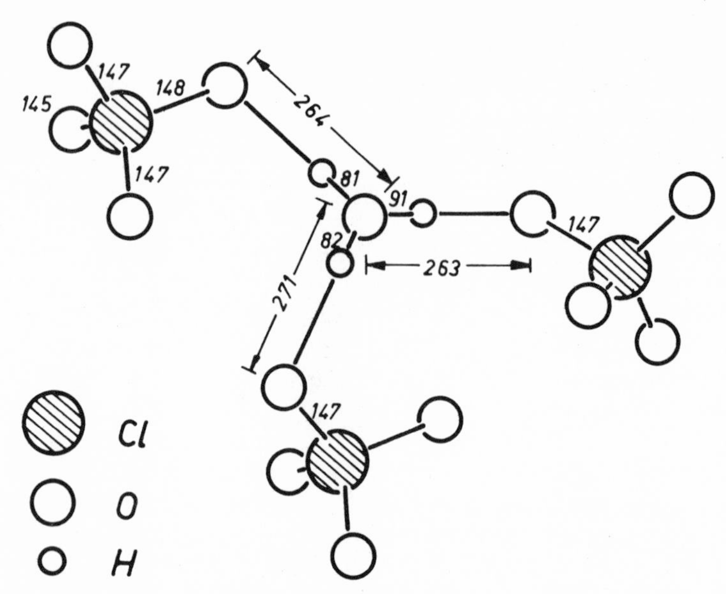

Fig. 7.4. Crystal structure of $HClO_4 \cdot H_2O$ (from Lundgren and Olavsson, 1976, courtesy of International Union of Crystallography).

Results of X-ray diffraction methods provide detailed information on internuclear distances and bond angles in crystalline hydrates of strong acids (Lundgren and Olavsson, 1976). These factors greatly depend on the molecular environment, e.g., the nature and relative position of anions as well as the number of water molecules. The structure of the H_3O^+ ion is symmetrical as long as each of the three hydrogen atoms is coordinated to the same group, e.g., as in $HClO_4 \cdot H_2O$ (Fig. 7.4), where the O···O bond distance is 266 pm (Nordman, 1962).

The effect of the counter ion may be seen by comparing the bond distances in $HClO_4 \cdot H_2O$ (Fig. 7.4) and in $HClO_4 \cdot 2H_2O$ (Fig. 7.5). The latter structure may be derived from the former by replacing a perchlorate ion linked to one of the three hydrogen positions by a water molecule. Since water is a stronger donor than the perchlorate ion the O···O bond distance is shortened and the O—H bond distance (within the H_3O^+ unit) is lengthened from 96 to 121 pm, so that a symmetrical hydrogen bond is provided within the newly established $H_5O_2^+$ unit (Olavsson, 1968). The bond-length variation rules are followed throughout the structure: The two adjacent O···O (perchlorate) bond distances have been lengthened from 264 to 279 pm and the O—Cl bond distances within each of the perchlorate groups have been shortened from 147 to about 143 pm (Fig. 7.5).

Deviations from the symmetrical O—H—O structure in the $H_5O_2^+$ ion are due to an asymmetrical environment such as in crystalline $HBr \cdot 2H_2O$, where the four Br—O bond distances are not equal (Lundgren, 1970). The neutron diffraction analysis (Attig and Williams, 1976) shows the bond-length relationships in accordance with the bond-length variation rules. The site with long Br—O bond distances of 324 and 326 pm provides an H—O distance of 97 pm and an O—H (bridge) bond distance of 122 pm. The stronger interaction on the other site with shorter Br—O bond distances of 320 pm and 325 pm increases the adjacent H—O bond distance to 99 pm and 98 pm, respectively, and decreases the O—H (bridge) bond distance to 117 pm (Fig. 7.6). The central proton has the greater vibrational amplitude along the bond axis, which indicates a broad and shallow potential minimum of the bridging bond that cannot be clearly distinguished from a symmetrical double minimum potential. In both cases the position of the bridging proton may be smeared.

The situation is analogous for the $H_5O_2^+$ unit in $H_2PtCl_6 \cdot 6H_2O$ (Peterson *et al.*, 1976). The site with long H—O bond distances of 99 and 95 pm induces an O—H (bridge) bond distance of 116 pm, and the site of the shorter H—O bond distances of 94 and 90 pm induces an O—H (bridge) bond distance of 122 pm (Fig. 7.7). An analogous situation is found in picrylsulfonic acid tetrahydrate $C_6H_2(NO_2)_3SO_3H \cdot 4H_2O$ (Lundgren and Tellgren, 1974).

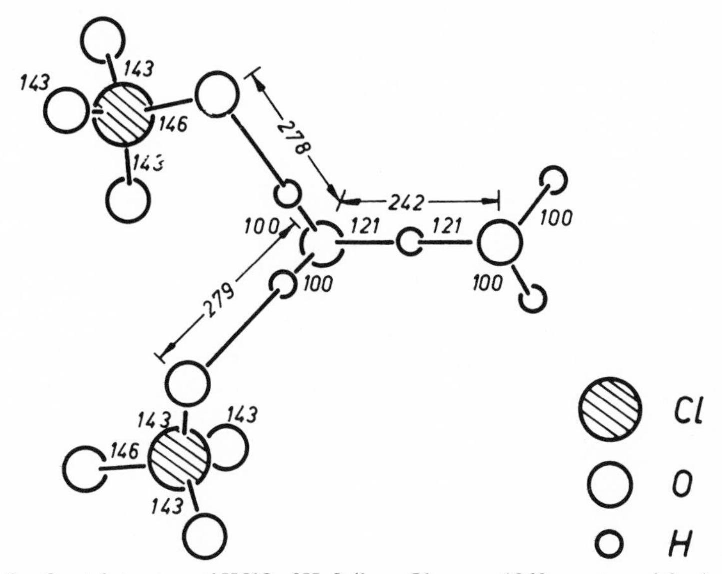

Fig. 7.5. Crystal structure of $HClO_4 \cdot 2H_2O$ (from Olavsson, 1968, courtesy of the American Institute of Physics).

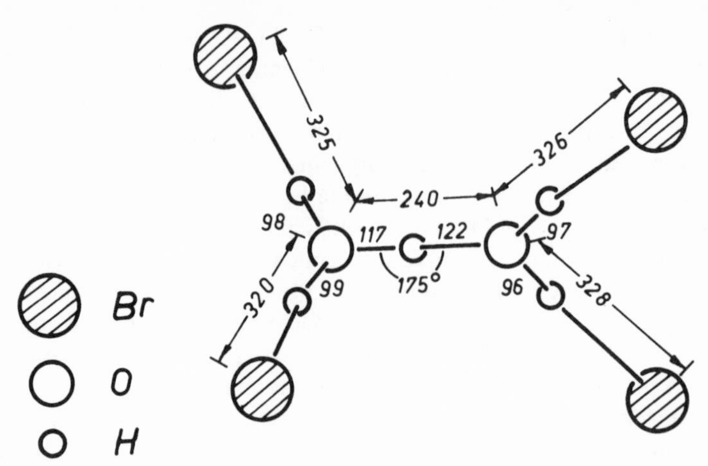

Fig. 7.6. Crystal structure of $HBr\cdot 2H_2O$ (from Attig and Williams, 1976, courtesy of Verlag Chemie, Weinheim).

The classical Eigen-type $H_9O_4^+$ ion is represented by a central H_3O^+ component pyramidically surrounded by three water molecules, with an average O···O bond distance of 257 pm and an average O···O···O bond angle of 111° (Lundgren and Olavsson, 1976). The H-bond distances and angles within the individual complexes deviate quite considerably in some cases due to the crystallographic environment of the water molecules surrounding the H_3O^+. The O···O distance within the $H_9O_4^+$ unit is increased by increasing coordination of the terminal oxygen atoms. For example, in $HClO_4\cdot 3\frac{1}{2}H_2O$ the O···O bond distance for a three-coordinate oxygen atom is 248 pm and hence shorter than that for a tetrahedrally coordinated oxygen atom (268 pm) (Almhöf, 1973).

The largest protonated unit structurally characterized is the $H_{13}O_6^+$ ion, which has a high symmetry [$2/m$ (C_{2h})] when symmetrically

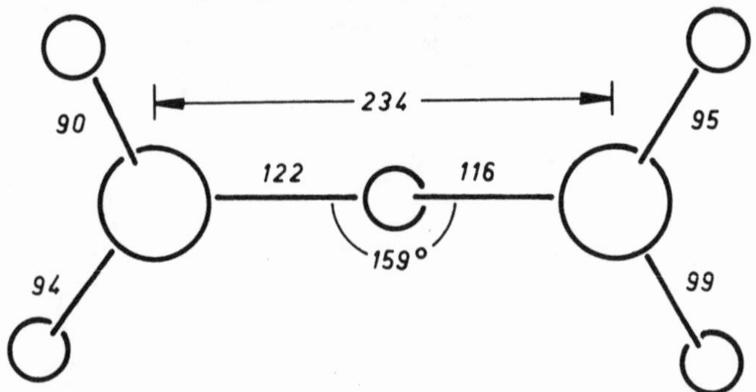

Fig. 7.7. Crystal structure of the $H_5O_2^+$ ion in $H_2PtCl_6\cdot 6H_2O$ (from Peterson *et al.*, 1976, courtesy of Hüthig-Verlag, Heidelberg).

surrounded by chloride ions (Bell *et al.*, 1975). The central O—H—O hydrogen bond lies across a center of symmetry with the O···O bond distance of 239 pm, which is among the shortest such distances observed to date. An electron-density-difference map supports the expectation of a symmetrical hydrogen bond with its potential function represented by a curve with a single minimum, although a certain degree of asymmetry can never be excluded (Hamilton and Ibers, 1968).

Cation solvation in different aprotic solvents may be characterized by the donor number of the solvent. A linear relationship exists between the latter and the chemical shift of the ^{23}Na nucleus in solutions of sodium perchlorate or sodium tetraphenylborate in various solvents (Erlich *et al.*, 1970; Erlich and Popov, 1971; Popov, 1975) (Fig. 7.8). However, no relationship is found between the ^{23}Na chemical shift and the dipole moment or the dielectric constant of the solvents. Thus the chemical shift of the ^{23}Na nucleus is an alternative measure for the solvating power of a donor solvent. Although the electrostatic contributions to solvate bonding dominate, the differences in charge densities in different solvents are related to the donor strength of the solvent toward the sodium ion. An analogous relationship has been found between the $^{205}Tl^+$ chemical shift in

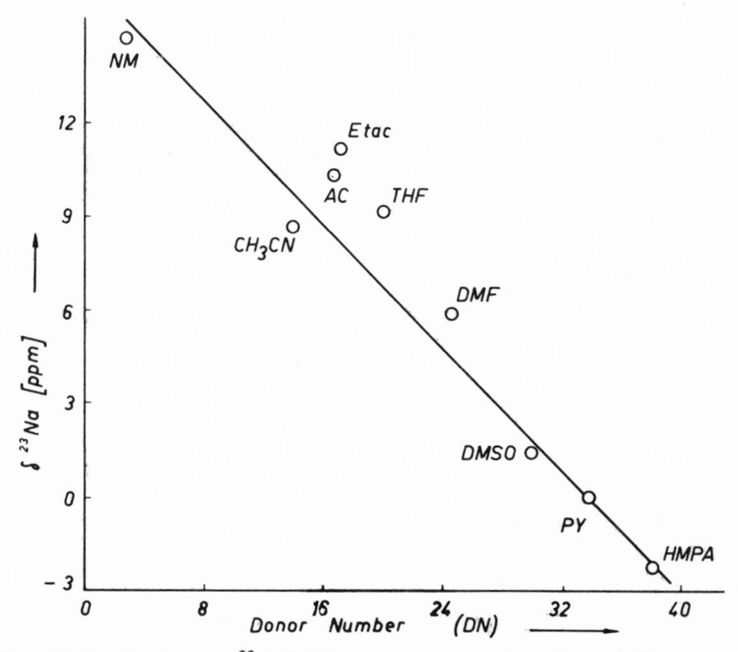

Fig. 7.8. Relationship between ^{23}Na NMR chemical shifts of $NaClO_4$ in aprotic solvents and the donor number of these solvents (from Erlich and Popov, 1971, courtesy of the American Chemical Society).

Table 7.2. Polarographic Half-Wave Potentials of $[Co(en)_3][ClO_4]_3$ in Various Solvents

Solvent	$E_{1/2}$ (volts)	DN	ε
NM	+0.61	2.7	35.9
BN	+0.44	11.9	25.2
CH_3CN	+0.46	14.1	38.0
Acetone	+0.23	17.0	20.7
DMF	+0.28	26.6	36.1
DMSO	+0.12	29.8	45.0

different solvents and the donor number of the solvent (Hinton and Briggs, 1975).

The outer-sphere effect for cations is well documented by the results of polarographic measurements on $Co(en)_3^{3+}$ in solvents of different donor number (Mayer *et al.*, 1977*a*). As the donor number of the outer-sphere solvent increases, the half-wave potential is shifted to more negative values (Table 7.2).

The effects are analogous when anions are involved in outer-sphere coordination, which has frequently been considered as ion-pair formation. The association constants of $Co(en)_2Cl_2{}^+Cl^-$ in different solvents (Fitzgerald *et al.*, 1968) are not related to the dielectric constant or the dipole moment of the solvent, but rather to its donor and acceptor properties. In the process,

$$Co(en)_2Cl^{+}_{2\,solv} + Cl^{-}_{solv} \rightleftharpoons [Co(en)_2Cl_2]Cl$$

solvent molecules are replaced at both the cationic species and at the chloride ion. The smaller the donor number of the solvent, the greater the ease of displacement of solvent molecules at the cation; the smaller the acceptor number of the solvent, the greater the donor property of the chloride ion and the more readily the latter is desolvated prior to outer-sphere coordination at the complex cation (Mayer, 1976). Hence the formation constant of the outer-sphere complex $Co(en)_2Cl_2{\cdot}Cl$ is greater the smaller the donor number and acceptor number of the solvent. The increase in formation constant in the order, MeOH < DMSO < DMF < DMA (Table 7.3) primarily reflects the decreasing acceptor properties of the solvent. If MeOH had a higher donor number, the formation constant would be even smaller in this solvent; the high value in TMS is due mainly to the low donor number of the solvent.

In concentrated salt solutions outer-sphere cation–anion interactions are due to mutual penetration of the outer-sphere hydration layers. This type of species is sometimes considered as a "solvent-separated" or "solvent-shared" ion pair (Schaschel and Day, 1968; Hammonds and Day, 1969) although they differ from the latter in spectroscopic and in redox

Table 7.3. Formation Conatants of the 1:1 Outer-Sphere Complex between $[Co(en)_2Cl_2]^+$ and Cl^-

Solvent	K_{form}	DN	AN	ε	μ (Debye)
Methanol	150	19.0	41.3	32.6	1.7
DMSO	400	29.8	19.3	46.7	3.9
DMF	8,000	26.6	16.0	36.7	3.8
DMA	20,000	27.8	13.6	37.8	3.8
TMS	42,000	14.8	19.0	43.3	4.8

properties as well as in dissociation behavior: While a solvent-separated ion pair in the sense of Bjerrum (1926) with tight solvation layers is separated by increasing the dielectric constant of the medium, solvent-shared ion pairs remain combined under these conditions. In order to avoid confusion, the species is better described as an "outer-sphere complex" involving ions of opposite charge (see Chapter 10).

7.5. *Outer-Sphere Effects in Anion Solvation*

The charge-density rearrangement due to the coordination of an anion by water molecules in the inner hydration sphere leads to a lengthening of the H—O bond and to an increase in donor property at the oxygen atom, which will interact with water molecules to build up a second hydration layer (outer-sphere solvation). The bond distances within the hydrated anion are shown below:

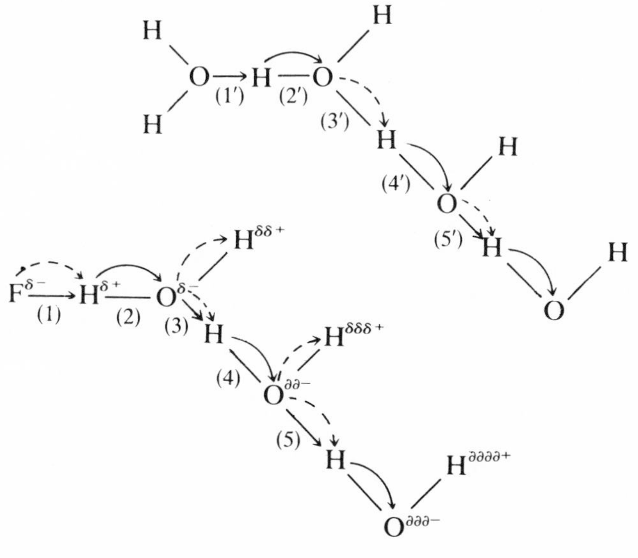

Bond (2) is longer than bond (2′), bond (3) is shorter than bond (3′), and bond (5) is longer than bond (5′). All of these effects are cooperative: Bond (1) is further shortened by the addition of hydration spheres, by which bond (2) is further lengthened, bond (3) is further shortened, and bond (4) is further lengthened. The strength of the inner-sphere bond (1) is the result of both the donor properties of the anion and the acceptor properties of the bulk solvent.

The outer-sphere effect of anions is analogous to that of cations, as outer-sphere coordination results in the shortening of the inner-sphere coordinate bonds. This has been referred to as the "outer-sphere" effect for acceptor ligands" (Gutmann and Schmid, 1974).

The results of quantum chemical calculations, indicating the charge-density rearrangement due to outer-sphere coordination, are given in Table 7.4.

Outer-sphere coordination of hexacyanoferrate ions is reflected in changes of half-wave potentials for the reduction of hexacyanoferrate(III) to hexacyanoferrate(II) (Gutmann *et al.*, 1976; Gritzner *et al.*, 1976*a*). The half-wave potentials are shifted to more positive potential values by an increase in the acceptor number of the solvent (see Section 8.1).

The vibrational spectrum of the cobalt tetracarbonylate anion $Co(CO)_4^-$ has been used as a probe of solution structure at ion sites in different solvents (Edgell *et al.*, 1971; Edgell and Lyford, 1971). The C—O stretching vibrations of the $Co(CO)_4^-$ anion in six different solvents showed a single absorption band ν_M. Extremely careful computer resolution of the band shape revealed also a very weak band ν_s at the low-frequency side, which was assigned to the ^{13}C species present (Edgell and Barbetta, 1974). This band showed the same trend as a function of the solvent, as does the main band ν_M. It has been concluded that "the solution

Table 7.4. Charge Distribution in the Hydrated Fluoride Ion of Tetrahedral Coordination

Number of water molecules	Net charge of central ion (electron)	Mean charge (electron) of water molecules		
		Sphere 1	Sphere 2	Sphere 3
1	−0.65	−0.35	—	—
2	−0.59	−0.21	—	—
4	−0.55	−0.11	—	—
7	−0.55	−0.09	−0.06	—
16	−0.55	−0.05	−0.02	—
17	−0.55	−0.05	−0.02	−0.01

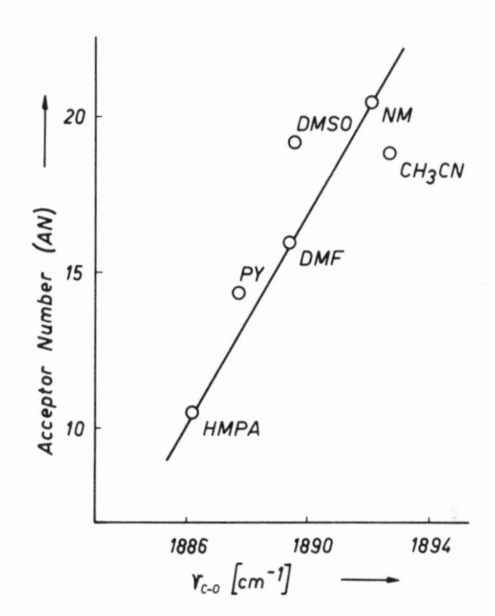

Fig. 7.9. Relationship between the C—O stretching frequency of $[Co(CO)_4]^-$ in various solvents (Edgell and Barbetta, 1974) and the acceptor number of these solvents (from Gutmann, 1977*b*, courtesy of Springer-Verlag, Vienna, New York).

environment of the anions consists of just one kind of site with forces coming from the medium which are large enough to influence the rotary–translatory motion and small relative to the forces required to distort the C—O bonds of the anion" (Edgell and Barbetta, 1974).

A correlation exists between the acceptor number of the solvent and ν_M (Fig. 7.9). However, an increase in acceptor number causes an increase in wave number corresponding to shortening of the C—O bonds, in violation of the first bond-length variation rule (Gutmann, 1977*b*). The tetracarbonylate ion contains cobalt with the oxidation number −1, and the Co—C π-bonding is stronger and the C—O bonds are longer than in states of higher oxidation number for cobalt. The electrophilic solvent attack at the oxygen atoms of the carbonyl groups within the $Co(CO)_4^-$ ion leads to a decrease in electron density in particular around the cobalt atom. This leads to weakening of Co—C π-bonding as well as to shortening of the C—O distances.

The outer-sphere effects on oxo anions by strong acceptors will be discussed in Section 15.1.

Chapter 8

Redox Properties

8.1. Solvent Effects

The relationships between thermodynamic properties of ions and empirical parameters of solvents have been exemplified in Section 7.2. We shall now examine in more detail the effects of solvents on redox properties.

The previously derived relationships between thermodynamic properties of single ions and empirical parameters of solvents can be used together with some chemical intuition to make semiquantitative predictions about solvent effects on redox properties. The standard redox potential E^0 of a redox active species is defined by

$$-\Delta G^0 = n \cdot F \cdot E^0$$

For a given redox system, in different solvents E^0 is determined by the interaction of the solvent with both the oxidized and reduced form of a redox couple. Thus knowledge of the free enthalpy of solvation of single ions (Sections 7.1 and 7.2) will yield information on E^0. Interaction of the solvent with only the oxidized form of a redox couple (e.g., Ag/Ag^+) causes a decrease in free enthalpy of this species and a shift of the E^0 to more negative values.

Interaction of a solvent with only the reduced form will lead to a shift of the E^0 to more positive values. In cases where both forms of a redox couple interact with the solvent (e.g., Fe^{3+}/Fe^{2+}), the direction of the shift will be determined by the relative strength of the interactions.

Solvent interactions with cations are generally stronger for the oxidized form. Since the σ-acceptor properties of cations increase with the oxidation state, a shift to negative values will generally be observed with increasing donor properties of the solvent. Stronger interaction with the reduced form of cations can be observed when the reduced form is capable of back donation and the solvent can act as a π-acceptor. Such interactions can exceed the interaction of the solvent with the higher-valent form, so that E^0 will shift to more positive values. Such interactions will often lead to stabilization of lower-valent forms of cations not generally observed in hard solvents. An example is the copper(II)–copper(I) couple in acetonitrile (Coetzee *et al.*, 1963; Duschek and Gutmann, 1972), dimethylthioformamide (Gutmann *et al.*, 1974*a*), and *N*-methylthiopyrrolidinone (Gritzner *et al.*, 1976*a*).

Solvent interactions with anions will generally stabilize the reduced form, which carries the higher negative charge. This causes a shift of E^0 values for anions to more positive values with increasing acceptor properties of the solvents (Gutmann, 1973).

Polarographic and voltammetric techniques in combination with suitable reference redox systems, as well as potentiometry employing a suitable salt bridge, have been most frequently used to obtain information about solvent effects on redox properties. The polarographic half-wave potential, as well as half the difference of the anodic and cathodic peak potentials derived in cyclic voltammetry, agree within a few millivolts with the standard redox potential for electrode reactions that are reversible, and both forms of the redox couple are dissolved (Heyrovsky and Kůta, 1965). In the case of amalgam formation the standard redox potential contains the free enthalpy of amalgam formation. Conversion to the standard redox potential of the metallic state can be made since free enthalpies of amalgam formation are available in the literature (Heyrovsky, 1941).

In order to compare half-wave potentials in various solvents, a proper reference redox system is needed. Such a reference redox system should consist of a redox couple, the standard redox potential of which is solvent independent (Pleskov, 1947; Koepp *et al.*, 1960; Strehlow, 1966; Gritzner *et al.*, 1968). These conditions are best approximated bv a redox couple where the interactions with the solvent are small and of the same magnitude for both the oxidized and reduced form. Rubidium was originally proposed (Pleskov, 1947), whereas ferrocene and *bis*-biphenylchromium are the most commonly used reference redox systems. A method of interconversion between these three systems has been published recently (Gritzner, 1977*a*). Unfortunately *bis*-biphenylchromium cannot be used as a reference redox system in aqueous solution due to the insolubility of the reduced form in water and to absorption phenomena of both the reduced and the oxidized form (Gritzner, 1975). The applicability of ferrocene in water has also been questioned (Kolthoff and Chantooni, 1971; Alfenaar, 1975). For these reasons water will be excluded when solvent effects on half-wave potentials are compared.

Potentiometric data in various nonaqueous solutions have been obtained by means of a salt bridge of 0.1 M tetraethylammoniumpicrate in one of the solvents under study (Diggle and Parker, 1973; Owensby *et al.*, 1974). It has been assumed that under such conditions there is a negligible liquid junction potential at the boundary of the bridge solvent and the other solvent. Potential values and free energies of transfer obtained on the basis of this assumption agree reasonably well with those obtained by use of a reference redox system (Gritzner, 1977*a*) as well as by the tetraphenylarsonium tetraphenylborate assumption (see Section 7.1).

For cations that are reduced to the metal or to the metal amalgam, the solvent interaction takes place with the oxidized form only. In this case the solvent effect on redox properties expresses itself most clearly. The redox potential of the system $M^{n+}_{solv} + n \cdot e \rightleftharpoons M^0$ will be a function of the donor properties of the solvent: The greater the donor number of the solvent, the more negative the standard redox potential. Correlations between half-wave potential and the donor number of the solvent have been obtained experimentally for a number of cations in a variety of solvents (Gutmann and Schmid, 1969; Duschek and Gutmann, 1973*a*, *b*; Duschek *et al.*, 1974). A graphical description for Na^+ and K^+ is given in Fig. 8.1. The correlations found for several other cations are given in Figs. 8.2 and 8.3.

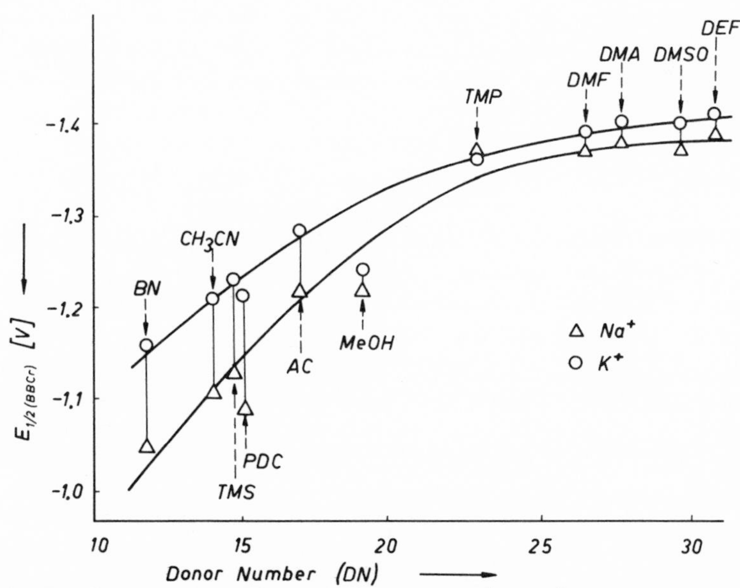

Fig. 8.1. Relationship between half-wave potentials of Na^+ and K^+ vs. that of *bis*-biphenylchromium in various solvents and the donor numbers of these solvents (from Duschek and Gutmann, 1973*a*, courtesy of Springer-Verlag, Vienna, New York).

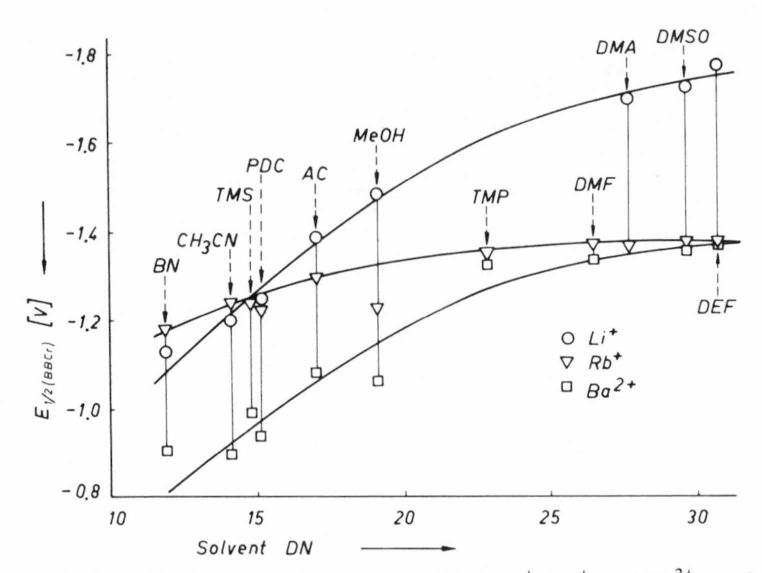

Fig. 8.2. Relationship between half-wave potentials of Li^+, Rb^+, and Ba^{2+} vs. that of *bis*-biphenylchromium in various solvents and the donor number of these solvents (from Duschek and Gutmann, 1973*b*, courtesy of Springer-Verlag, Vienna, New York).

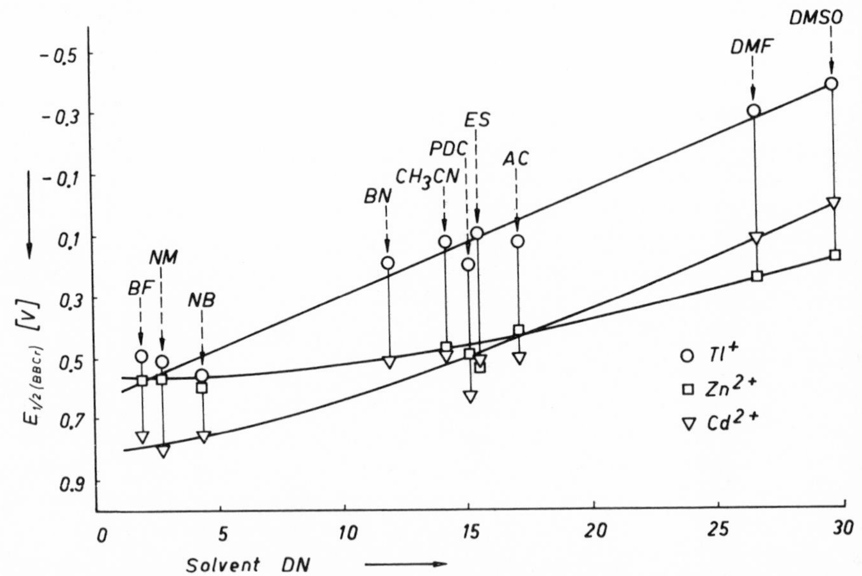

Fig. 8.3. Relationship between half-wave potentials of Tl^+, Zn^{2+}, and Cd^{2+} vs. that of *bis*-biphenylchromium in various solvents and the donor numbers of these solvents (from Gutmann, 1971*a*, courtesy of Springer-Verlag, Vienna, New York).

A correlation between the $E_{1/2(\text{BBCr})}$ and the donor number of the solvents studied has also been found for the reduction of *bis*-triphenylphosphine mercury(II) perchlorate to metallic mercury (Gritzner *et al.*, 1976*b*). Although two coordination sites are taken by triphenylphosphine ligands, there are at least two more sites for solvent coordination. Their occupation by solvent molecules leads to a shift of the redox potential to more negative values as the donor number of the solvent increases.

The examples given so far deal with cases where only the oxidized form interacted with the solvent. In cases where both the oxidized and reduced form remain in solution, the solvent effect on both forms of the redox couple has to be considered.

The coordination of solvent molecules with only the higher-valent form of a cation in a redox couple would lead to a shift of the redox potential to more negative values. A coordination of the solvent with only the lower-valent form would lead to a shift to more positive values. For hard cations the solvent will interact with both forms of the redox couple. The interaction with the higher-valent form of hard cations will generally be stronger than that with the lower-valent form. These differences are reflected in the shift of the redox potentials to more negative values. The $E_{1/2(\text{BBCr})}$ of the reactions

$$\text{Sm}^{3+}_{\text{solv}} + \text{e} \longrightarrow \text{Sm}^{2+}_{\text{solv}}, \qquad \text{Yb}^{3+}_{\text{solv}} + \text{e} \longrightarrow \text{Yb}^{2+}_{\text{solv}}$$

$$\text{Eu}^{3+}_{\text{solv}} + \text{e} \longrightarrow \text{Eu}^{2+}$$

as a function of the donor number of the solvents are given in Fig. 8.4 (Gutmann and Peychal-Heiling, 1969).

Such correlations between donor properties of solvents and cations are also applicable to complex cations. In the case of the redox couple $\text{Co(en)}_3{}^{3+}/\text{Co(en)}_3{}^{2+}$ a donor solvent can interact with the hydrogen atoms of the ligand. Such an interaction increases the stability of the oxidized form more than the stability of the reduced form. Thus a shift to more negative values of the half-wave potentials with increasing donor properties of the solvent has been observed (Mayer *et al.*, 1977*a*) (Table 7.2).

Recent studies of *N*-methylthiopyrrolidinone (Gritzner *et al.*, 1977) and dimethylthioformamide (Gutmann *et al.*, 1974*a*; Alexander *et al.*, 1974) have shown that the solvent effects in these solvents differ remarkably for hard and soft cations. Figure 8.5 shows a comparison of $E_{1/2(\text{BBCr})}$ of several cations in *N*-methylpyrrolidinone(2) and *N*-methylthiopyrrolidinone(2). Typically hard cations such as Na^+, K^+, and Rb^+ undergo much stronger interactions with the O-donor atom in *N*-methylpyrrolidinone than with the S-donor atom in *N*-methylthiopyrrolidinone. For typically soft cations such as Ag^+ and Cu^+

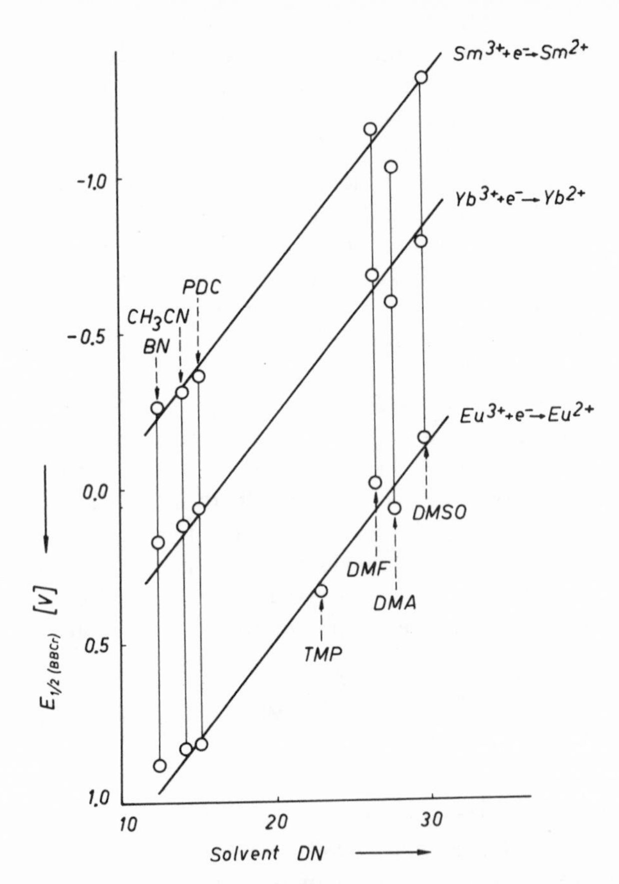

Fig. 8.4. Relationship between half-wave potentials of the reductions Sm^{3+}/Sm^{2+}, Eu^{3+}/Eu^{2+}, and Yb^{3+}/Yb^{2+} vs. that of *bis*-biphenylchromium in various solvents and the donor numbers of these solvents (from Gutmann, 1971*a*, courtesy of Springer-Verlag, Vienna, New York).

the effect is reversed. Since empirical parameters to describe the nature of soft–soft interactions are not available, the interpretation of the effects of soft solvents is limited to a qualitative description. The redox potential of a soft cation will be shifted to more negative potential values in a solvent with a soft donor atom as compared to a solvent with a hard donor atom. The reverse will hold true for hard cations. Soft cations very often have the ability to undergo π-back-bonding. Such interactions are responsible for the negative redox potential of Ag^+ and Cu^+ in nitrile solvents such as acetonitrile or benzonitrile (Duschek and Gutmann, 1972).

Soft solvents have the ability to stabilize the low-valent forms of certain cations. Thus the reduction of Cu^{2+}—in most hard solvents a

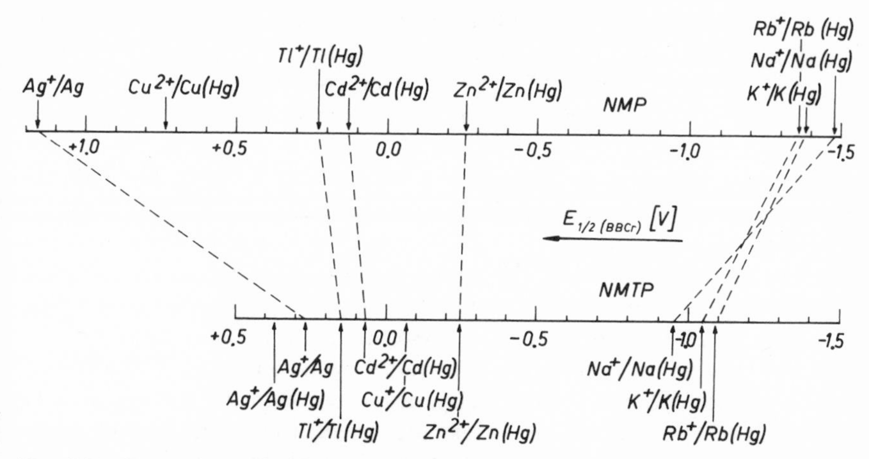

Fig. 8.5. Comparison of half-wave potentials vs that of *bis*-biphenylchromium in *N*-methylpyrrolidinone and *N*-methylthiopyrrolidinone (from Gritzner *et al.*, 1977, courtesy of Elsevier Sequoia, Lausanne).

one-step two-electron process—is a two-step process via Cu^+ in acetonitrile. Studies on the redox system $Mn(CNCH_3)_6^{2+}/Mn(CNCH_3)_6^+$ have shown that in this redox system, interaction with donor solvents leads to an increase in stability of the reduced form over the oxidized form with increasing donor properties of the solvent (Gritzner, 1976, 1977*b*) (Fig. 8.6).

In studies using polarographic or voltammetric methods, the redox behavior of single-atom anions was found to be irreversible. Complex

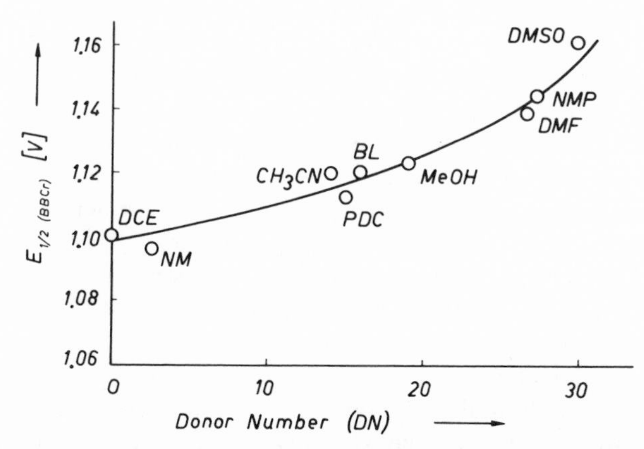

Fig. 8.6. Relationship between half-wave potentials of the oxidation $Mn(CNCH_3)_6^+/Mn(CNCH_3)_6^{2+}$ vs. that of *bis*-biphenylchromium in various solvents and the donor numbers of these solvents (from Gritzner, 1976, courtesy of Springer-Verlag, Vienna, New York).

anions had to be selected to study the solvent effects on anions. The system $Fe(CN)_6^{3-}/Fe(CN)_6^{4-}$ has been investigated in 12 solvents (Gutmann *et al.*, 1976; Gritzner *et al.*, 1976*a*). The interaction of the solvent with the reduced form, namely, $Fe(CN)_6^{4-}$, is much stronger than with the oxidized form. For this redox system, two effects can be held responsible for the stronger interaction of the solvent with the reduced form: (i) The reduced form carries a higher negative charge, (ii) the change in the formal oxidation state of iron from +3 to +2 enhances the donor properties of the nitrogen atoms on each of the cyano groups, since Fe(II) is a weaker σ-electron-pair acceptor and a stronger π-electron-pair donor than Fe(III). The solvent effects on this redox system therefore lead to a shift of half-wave potentials to more positive values with increasing acceptor properties of the solvent. A linear relationship between the solvent activity coefficient of the chloride ion (Alexander *et al.*, 1974) and the half-wave potentials of this redox system exists (Fig. 8.7) as well as a relationship with the acceptor number of the solvent (Mayer *et al.*, 1975; Gutmann, 1976*b*) (Fig. 8.8).

8.2. Ligand Effects

In a solution of a complex compound, competition takes place between the ligands and solvent molecules for coordination sites on the redox active species. This discussion will be restricted to cases where the solvent molecules are in large excess over the ligand molecules. Stable complexes with ligands will only occur when the donor or acceptor properties, respectively, of the ligands exceed those of the solvent. The reference state in a given solvent is the standard redox potential of the solvated species. In water the standard redox potential vs. the standard hydrogen electrode is reported. Standard hydrogen electrodes as well as other suitable reference electrodes have been used in nonaqueous solvents (Bauer and Foucault, 1976; Owensby *et al.*, 1974) (see Fig. 7.2).

Ligand effects on the redox behavior of a redox active species are analogous to solvent effects. For cations the reduced form of which is independent of the solvent and of complex formation, e.g., the metal state, coordination of the cation by ligands instead of solvent molecules will shift the redox potential to more negative values. For coordination of ligands with anions where the oxidized form is independent of solvent and ligand, a shift to more positive potentials will be observed. These rules have long been employed to measure stability constants. The change in the redox potential due to coordination by ligands has been expressed by a decrease in the concentration of "free," that is, solvated, ions as employed in the Nernst

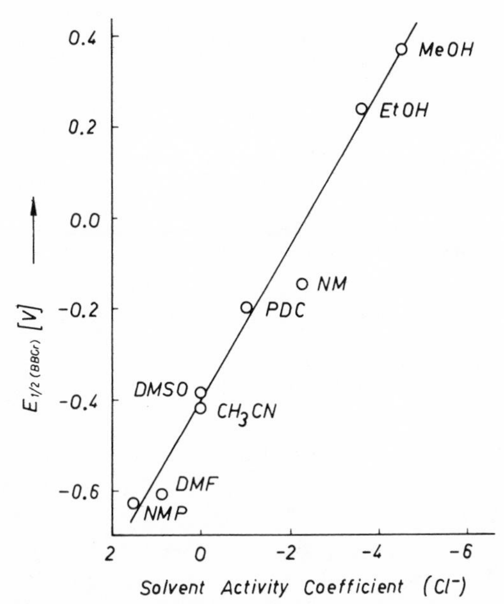

Fig. 8.7. Relationship between half-wave potentials of the reduction $Fe(CN)_6^{3-}/Fe(CN)_6^{4-}$ vs. that of *bis*-biphenylchromium in various solvents and the solvent activity coefficient of the chloride ion.

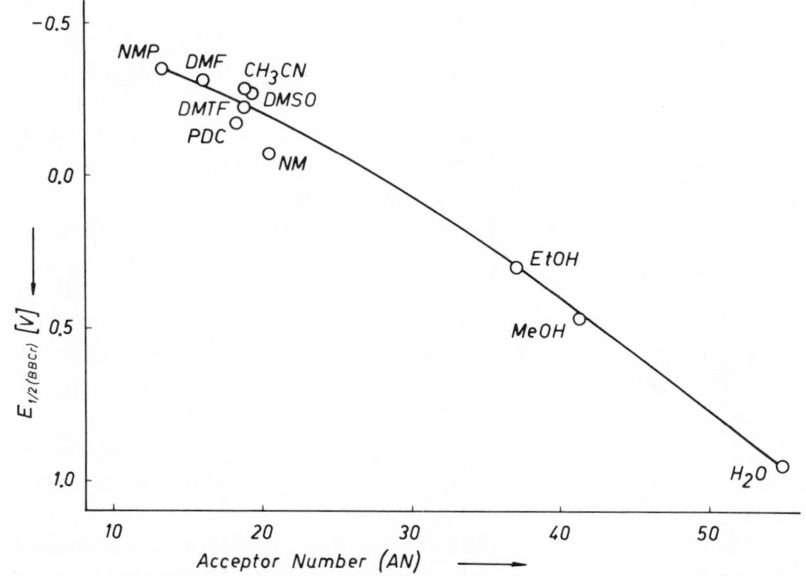

Fig. 8.8. Relationship between half-wave potentials of the reduction $Fe(CN)_6^{3-}/Fe(CN)_6^{4-}$ vs. that of *bis*-biphenylchromium in various solvents and the acceptor numbers of these solvents.

equation. For the reaction, $MS_n^{m+} + me \longrightarrow M^0 + nS_{(l)}$, where M^{m+} represents the metal ion and S represents solvent molecules, the Nernst equation yields

$$E = E^0 + \frac{RT}{nF} \ln \frac{a_{MS_n^{m+}}}{a_{M^0}} \tag{1}$$

The effect of complex formation will be exemplified by the reaction

$$MS_n^{m+} + nL = ML_n^{m+} + nS \tag{2}$$

The ligands used in these equations are uncharged and the coordination number is the same for ligand and solvent, for reasons of simplicity. It should be noted that many other ligands which coordinate to cations are negatively charged. For treatment of the redox behavior it is now generally assumed that only the "free" ions can undergo the redox processes. The concentration of "free" ions can be obtained from the equilibrium constant of reaction (2). This yields the Nernst equation in the form

$$E = E^0 + \frac{RT}{nF} \ln \frac{a_{ML_n^{m+}}}{K \cdot a_L^n \cdot a_{M^0}}$$

where K is the equilibrium constant of reaction (2). It becomes obvious that for a given ligand concentration, E becomes more negative with increasing values of K. For cases where both the oxidized and reduced form are subject to complex formation, the change in redox properties will be determined by the form that undergoes stronger interaction with the ligands. Extended forms of the Nernst equation have been derived and are accessible (Heyrovsky and Kůta, 1965). If complex formation with the oxidized form is stronger than complex formation with the reduced form of a redox couple the redox potential will shift to more negative values. This will be generally the case for complexes between hard cations and hard ligands. If the reduced species forms a stronger complex than the oxidized species the presence of the ligand will shift the redox potential to more positive values. Such behavior may occur for cations whose reduced form is soft, e.g., Cu^+, and whose ligand is also soft, such as the cyanide ion or ligands containing sulfur as the donor atom.

Soft ligands can often stabilize oxidation states of a cation, e.g., Cu^+, which are not generally observed in hard solvents, e.g., water. Another example is the reduction of $Mn(CNCH_3)_6^{2+}$ to the metal, which proceeds via a very stable $Mn(CNCH_3)_6^+$ complex, whereas a reduction of Mn(II) in the absence of soft ligands proceeds directly to the metal. This unusual stability of the $Mn(CNCH_3)_6^+$ complex is due to the π-donor properties of Mn(I) and the π-acceptor properties of methylisonitrile. π-donation by the metal ion and π-acceptance by the ligand play an important part in soft–soft interactions.

Chapter 9

Solvation in Solvent Mixtures

9.1. *Preferential Solvation*

The common feature of ionic solutions in solvent mixtures is the presence of mixed solvates (Kortüm and Weller, 1950).

Ion solvation in solvent mixtures is usually interpreted in terms of "preferential solvation" by one of the solvent components: For example, in a mixture of water and methanol, nickel(II) ions are preferentially solvated by water molecules (MacKellar and Rorabacher, 1971). Likewise, most cations in mixtures of water and dioxane, THF, or PDC are preferentially solvated by water, while in pyridine–water mixtures pyridine is preferred by the cations (Fratiello and Douglas, 1963*a*, *b*; Fratiello and Christie, 1964; Fratiello and Miller, 1965; Maciel *et al.*, 1966; Hainton and Amis, 1967; Cogley *et al.*, 1971). Preferential solvation of alkali metal ions by DMSO occurs in its mixtures with 1-pentanol (Wong *et al.*, 1971) and by acetone in its mixtures with nitromethane (Maxey and Popov, 1968). The extent of preferential solvation may be characterized by the position of the "isosolvation" point, i.e., the solvent composition at which the coordinating properties of the mixture toward a given solute lie midway between those of the pure solvent components (Frankel *et al.*, 1965).

Preferential solvation of cations has been studied by the NMR technique. For example, the ^{23}Na chemical shift of $NaClO_4$ in various binary solvent mixtures as a function of solvent composition yields a smooth transition in proceeding from one pure solvent to the other (Frankel *et al.*, 1965, 1970), indicating preferential solvation by one of the solvents. Binary solvent mixtures of nitromethane with the donor solvents DMSO, pyridine, and HMPA exhibit isosolvation points 0.05, 0.12, and 0.15 mole fraction, respectively, of the nitromethane (Greenberg and Popov, 1975)

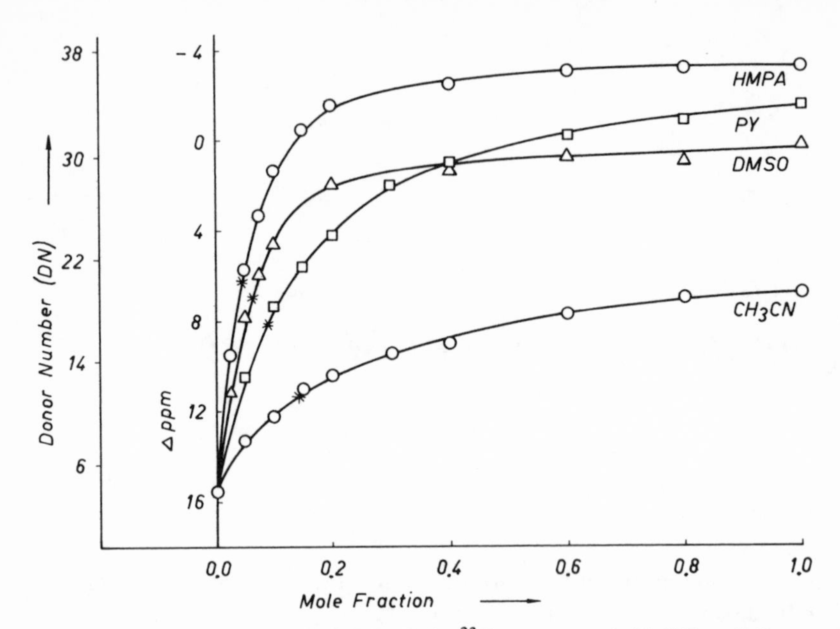

Fig. 9.1. Variation of the chemical shift of the ^{23}Na resonance in $NaClO_4$ solutions, and donor numbers estimated from the Popov correlation as a function of solvent composition for binary solvent mixtures of nitromethane with acetonitrile, dimethylsulfoxide, pyridine, and hexamethylphosphoric amide; * represent isosolvation points.

(Fig. 9.1). Na^+ is to a large degree preferentially solvated by these solvents relative to nitromethane.

In solutions of sodium tetraphenylborate in DMSO–CH_3CN mixtures the isosolvation point occurs at 15 mole percent DMSO, whereas for its solutions in pyridine–acetonitrile mixtures the isosolvation point is found at 40 mole percent pyridine (Fig. 9.2) (Erlich *et al.*, 1973). Although pyridine is a stronger donor solvent than DMSO, the latter is more strongly coordinated to Na^+ in acetonitrile mixtures below $x = 0.7$.

Apart from preferential solvation specific solvent–solvent interactions must be considered. Each particular solvent mixture is characterized by structural features which are absent in either of the pure solvents.

9.2. *Solvent–Solvent Interactions*

Regarding the DMSO–pyridine mixtures, DMSO is an associated liquid with a chainlike structure (Lindberg *et al.*, 1961), the sulfur atom acting as an acceptor atom toward the oxygen atom of another DMSO molecule (Szmant, 1972). The outer-solvation spheres around a given ion are more highly structured for DMSO than for pyridine molecules. In the presence of both DMSO and pyridine the original structures of the pure liquids have been replaced by new structural characteristics, resulting in an enhancement of the donor properties of DMSO in the mixtures (Greenberg and Popov, 1975). The structural features of the mixtures are hardly understood, although indications are available from the results of Brillouin scattering and infrared studies. The addition of small amounts of pyridine to DMSO results in a sharp increase in the ν_{S-O} frequency up to 5 mole percent pyridine; strong interactions of the type

$$\begin{array}{c} CH_3 \\ | \\ O{=}S{\leftarrow}PY \\ | \\ CH_3 \end{array}$$

may be involved which result in increasing donor properties of DMSO at its oxygen atoms (Gutmann and Schmid, 1974).

The structural features due to association between two types of solvent molecules can be seen by comparing the spectroscopic properties of the solvent mixture with its solution in an "inert" solvent. Schneider and Schulz (1976) have shown that the multiple spectral pattern of Al^{3+} NMR spectra in mixtures of DMSO and DMF that have similar donor numbers follows a statistical distribution when a fairly inert diluent, such as

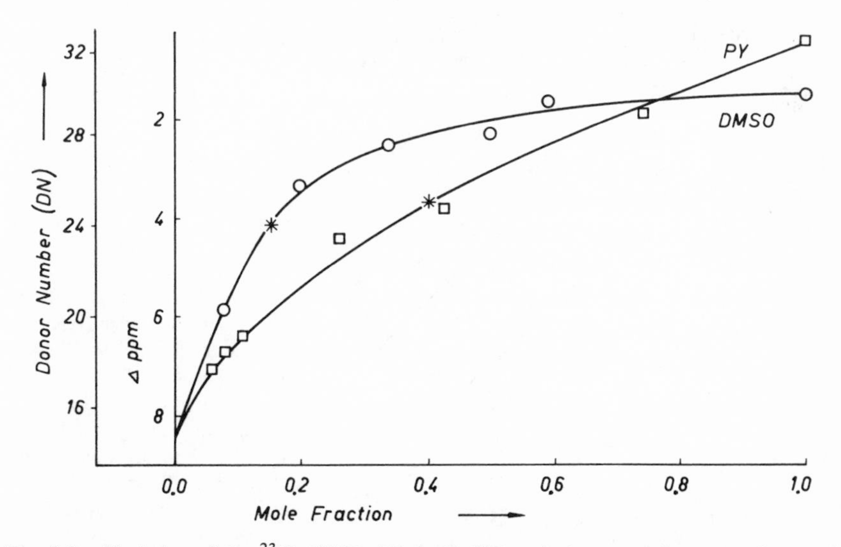

Fig. 9.2. Variation of the ^{23}Na NMR shift in $NaClO_4$ solutions, and donor numbers estimated from the Popov correlation as a function of solvent composition for binary solvent mixtures of acetonitrile with dimethylsulfoxide and pyridine; * represent isosolvation points.

nitromethane, is predominant in the solution. In the absence of an "inert" diluent the resonance lines of coordinated molecules are multiply split and these lines represent the various Al^{3+} species of different coordination shells that exist simultaneously in solution. This again supports the view that associated "pseudomacromolecules" are responsible for the structural features in a solvent mixture.

Cox *et al.* (1974), in studying solvation of ions in water–DMSO mixtures, concluded strong interactions between the components of the solvent mixture. The variation of the free enthalpies of transfer is not a function of a change in some bulk physical property of the system, such as dielectric constant, but rather is related to the changes in concentration of the component solvents. In $CH_3CN—H_2O$ mixtures Cu^+ and Ag^+ ions have favorable values for the free enthalpies of transfer from H_2O to CH_3CN (they are preferentially solvated by CH_3CN), while other ions such as Cu^{2+}, Fe^{3+}, Fe^{2+}, Na^+, H^+, and Cl^- all have unfavorable values for the free enthalpies of transfer from H_2O to CH_3CN (see Section 9.3).

The acceptor properties of the solvent mixture are highly influenced by solvent–solvent interactions. In mixtures consisting of water and a co-solvent, such as HMPA, EtOH, or CH_3CN, a positive deviation from the straight line should be expected for preferential interaction of the

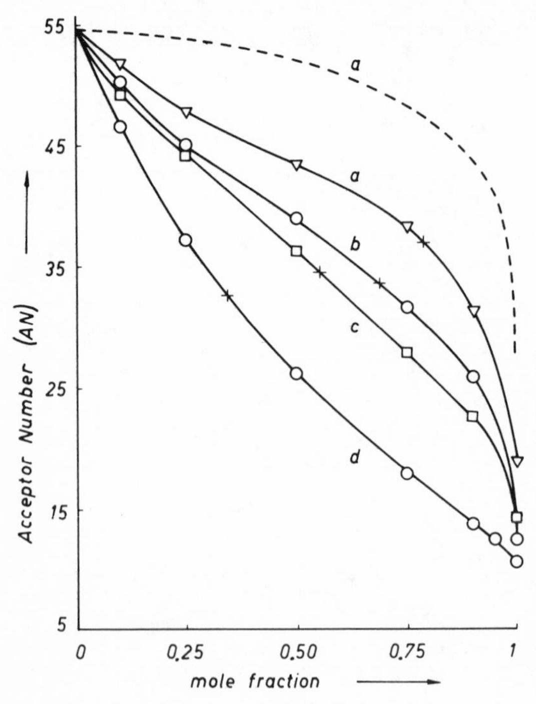

Fig. 9.3. Variation of acceptor numbers in binary mixtures of water and an aprotic solvent; curve a = acetonitrile, curve b = acetone, curve c = pyridine, curve d = HMPA; × represent the corresponding isosolvation points, and curve a′ allows for the polymeric nature of water by replacing x_{H_2O} by $\bar{x}_{H_2O}$ assuming a cluster size of 10 (from Mayer *et al.*, 1977*b*, courtesy of Springer-Verlag, Vienna, New York).

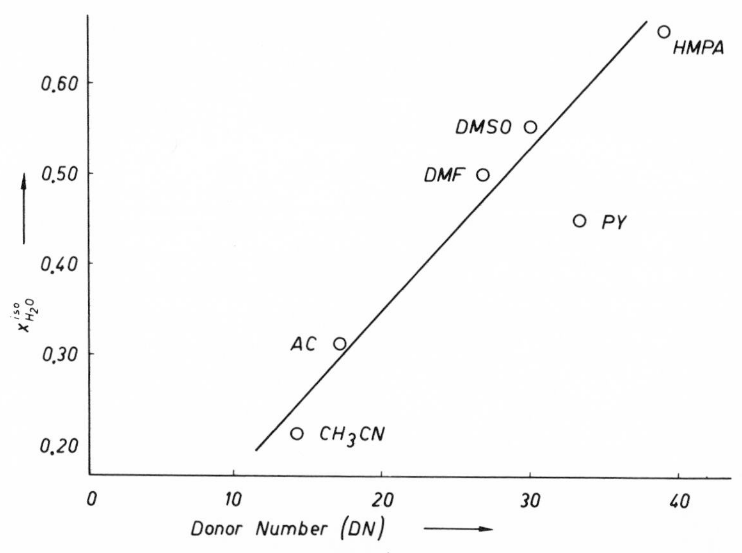

Fig. 9.4. Variation of the isosolvation point $X^{iso}_{H_2O}$ for binary mixtures of water with various aprotic solvents, as a function of the donor number of the aprotic solvent component (from Mayer *et al.*, 1977*b*, courtesy of Springer-Verlag, Vienna, New York).

stronger acceptor solvent, e.g., water with the solute Et_3PO (curve a′ in Fig. 9.3). Since HMPA has a smaller acceptor number than CH_3CN, Et_3PO would be expected to be hydrated more strongly in a water–HMPA mixture than in a water–acetonitrile mixture of equal mole ratio. The observed behavior is opposite to this expectation. The strength of solvent–solvent interactions increases with increasing donor number of the co-solvent, as evidenced by the positions of the isosolvation points. A linear relationship exists between the position of the isosolvation point and the donor number of the aprotic co-solvent rather than the acceptor number (Mayer *et al.*, 1977*b*) (Fig. 9.4). This shows that there is a decrease in acceptor properties with increasing donor number of the co-solvent, resulting in new structural features of the mixture, such as

$$\overset{\delta^-}{\text{O}}\text{—H}\leftarrow\text{O}\text{=P(NR}_2)_3 \quad (\text{O bonded to H below})$$

The same conclusion has been reached for water–HMPA mixtures from studies of the behavior of R_2NO in such mixtures (Symons, 1976): Water is taken away by HMPA, leaving R_2NO hardly solvated:

$$m\text{HMPA} + (\text{HOH})_n + \text{R}_2\text{NO} \longrightarrow (\text{HMPA})_m(\text{HOH})_n + \text{R}_2\text{NO}$$

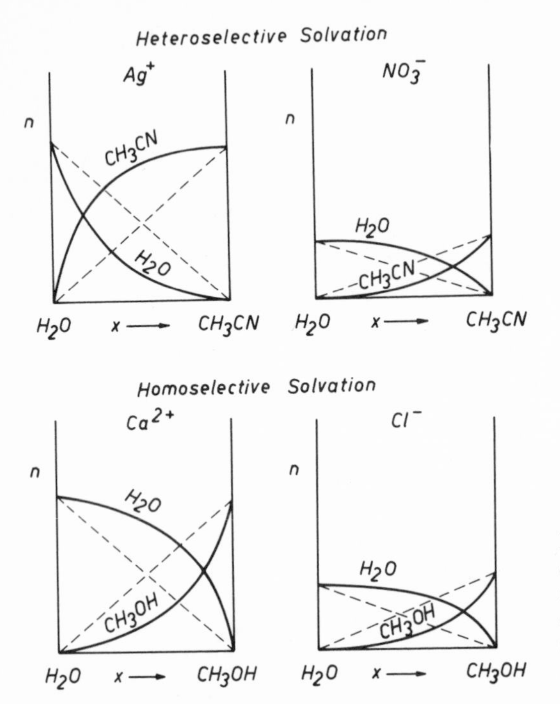

Fig. 9.5. Solvation numbers as a function of solvent composition for homo- and heteroselective solvation (from Strehlow and Schneider, 1969, courtesy of the Societé de Chimie Physique).

Thus the strength of the specific solvent–solvent interactions between water and the co-solvent is mainly regulated by the donor properties of the latter. The situation is similar for water–alcohol mixtures, when the water structure is altered by the alcoholic solvent.

It should be expected that the donor properties of a solvent, after mixing with another solvent, should be decreased as the acceptor number of the co-solvent increases. For such experiments a suitable cation should be used as a probe.

Neither the donor number nor the acceptor number of a given solvent mixture can be calculated from the donor and acceptor numbers, respectively, of the pure solvents. It is therefore necessary to investigate the donor and acceptor properties for each of the solvent systems in order to allow further progress in the application of solvent mixtures. The situation is similar to that in material science several decades ago, when the systematic accumulation of experimental data on two-component and three-component systems started. The specific properties of solvent mixtures will offer a large variety of new solvent systems.

9.3. *Homoselective and Heteroselective Solvation*

In solutions of a binary electrolyte in a mixture of two solvents the following cases may be encountered: Either both anion and cation are preferentially solvated by the same solvent component, or, with more interesting consequences, the cation is bound more strongly by one solvent component and the anion prefers the other solvent (Schneider and Strehlow, 1965). The term "homoselective solvation" has been suggested for the former and "heteroselective solvation" for the latter case (Strehlow and Schneider, 1969).

Figure 9.5 schematically shows two such cases, namely, silver nitrate in water–acetonitrile mixtures (Strehlow and Koepp, 1958) and calcium chloride in water–methanol mixtures (Strehlow and Schneider, 1969). Silver ion is preferentially solvated by acetonitrile in a 1 : 1 admixture with water, whereas the nitrate ion is considerably stronger solvated by water

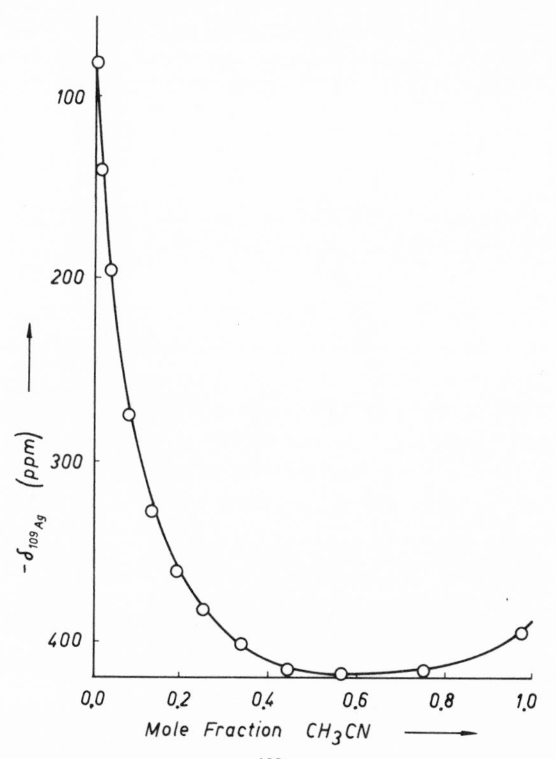

Fig. 9.6. Variation of the chemical shift of ^{109}Ag resonance in 2 M $AgNO_3$ as a function of solvent composition for binary water–acetonitrile mixtures (from Rahimi and Popov, 1976, courtesy of Pergamon Press).

molecules in the same mixture (heteroselective solvation). On the other hand, both calcium ions and chloride ions are preferentially solvated by water molecules in the 1 : 1 mixture of water and methanol (homoselective solvation). Solutions of $AgNO_3$ in water–acetonitrile mixtures have also been studied by $^{109}Ag^+$ NMR spectroscopy (Rahimi and Popov, 1976). The isosolvation point was found at 0.075 mole fraction of acetonitrile, indicating a strong preferential solvation by this solvent. It is interesting to note that the plot is not monotonic, but shows a definite and reproducible minimum of ≈0.6 mole fraction of acetonitrile (Fig. 9.6). This is an indication of strong cation–anion interactions in this region.

The consequences of heteroselective solvation are as follows: The solubility in such systems is generally higher in mixtures than in the pure solvents. For example, the solubility of Ag_2SO_4 is considerably greater in water–acetonitrile mixtures than in either of the pure solvents (Driessen and Groeneveld, 1968, 1969). Likewise, magnesium chloride is insoluble in either $POCl_3$ or $SbCl_5$, but it is readily dissolved in $POCl_3$ containing $SbCl_5$ (see Section 10.1). This behavior is expected, since both ions can solvate in the mixture with that solvent for which they exhibit the greater

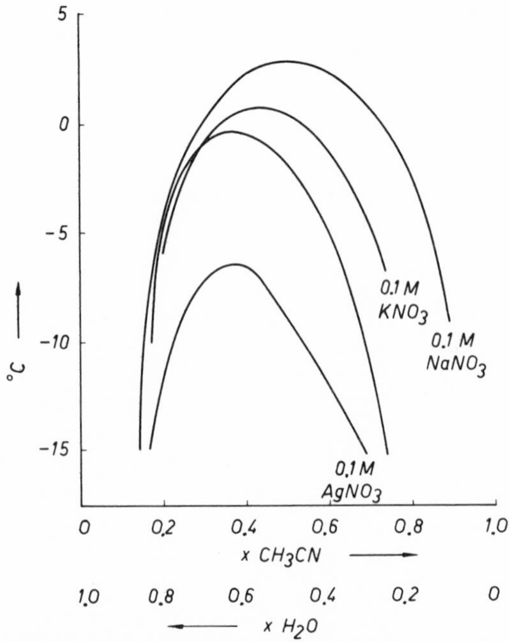

Fig. 9.7. The T–x phase diagram of $H_2O—CH_3CN$. Effect of addition of homoselective ($NaNO_3$) and heteroselective ($AgNO_3$) electrolytes (from Strehlow and Schneider, 1969, courtesy of the Societé de Chimie Physique).

affinity. An unusual behavior of heteroselective systems is found with phase-separation experiments. Water and acetonitrile form a strongly nonideal mixture. Below about 0°C the two solvents are no longer miscible in all proportions (Fig. 9.7). Addition of homoselective electrolytes, e.g., $NaNO_3$, widens the range of immiscibility. Addition of $AgNO_3$, however, considerably decreases the range of immiscibility (Strehlow and Schneider, 1971). The phase separation is opposed by the heteroselective electrolyte (Fig. 9.7). Rather small concentrations of electrolyte produce large effects, and this shows that physical and hence structural properties of the solvents are characteristically different from those of the solutions, the latter being highly influenced by the individual interactions between solvent and solute.

Chapter 10
Ionization Equilibria

10.1. Heterolysis by Coordinating Interactions

A coordinating interaction of a given substrate A—B leads to the formation of a molecular adduct, in which the bond A—B is longer than in the uncoordinated species (first and third bond-length variation rule). The lengthening of this bond is a function of the extent of the coordinating interactions: Strong coordinating interactions may lead to heterolysis (ionization) (see Section 3.2). The coordinating species is frequently a solvent, which attacks at A when acting as a donor and at B when acting as an acceptor (Gutmann, 1969, 1970, 1972). Each of these interactions results in a shift in electron charge from A toward B. In the first case the cation A^+ becomes coordinated by donor molecules, and in the second case the anion B^- is coordinated by acceptor molecules. When the solvent has amphoteric properties, some of its molecules may act as donors solvating the produced cations A^+, while others act as acceptors with solvation of the produced anions B^-. This is also known as "push–pull effect" (Swain, 1948).

$$n\mathrm{D}\rightarrow\overset{\delta+}{\mathrm{A}}\!-\!\overset{\delta-}{\mathrm{B}}\rightarrow m\mathrm{Ac} \rightleftharpoons \mathrm{AD}_n^+\cdot\mathrm{BAc}_m^-$$

The equilibrium will depend on:

(i) the bond strength A—B or rather on its heterolytic dissociation enthalpy (see Section 10.3);
(ii) the acceptor property at A within the compound;
(iii) the donor property at B within the compound;
(iv) the donor property of D, attacking at A;
(v) the acceptor property of Ac, attacking at B.

Thus for a given substrate, (i) to (iii) are constant and the equilibrium depends on (iv) and (v): Both donor and acceptor actions of the solvent—and not their polarity in a physical sense—must be considered in order to account for the ionizing effects of the solvent (Mayer *et al.*, 1975; Gutmann, 1976*b*).

If the solvent acts as a strong donor and as a very weak acceptor the produced cations will be solvated, whereas the anions will be poorly solvated and hence rather reactive:

Action of a donor solvent (D):

$$n\text{D}\rightarrow\overset{\frown}{\text{A—B}} \rightleftharpoons \underset{\text{Stabilized cation}}{\text{AD}_n^+} + \text{B}^-$$

Example:

$$4\text{HMPA} + \text{CoCl}_2 \rightleftharpoons \text{Co(HMPA)}_4^{2+} + 2\text{Cl}^-$$

Ionization of KF is achieved even in benzene, if a strong donor is present, for example, a so-called "crown ether," which stabilizes the potassium ion (Liotta and Harris, 1974). At the same time "naked," reactive fluoride ions are produced. Thus reactive anions can be made available by strong donors in the absence of acceptors.

On the other hand, reactive cations can be produced in an environment of strong acceptors, which are good anion solvators (Olah, 1974; Gillespie and Morton, 1971; Gillespie and Passmore, 1975):

Action of an acceptor solvent (Ac):

$$\overset{\frown}{\text{A—B}}\rightarrow n\text{Ac} \rightleftharpoons \text{A}^+ + \underset{\text{Stabilized anion}}{\text{BAc}_n^-}$$

Example:

$$\text{Ph}_3\overset{\frown}{\text{C—Cl}}\rightarrow\text{AsCl}_3 \rightleftharpoons \text{Ph}_3\text{C}^+ + \text{AsCl}_4^-$$

Thus carbonium ions are produced in highly acidic media, for example, in mixtures of HF and SbF_5 due to extensive anion solvation (Olah *et al.*, 1964; Olah and Lukas, 1967).

Simultaneous stabilization of both cations and anions may be effected in an amphoteric solvent such as water which is an ideal solvent for interactions between solvated ions. Alternatively a suitable combination of a donor solvent and of an acceptor solvent may be used for this purpose. For example, phosphorus oxychloride functions as a donor solvent, but $CoCl_2$ remains virtually insoluble in it. Addition of a strong acceptor, such as $SbCl_5$ to the suspension dissolves the cobalt chloride as simultaneous attack by $POCl_3$ at the cobalt and by $SbCl_5$ at the chlorine atoms heterolyses the Co—Cl bonds with subsequent stabilization of both cationic and anionic species by coordination (Driessen and Groeneveld, 1968; Dragulescu *et al.*, 1974):

$$n\text{POCl}_3 + \text{CoCl}_2 \qquad \text{insoluble, no ionization}$$

$$\text{CoCl}_2 + 2\text{SbCl}_5 \qquad \text{insoluble, no ionization}$$

$$\underset{\text{Donor}}{n\text{POCl}_3} + \underset{\text{Substrate}}{\text{CoCl}_2} + \underset{\text{Acceptor}}{2\text{SbCl}_5} \rightleftharpoons \text{Co(OPCl}_3)_n^{2+} + 2\text{SbCl}_6^-$$

Likewise, magnesium chloride is insoluble in nitromethane, but addition of ferric chloride leads to dissolution with ionization:

$$n\text{NM} + \text{MgCl}_2 \qquad \text{insoluble, no ionization}$$

$$\text{MgCl}_2 + 2\text{FeCl}_3 \qquad \text{insoluble, no ionization}$$

$$n\text{NM} + \text{MgCl}_2 + 2\text{FeCl}_3 \rightleftharpoons \text{Mg(NM)}_n^{2+} + 2\text{FeCl}_4^-$$

The ionization process for triethylammonium picrate in nitrobenzene is altered by addition of imidazole derivatives (Pirson and Huyskens, 1974; Haulait and Huyskens, 1975). While a single complex species is formed with 1-methyl imidazole, more than one azole molecule is involved in complex formation with imidazole or benzimidazole. Whereas the tertiary nitrogen has basic properties in every case, the imino group in imidazoles is weakly acidic and according to the second bond-length variation rule their acidity is enhanced by coordination of the tertiary nitrogen atoms, thus allowing coordination of a further imidazole molecule or of a picrate ion:

pi→H—N⌒N→H—N⌒N→Et_3NH

$$\text{pi}^-\left[\text{CH}_3\text{—N}\frown\text{N}\rightarrow\text{Et}_3\text{NH}\right]^+$$

The relationship between thermodynamic properties, e.g., ionization constants K or solubility products K_s of a compound A—B and the coordinating properties of solvents can be deduced from the following free-energy cycle:

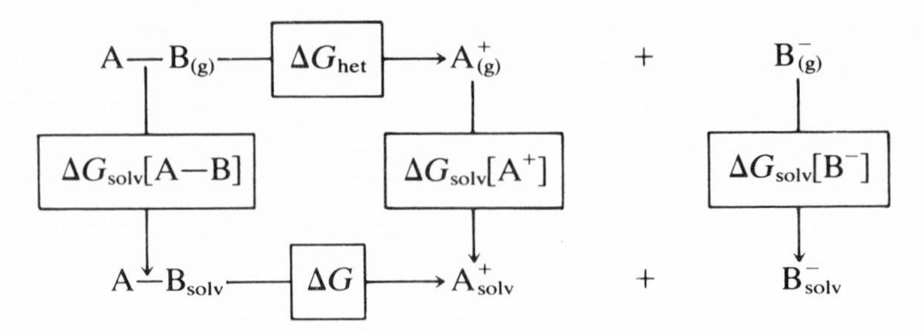

where ΔG_{het} denotes the free enthalpy for heterolytic bond cleavage (Section 10.3) of the gaseous molecules $A—B_{(g)}$ into the gaseous ions $A^+_{(g)}$ and $B^-_{(g)}$ and ΔG the free enthalpy of ionization in the respective solvent:

$$\Delta G = -RT \ln K$$

$$K = \frac{[A^+_{solv}][B^-_{solv}]}{[A—B_{solv}]}$$

$$\Delta G = \Delta G_{het} - \Delta G_{solv}[A—B] + \Delta G_{solv}[A^+] + \Delta G_{solv}[B^-]$$

The ionization constant of a given compound A—B in different solvents is therefore a function of the free enthalpies of solvation of A—B, A^+, and B^-, and these represent a measure of the solvating properties of the solvents. According to

$$\Delta G_{solv} = \Delta H_{solv} - T\,\Delta S_{solv}$$

a relationship exists between ionization constant and the enthalpies of solvation ΔH_{solv} of the species present in solution.

We have seen that the free enthalpy of solvation of a given cation is lower the higher the solvent donor number, whereas the free enthalpy of solvation of a given anion is decreased as the acceptor number of the solvent is increased. From that it follows that the free enthalpy of ionization ΔG for a given substrate will become more negative the greater the donor number and the greater the acceptor number of the solvent.

The role of the donor effect on the ionization equilibrium

$$n\text{D} + \text{Sn(CH}_3)_3\text{—I} \rightleftharpoons \text{D}_n\text{Sn(CH}_3)_3^+ + \text{I}^-$$

has been shown by conductometric techniques in nitrobenzene, a weakly coordinating medium with a dielectric constant of 34.8. The interference of the acceptor properties of the solvent has been minimized by investigating aprotic donor molecules D with nearly identical acceptor numbers so that the iodide ion is similarly solvated in all systems under investigation. When the donor solvent is added to the solute A—B, ions are produced by the

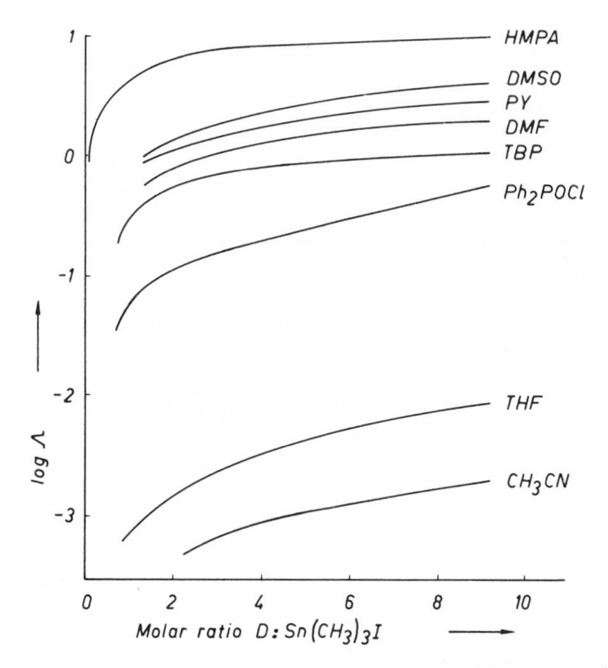

Fig. 10.1. Conductometric titrations of $Sn(CH_3)_3I$ in nitrobenzene ($c \approx 10^{-2}$ mol·liter^{-1}) with various neutral donor molecules (from Gutmann and Mayer, 1969, courtesy of Academic Press, London, New York).

donor action of the added donor in the medium of given dielectric constant. Thus the extent of ion-pair formation (see below) between cationic and anionic species remains essentially constant* in the presence of various donors (Gutmann and Mayer, 1969). The conductivities in nitrobenzene in the presence of added amounts of stronger donors correspond to the relative donor effects at the tin atom.

Figure 10.1 shows the increase in molar conductivities of $Sn(CH_3)_3I$ dissolved at $c \approx 10^{-2}$ mol/liter in nitrobenzene as a function of the molar ratios donor: $(CH_3)_3SnI$ (Gutmann and Mayer, 1969). In the absence of a donor, $Sn(CH_3)_3I$ in nitrobenzene is un-ionized (tetrahedral structure). This is found in all weakly coordinating solvents such as hexane, carbon tetrachloride, 1,2-dichloromethane, or nitromethane, irrespective of their dielectric constants. Addition of a donor solvent to this solution gives rise to conductivity, due to ionization presumably with formation of trigonal bipyramidal cations $D_2Sn(CH_3)_3^+$ and iodide ions. The molar conductivities of a given molar ratio donor: $Sn(CH_3)_3I$ are (with the exception of

*This is strictly true only if the coordinated cations AD_n^+ are very large and fully coordinated.

pyridine) related to the relative donor numbers, whereas there is no relationship between the conductivities of $Sn(CH_3)_3I$ in nitrobenzene and the dipole moments μ of the donor molecules added. A relationship exists also between donor number of the solvent and the coupling constants J of the ionized substrate (Gutmann and Mayer, 1969).

The role of the acceptor effect on an ionization equilibrium has been demonstrated by conductometric titration of triphenylchloromethane dissolved in nitrobenzene with sulfur dioxide or arsenic(III) chloride (Baaz *et al.*, 1962; Gutmann and Kunze, 1963); this role has also been demonstrated in relation to the acceptor number of the solvent by investigating the ionization equilibrium of $VO(acac)_2Cl^-$ in various solvents (Mayer and Gutman, 1970):

$$VO(acac)_2Cl^- + S \rightleftharpoons VO(acac)_2^+ + Cl^-_{solv}$$

The equilibrium constants are found to increase as the acceptor number of the solvent is increased (see Section 2.3).

In most cases both the donor and acceptor number of the solvent must be taken into consideration. For example, the equilibrium constant for the reaction

$$R_3NHCl + (n+1)S \rightleftharpoons R_3NHS^+ + ClS_n^-$$

is increased both by increasing donor number of the solvent (stabilization of the cation) *and* by increasing acceptor number of the solvent (stabilization of the anion) (Mayer *et al.*, 1976) (Table 10.1). The equilibrium constant is smaller in nitrobenzene than in nitromethane because of the differences in acceptor properties (both solvents have nearly identical donor numbers and dielectric constants). Likewise for the pair of solvents DMF and DMA, donor numbers and dielectric constants are very similar but the former has a higher acceptor number and hence the value for the equilibrium constant is higher in DMF than in DMA. Acetonitrile and

Table 10.1. Equilibrium Constants K for the Ionization of Chinuclidinhydrochloride in Various Solvents at 25°C

Solvent	K	DN	AN	ε	μ (Debye)
NB	$\approx 1.0\times10^{-6}$	4.4	14.8	34.7	4.0
NM	3.8×10^{-5}	2.7	20.5	36.7	3.6
CH_3CN	6.0×10^{-5}	14.1	19.3	36.0	4.0
PDC	2.74×10^{-4}	15.1	18.3	65.0	5.0
DMF	4.1×10^{-4}	26.6	16.0	36.7	3.9
DMA	1.8×10^{-4}	27.8	13.6	37.8	3.8
DMSO	1.2×10^{-2}	29.8	19.3	46.7	3.9

DMSO have practically the same acceptor number but the greater donor number of DMSO leads to a higher value for the equilibrium constant in DMSO. The differences found for acetonitrile and PDC are accounted for by the different dielectric constants, as will be discussed in Section 10.4.

10.2. Heterolysis by Redox Interactions

Redox interactions may also lead to the production of ions. Reductive bond cleavage produces anions and radicals:

$$\underset{\text{(ED)}}{\underset{\text{Donor}}{Na}} + \underset{\text{(EA)}}{\underset{\text{Acceptor}}{H{-}I}} \rightleftharpoons Na^{+} + \underset{\text{Radical}}{\cdot H} + \underset{\text{Anion}}{I^{-}}$$

This type of interaction may be applied for the production of radicals, such as ·OH:

$$\underset{\text{(ED)}}{\underset{\text{Donor}}{Fe^{2+}}} + \underset{\text{(EA)}}{\underset{\text{Acceptor}}{HO{-}OH}} \rightleftharpoons Fe^{3+} + \underset{\text{Radical}}{\cdot OH} + \underset{\text{Anion}}{OH^{-}}$$

Oxidative bond cleavage leads to cations and radicals, for example:

$$\underset{\text{(ED)}}{\underset{\text{Donor}}{H{-}I}} + \underset{\text{(EA)}}{\underset{\text{Acceptor}}{F}} \rightleftharpoons \underset{\text{Cation}}{H^{+}} + \underset{\text{Radical}}{\cdot I} + F^{-}$$

$$Ph{-}\overset{\overset{\displaystyle O}{\|}}{C}{-}H + Fe^{3+} \rightleftharpoons H^{+} + Ph{-}\dot{C}{=}O + Fe^{2+}$$

A simple representation for bond cleavage is given in Fig. 10.2. None of the donor–acceptor functions leads to homolytic bond fission, since the transfer of negative charge from $B^{\partial-}$ to $A^{\partial+}$ would require attack either by a donor at $B^{\delta-}$ or by an acceptor at $A^{\delta+}$, which is very unlikely.

10.3. Homolytic and Heterolytic Dissociation Energy

A simple approach, which is frequently used to estimate the ionizability of a bond, is based on a comparison of bond energies. The term "bond energy" refers to the dissociation energy of the gas-phase reaction and this may be more precisely termed the homolytic dissociation energy D_{hom}:

$$A{-}B_{(g)} \rightleftharpoons \dot{A}_{(g)} + \cdot B_{(g)} \qquad D_{\text{hom}}$$

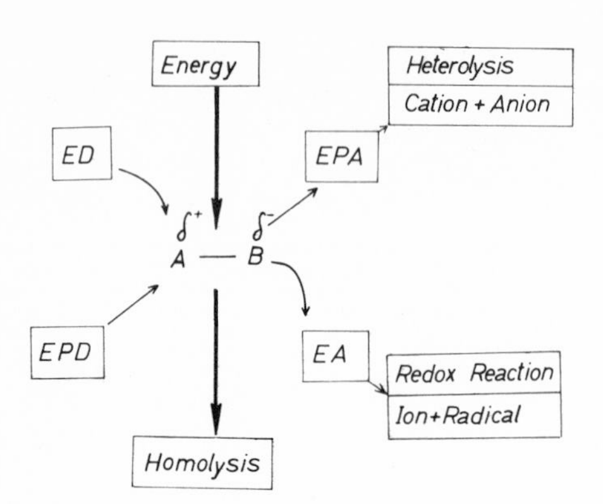

Fig. 10.2. Simple representation for bond cleavage for a molecule A—B.

In Section 10.1 it was stated that for ionization equilibria the values for the heterolytic dissociation energies D_{het} should be consulted (Gutmann and Mayer, 1971). D_{het} refers to the gas-phase reaction

$$A{-}B_{(g)} \rightleftharpoons A^{+}_{(g)} + B^{-}_{(g)} \qquad D_{het}$$

which involves a heterolytic cleavage of the bond A—B. Since a second mode of ionization is possible, namely, $A{-}B \rightleftharpoons A^{-} + B^{+}$, D_{het} always refers to the process that requires less energy. The relationship between D_{hom} and D_{het} is provided by

$$D_{het} = D_{hom} + I_A + E_B$$

where I_A denotes the ionization energy of A and E_B the electron affinity of B. Cesium has the lowest ionization potential (+3.89 eV) and chlorine the highest electron affinity of any atom (−3.61 eV). This means that D_{het} is always greater than D_{hom}.

10.4. *Solvent-Separated Ion Pairs: The Role of the Dielectric Constant*

So far the existence of ion-pair equilibria has been excluded from the considerations and ionization of a covalent compound has been described as the process leading to solvated ions (independent of their presence as associated ions or as free entities). In a coordinating solvent of low dielectric constant, heterolysis of A—B leads to ions with tight solvation spheres, which are associated to nonconducting ion pairs (Bjerrum, 1926), also known as "solvent-separated ion pairs." The equilibrium

$$AB \rightleftharpoons A^{+}B^{-}$$

may be characterized by the equilibrium constant K_{form}, which may be termed "formation constant of the (solvent-separated) ion pair* from the covalent substrate"†:

$$K_{form} = \frac{c_{A^+ \cdot B^-}}{c_{AB}}$$

In the above, as well as in the following equations, the indices for the solvation of any of the dissolved species have been omitted. In a medium of reasonably high dielectric constant the ion pair undergoes electrolytic dissociation:

$$A^+B^- \rightleftharpoons A^+ + B^-$$

This process may be characterized by the "ion-pair separation constant":

$$K_{sep} = \frac{c_{A^+} c_{B^-}}{c_{A^+ \cdot B^-}}$$

K_{form} is primarily influenced by the coordinating properties of the solvent; K_{sep} (according to Coulomb's law) is a function of the dielectric constant of the medium.

The equilibrium constant K_{ion} for the overall process (Gutmann, 1969, 1970, 1971*a*,*b*, 1972) is expressed by:

$$A{-}B \rightleftharpoons A^+ + B^- \qquad K_{ion}$$

$$K_{ion} = K_{form} \cdot K_{sep}$$

K_{ion} is identical with the ionization or dissociation constant in the classical sense. In a medium of high dielectric constant such as water the concentration of associated ions is negligibly small: K_{form} and K_{sep} cannot be measured separately, and this is the reason why in water and other solvents of high dielectric constant only the overall (classical) equilibrium constant K is meaningful. On the other hand, in solvents of low dielectric constant such as tributylphosphate there is almost no electrolytic dissociation, since an ionized substrate will be present mainly as associated ions. Consequently the ionization process is best characterized by K_{form} (Mayer and Gutmann, 1975).

In solvents of medium dielectric constant, K_{form} and K_{sep} may be determined separately by combination of appropriate experimental

*The term "loose ion pair" is also in use.

†This term is not identical with the formation constant of the ion pair from free ions, frequently referred to as association constant K_{assoc} and sometimes confused with outer-sphere complexation involving ions of opposite charge (Section 10.5), which according to this presentation is equal to $1/K_{sep}$.

techniques. For example, spectrophotometric or NMR techniques may be applied to obtain the total concentration c_i of ionized substrate, that is,

$$c_i = c_{A^+ \cdot B^-} + c_{A^+} = c_{A^+ \cdot B^-} + c_{B^-}$$

and the concentration of free ions $c_{A^+} = c_{B^-}$ may be determined by conductometric measurements. With c_0 denoting the analytical concentration of the solute one obtains:

$$c_0 = c_{AB} + c_{A^+ \cdot B^-} + c_{A^+} = c_{AB} + c_i$$

$$K_{form} = \frac{c_i - c_{A^+}}{c_0 - c_i}, \qquad K_{sep} = \frac{c_{A^+}}{c_i - c_{A^+}}$$

If K_{form} and K_{sep} cannot be determined directly as shown above, approximate values of K_{sep} may be obtained by comparison with the electrolytic dissociation behavior of suitable model electrolytes, such as quaternary ammonium salts. K_{sep} is approximately equal to the dissociation constant of Bu_4NClO_4. Under such conditions K_{sep} is inversely proportional to the association constant K_{assoc} or ion-pairing constant:

$$K_{sep} \approx K_{dissoc(R_4N^+X^-)} = \frac{1}{K_{assoc(R_4N^+X^-)}}$$

Hence the formation of free ions from a covalent substrate is a function of both the coordinating and the dielectric properties of the solvent, as is illustrated by a simple model calculation (Gutmann, 1970, 1972). Consider two donor solvents S_1 and S_2, which have the same coordinating properties and hence the same values K_{form}. In the present case we shall assume $K_{form} = 1$. S_3 is assumed to be a slightly stronger donor solvent, giving $K_{form} = 3$. S_1 and S_3 may have similar dielectric constants, namely, of about 10, while S_2 is assumed to have a higher dielectric constant of about 35; this may correspond to $K_{sep} \approx 10^{-4}$ (S_1, S_3) and $K_{sep} \approx 10^{-2}$ (S_2). The extent of ionization, that is, the percentage of ionized substrate (ion pairs + free ions) of total solute concentration, has been calculated and the results are presented in Table 10.2. Thus, AB is

Table 10.2. Extent of Ionization of the Covalent Substrate A—B at $c_0 \approx 10^{-2}$ mol/liter under Arbitrarily Assumed Conditions (Gutmann, 1969)

Solvent	Assumed ε	Assumed K_{sep}	Assumed K_{form}	Ionized (%)
S_1	10	1×10^{-4}	1	53
S_2	35	1×10^{-2}	1	75
S_3	10	1×10^{-4}	3	77

more strongly ionized in solvent S_2, which has the same coordinating properties as S_1 but a considerably higher dielectric constant. On the other hand, AB is even more strongly ionized in S_3 although its dielectric constant is low compared to S_2, but its coordinating properties are somewhat greater. However, solutions of AB are much better conductors in S_2 than in S_3 because in the latter solvent the ionized substrate is mainly present as ion pairs, which do not contribute to the conductivity of the solution.

Comparison of dissociation constants of acids and bases derived from electrochemical or conductivity measurements in solvents of different dielectric constants are meaningless (Gutmann, 1968), since equilibrium constants determined in this way always represent the overall equilibrium constant $K_{ion} = K_{form} \cdot K_{sep}$. Liquid ammonia is known to have considerably stronger donor properties than water, but the acidity constant K_a of HCl is much smaller in ammonia ($K_a \approx 10^{-4}$). Ionization of HCl is essentially complete in both solvents, but the observed differences in dissociation constants are mainly due to the differences in dielectric constants (water: $\varepsilon = 81$; ammonia: $\varepsilon = 17$). On the other hand the acidity constant of acetic acid is higher in liquid ammonia ($K_a \approx 10^{-4}$) than in water ($K_a \approx 10^{-5}$). Acetic acid is moderately ionized in water, while in liquid ammonia (due to its stronger donor properties) ionization is nearly complete: The influence of K_{sep} for acetic acid in liquid ammonia is overcompensated by K_{form}.

The situation may be further illustrated by comparing the electrolytic dissociation behavior of lithium halides and tetrabutylammonium halides in the solvents propanediol-1,2-carbonate (PDC: DN = 15.1, $\varepsilon = 65.0$) and hexamethylphosphorictriamide (HMPA: DN = 38.8, $\varepsilon = 29.6$) (Mayer *et al.*, 1973). LiBr and in particular LiCl are associated in PDC, but they are fully dissociated in HMPA. This cannot be explained by simple electrostatic considerations, since HMPA has a lower dielectric constant than PDC and the dipole moments of the two solvents are similar ($\mu_{HMPA} = 5.37$, $\mu_{PDC} = 4.98$). The behavior is readily interpreted from the point of view of donor–acceptor interactions: HMPA acts as a very strong donor and consequently lithium halides are completely ionized with formation of tightly solvated Li^+ ions. Due to the large effective radius of the HMPA-solvated lithium ion the center-to-center distance in the ion pair $Li(HMPA)_n^+ \cdot X^-$ is so great that dissociation is easily effectuated. In PDC, which has a considerably smaller donor number, LiCl and LiBr exist as un-ionized molecules LiX (*contact ion pairs*) in equilibrium with the free solvated ions Li^+ and X^- (the concentration of ion pairs $Li(PDC)_n^+ \cdot X^-$ is small due to the high dielectric constant of this solvent). The behavior of lithium halides in PDC is analogous to that of acetic acid in water. The different behavior of lithium halides in PDC and HMPA is also in agreement with semiquantitative data for the relative donor properties of the halide ions in the respective solvents: HMPA has a higher donor number than both Cl^- and Br^- and hence these ligands are easily replaced from the lithium by HMPA. On the other hand, PDC is a weaker donor than Cl^- and Br^- and hence these ligands are only partly replaced despite the large excess of solvent molecules present. Contrary to lithium halides, tetrabutylammonium halides are fully dissociated in PDC, while they are associated in HMPA in agreement with simple electrostatic considerations. Similar results have been found in DMF, where alkali metal halides are fully dissociated while tetraalkylammonium halides are associated (Ames and Sears, 1955; Sears *et al.*, 1955).

10.5. Outer-Sphere Ion Association

Formation of ion pairs or higher ionic aggregates (triple ions, etc.) is a very common and characteristic phenomenon in nonaqueous media. In Section 10.4 the separation of pairs of ions with tight solvation shells, called solvent-separated ion pairs (in the original sense of Bjerrum), was discussed as a dielectric effect: The interaction between tightly solvated ions of opposite charge leads to solvent-separated ion pairs, which do not contribute to the conductivity of the solution. This process is also called "ion association" and is characterized by the "association constant" K_{assoc}, which is the reciprocal value of K_{sep} (Section 10.4).

The elementary electrostatic model may be applied only when the ions are either coordinatively saturated or do not undergo strong specific interactions with the solvent molecules (Mayer, 1976; Gutmann, 1977*c*). This is illustrated by the variation in the association constant for tetrabutylammonium iodide as a function of the dielectric constant, corresponding roughly to the trends predicted by the Bjerrum theory.

With increasing donor strength of the anion, "specific" anion solvation effects become increasingly important. Tetraalkylammonium chlorides are less extensively associated in methanol ($\mu = 1.71$ D, $\varepsilon = 32.6$) than they are in acetonitrile ($\mu = 3.37$ D, $\varepsilon = 36$) or nitromethane ($\mu = 3.57$ D, $\varepsilon = 35.9$), although both the dipole moments and the dielectric constants are greater for the latter than for methanol, since the chloride ion is strongly solvated in methanol via hydrogen bonds: Methanol (AN = 41.3) has a greater anion-solvating ability than either acetonitrile (AN = 18.9) or nitromethane (AN = 20.5) (Mayer *et al.*, 1975).

In the alcohols the association constants decrease in the order I > Br > Cl and this is in contrast with expectations from elementary electrostatic considerations. If this was due either to the numerical increase in Born solvation energies of the halide ions or to increasing ion–dipole interactions in the series, the same trend should be found in aprotic solvents of similar or higher dielectric constants. However, the stability order in CH_3CN, NB, or DMF is the reverse of that found in the alcohols (Table 10.3). In the gas phase the stability of the ion pairs with a given cation increases in the series $I < Br < Cl$; the free energies of halide ions increases in the same order. The differences in solvation energies are greater the stronger the acceptor properties of the solvent. For this reason, in the alcohols the chloride ion is more greatly weakened in donor properties than the bromide ion and hence the stability order of the gas phase is reversed. In acetonitrile the tetrapropylammonium and tetrabutylammonium salts have nearly the same association constants due to the leveling effect of acetonitrile. In solvents of low acceptor number and consequently of weak anion-solvating properties, the gas-phase conditions may be approached.

Substitution of solvent molecules in the solvation shell of Li^+ is facilitated by increasing donor properties of the anions, e.g., $I^- < Br < Cl^-$. In DMSO, LiI and LiBr are almost completely dissociated, whereas LiCl is weakly associated, indicating comparable donor properties of both DMSO and chloride ions in this solvent.

It has been stated that the enhanced donor properties of liquid water as compared to those under "gaslike" conditions are due to outer-sphere

Table 10.3. Ion-Pair Association Constants of Tetraalkylammonium Halides in Various Solvents at 25°C

Tetraalkylammonium halide	Solvent		
	EtOH[a]	PrOH[b]	NB[c]
Me_4NCl	122	456	—
Me_4NBr	146	638	—
Et_4NCl	—	—	80
Et_4NBr	99	373	62
Et_4NI	133	466	29
Pr_4NBr	78	270	—
Pr_4NI	120	391	—
Bu_4NCl	39	149	—
Bu_4NBr	75	266	56
Bu_4NI	123	415	27

[a] EtOH: $\varepsilon = 24.3$; $\mu = 1.7$; AN = 37.1.
[b] PrOH: $\varepsilon = 20.5$; $\mu = 1.6$; AN, approximately 34.
[c] NB: $\varepsilon = 34.7$; $\mu = 4.0$; AN = 14.8.

interactions between water molecules between the inner- and outer-hydration shells (Mayer and Gutmann, 1972). These donor–acceptor interactions are, of course, not restricted to interactions between solvent molecules. Generally the term "outer-sphere coordination," may be applied to all coordinative intermolecular actions between coordinated ligands and either neutral or charged ligands.

The term "outer-sphere complex" has been introduced by Werner (1913) for the interactions of coordinatively saturated complex ions with further ligands in order to account for the stereochemical course of substitution reactions (Taube and Posey, 1953). It was later postulated and subsequently experimentally confirmed by kineticists in order to explain certain features of rate laws observed in reactions that lead to transition metal–ammine complexes (Basolo *et al.*, 1957). The importance of outer-sphere interactions in substitution reactions has been emphasized recently. Various examples of outer-sphere coordination have been discussed in Chapter 7 and others will be presented in Chapters 12 and 13.

10.6. The Concept of Contact Ion Pairs

The concept of "contact ion pair" or "intimate ion pair" has been introduced in order to explain the stereochemical course of solvolysis and electrophilic substitution reactions (Winstein *et al.*, 1954). Since then the existence of both types of ion pairs has been widely accepted and supported by various experimental results (Winstein and Robinson, 1958; Cram *et al.*, 1959; Cram, 1965; Hogen-Esch and Smid, 1966; Szwarc, 1972). In contact ion pairs, cations and anions are considered to be in immediate contact with each other, while in solvent-separated ion pairs, ions are separated by solvent molecules.

The properties of a bond between two atoms do not depend on the mode of formation. For example, the properties of the H—F molecule will be the same whether it has been formed from the free atoms or from a hydrogen ion and a fluoride ion. Solvation in a particular solvent is also independent from its history and it will always lead to an increase in bond length compared to that in the gaseous state. In such cases it is impossible to distinguish by experiment a solvated contact ion pair from the un-ionized solvated species (Gutmann, 1971*a*, 1976*a*):

$$A_{(g)} + B_{(g)} \rightleftharpoons \overset{\delta^+}{A}-\overset{\delta^-}{B}_{(g)} \rightleftharpoons A^+_{(g)} + B^-_{(g)}$$

$$\overset{\delta^+}{A}-\overset{\delta^-}{B}_{(g)} \rightleftharpoons \overset{\delta^+}{A}-\overset{\delta^-}{B} \text{ (solvated)}$$

"contact ion pair"
= un-ionized species

The formation of such species from the free ions or from the solvent-separated ion pairs may be regarded as the inverse type of the ionization reaction discussed in Section 10.1 and the intimate ion pair may therefore be considered as solvated "un-ionized" species.

The contact ion pair is limited to salt solutions to which the "sphere-in-continuum model," e.g., where neither cation nor anion are solvated such as for $Bu_4N^+BPh_4^-$ in water (Mayer and Gutmann, 1972) or K(crown ether)$^+$F$^-$ in benzene solution (Liotta and Harris, 1974), can be applied to equilibria involving asymmetric carbonium ions, to highly concentrated salt solutions, and to fused-salt media.

In all other cases a contact ion pair cannot be distinguished from the un-ionized species on spectroscopic or on chemical grounds. This can be seen from the following examples, which have originally been presented in support of the contact-ion-pair concept. Hogen-Esch and Smid (1966) have investigated the formation of contact ion pairs of alkali metal salts of various carbanions and radical anions in a number of solvents. The stabilities of contact ion pairs, however, are found to decrease as the donor numbers of the solvents increase, this represents the typical behavior of covalent substrates. In a solvent of high donor number such as pyridine or DMSO only solvent-separated ion pairs and free solvated ions are present, while the un-ionized species (contact ion pair) is found in a weak donor solvent such as toluene. The extent of ionization is governed by the donor property of the solvent and not by its dipole moment or dielectric constant. The lithium compounds are more easily ionized than the sodium compounds because of the greater acceptor strength of the lithium ion toward donor solvent molecules (Table 10.4).

Structural investigations have established that in the solid lithium-fluorenyl–THF adduct the carbanionoid electron pair is bonded to the

Table 10.4. Amount of Solvent-Separated Ion Pairs (+Free Ions) of Alkali Metal Salts of 9-Fluorenyl (F) in Various Solvents at 25°C

			Solvent-separated ion pairs (+free ions) (%)		
Solvent	DN	ε	Li^+F^-	Na^+F^-	Cs^+F^-
Toluene	0.1	2.4	0	0	0
Dioxane	14.8	2.2	0	0	0
2-Me–THF	18	6.3	25	0	0
THF	20	7.6	75	5	0
DME	≈24	7.2	100	95	0
DMSO	29.8	45.0	100	100	—
Pyridine	33.1	12.3	100	100	—

lithium despite the delocalization in the aromatic system. Three tetrahydrofuran molecules are bonded to each lithium, while the fourth ligand position is occupied by the carbanionoid ligand (Dixon *et al.*, 1965). The ionization of naphthalene sodium is drastically increased by replacing THF by glycoldimethylether (Zandstra and Weissman, 1962; Bhattacharya *et al.*, 1963).

Styrene lithium gives solvent-separated ion pairs in THF. The polymerization of styrene is considerably faster in THF (DN = 20) than in dioxane (DN = 14.8), although the dielectric constants of the two solvents are nearly identical. In dioxane the pattern is reversed, the cesium compound being more reactive than the lithium compound (Hogen-Esch and Smid, 1967; Waack and Doran, 1962). This shows that the Cs—C bond is more polar than the Li—C bond in the weakly solvated state, while a stronger solvating solvent mobilizes the carbanion from the lithium compound to a greater extent due to the greater acceptor strength of the Li^+ ion and the greater ease of increasing the polarity of the Li—C bond. The absorption maximum of 1,1-diphenylhexyl lithium is shifted to longer wavelengths with increasing donor properties of the solvent (Waack and Doran, 1962, 1963, 1964; Waack *et al.*, 1966).

In solvents of medium donor strength, such as THF (Edgell *et al.*, 1966, 1970) or acetone (Wong *et al.*, 1971), the IR cage frequencies of lithium, sodium, and potassium ions are dependent on the anion; in contrast, in strong donor solvents, such as DMSO (Maxey and Popov, 1967, 1969) 1-methyl-2-pyrrolidone (Wuepper and Popov, 1969, 1970), and pyridine (McKinney and Popov, 1970), the frequencies are nearly independent of the anion. This was ascribed to the presence of solvent-separated ion pairs and free ions in the strong donor solvents and the presence of contact ion pairs in THF and acetone.

Variation of band frequency upon changing the anion, as observed in THF and acetone, cannot be ascribed to the rather low dielectric constants of these solvents, but are due to their moderate solvating power (Maxey and Popov, 1969; Wong *et al.*, 1971). For example, nearly constant frequencies are observed for lithium salts independent from the anion in mixtures of benzene and DMSO with dielectric constants ranging from 7 to 46, in which the lithium ions are fully coordinated by DMSO molecules. Wuepper and Popov (1970) dissolved alkali metal salts in dioxane, added DMSO to the solution, and found the same characteristics as in pure DMSO. Despite the low dielectric constant of the dioxane solution the high donor number of the added DMSO gives DMSO solvates of the alkali metal ions, and the solvation sphere is no longer penetrable by anions competing for coordination.

Valuable information is further provided by studies of the lithium fluorenyl in mixtures of acetone and nitromethane (Wong *et al.*, 1971). In these mixtures, the intensity of the band observed at 425 cm^{-1} and assigned to the stretching motions of the Li—O coordinate bonds is a linear function of the mole ratio, acetone/$LiClO_4$ provided that the mole ratio is greater than 4. It was concluded that under these conditions Li^+ is coordinated by four molecules of acetone. Preferential solvation of the lithium ion by acetone is readily understood by comparing the donor numbers of the two solvents: $DN_{acetone} = 17.0$, $DN_{NM} = 2, 7$, whereas elementary electrostatic considerations are not helpful: NM has a higher dipole moment ($\mu = 3.57$ D) than acetone ($\mu = 2.88$ D). Raman and NMR results were explained by the formation of contact ion pairs at mole ratio acetone/$LiClO_4 < 4$ with $[ClO_4]^-$ ions entering the coordination sphere of the lithium ions; in other words they indicate the presence of solvated un-ionized lithium perchlorate (the perchlorate ion acts as an inner-sphere ligand in nickel and cobalt complexes only in solvents of low or moderate donor number) (Gutmann and Schmidt, 1974).

The influence of various anions on the position of the Li^+ and Na^+ solvation bands was investigated in acetone (Wong *et al.*, 1971): With perchlorates, tetraphenylborates, thiocyanates, nitrates, and iodides, band frequencies are independent of the nature of the anion. For bromides and chlorides a shift to lower frequencies is observed. Addition of $Bu_4N^+ClO_4^-$ or $Bu_4N^+I^-$ to solutions of $LiClO_4$ or LiI does not change the bands, but a shift to lower frequencies occurs on addition of $Bu_4N^+Br^-$ or $Bu_4N^+Cl^-$.

It has been shown that outer-sphere acceptor solvent effects shift the half-wave potential of the redox equilibrium $Fe(CN)_6^{4-} \rightleftharpoons Fe(CN)_6^{3-} + e$ to positive potential values (Chapter 8). Even more remarkable is the change in half-wave potential by varying the cation of the supporting electrolyte from the tetraethylammonium to the tetrabutylammonium ion. The differences in half-wave potentials are greater in a weak acceptor solvent, such as DMF, than in a strong acceptor solvent, such as methanol (Gritzner *et al.*, 1976*a*), since competition between cation and solvent molecules for outer-sphere coordination is taking place (Gutmann, 1977*c*) (Table 10.5).

For ion-pair formation in water between Me_4N^+ and $Fe(CN)_6^{3-}$ and $Fe(CN)_6^{4-}$, respectively, a stronger interaction is expected with the anion of higher charge, while NMR measurements reveal the opposite behavior (Larsen, 1966). This is in agreement with the donor–acceptor concept since the donor properties of the ligand nitrogen atoms are stronger in the reduced form, in which coordinated water is less readily replaced by the

Table 10.5. Half-Wave Potentials for the Reduction $[Fe(CN)_6]^{3-}/[Fe(CN)_6]^{4-}$ in 0.1 M Solutions of Tetrabutylammoniumperchlorate (TBAP) and Tetramethylammonium Perchlorate (TEAP), Respectively, in Various Solvents at 25°C

Solvent	$E_{1/2}$ (volts)		$\Delta E_{1/2}$ (volts)	AN	ε
	TEAP	TBAP			
DMF	−0.31	−0.61	0.30	16	36.7
CH_3CN	−0.27	−0.42	0.14	18.9	36.0
DMSO	−0.27	−0.39	0.12	19.3	46.7
PDC	−0.17	−0.20	0.03	18.3	65.0
NM	−0.07	−0.14	0.07	20.5	36.7
EtOH	+0.30	+0.24	0.06	37.1	24.3
MeOH	+0.47	+0.38	0.09	41.3	32.6

weakly acidic R_4N^+ ions than in the oxidized form, where water molecules are more easily replaced.

The relationship between the donor number of the solvent and the chemical shift of the ^{23}Na nucleus in solutions of sodium perchlorate or sodium tetraphenylborate in various donor solvents (Section 7.4) does not hold for solutions of sodium iodide. The chemical shift depends on the salt concentration, and this has been regarded as due to the presence of contact ion pairs. The fact that the effect is greater the lower the donor number of the solvent, indicates replacement of solvent molecules by iodide ions in the inner-solvation core around the sodium ion. Hence the species described as the contact ion pair is identical with solvated sodium-iodide molecules in the respective solution, where the Na—I bonds are more strongly polar than in the gas phase the greater the donor and acceptor numbers of the solvent.

It has been stated frequently that the formation of solvent-separated ion pairs is facilitated by increased solvent "polarity." Often, small changes in solvent structure can also drastically influence the equilibria (Szwarc, 1972). It has further been recognized that, contrary to previous belief, solvent polarity *per se* does not affect the spectrum of the ion pair if its structure is retained. The spectra of un-ionized species in solution are virtually indistinguishable from the spectra of the contact ion pairs, which are different from the spectra of solvent-separated ion pairs. It has been observed in dioxane that increase in size of the counterions leads to a bathochromic shift in the spectra of negative radical ions derived from aromatic hydrocarbons and ketones. For alkali metal salts the increase in the absorption frequency corresponding to λ_{max} was found to have a linear relationship with $(r + \text{const})^{-1}$, r denoting the radius of the cation. Increase in r of alkali metal ions corresponds to a decrease in acceptor strength and hence the bathochromic shift increases with increasing acceptor strength of the metal ion, indicating the formation of outer-sphere complexes between ions of opposite charge (Carter *et al.*, 1960; Mathias and Warhurst, 1962; McClelland, 1964).

Association of ions of opposite charge with the formation of hydrogen bonding has also been described as contact ion pairing, although considerable charge transfer is known to be involved. In nitrobenzene, tetrabutylammonium picrate is fully dissociated, whereas the association constant of the tributylammonium salt in the same solvent is 526 due to hydrogen bonding (Witschonke and Kraus, 1947). Likewise, the formation of carbanions by deprotonation of an acid is considered as due to stabilization of the hydrogen ion by coordination of the donor D:

$$D \longrightarrow H \underset{\text{Hetero-lysis}}{\frown} CR_3 \rightleftharpoons BH^+ + CR_3^-$$

The application of the concept of ion pairing for the reversed reaction is rejected by the conductometric results on the substitution equilibria of quinuclidine hydrochloride and *n*-butylammonium hydrochloride in different solvents (Table 10.1). In these cations the H atom is sufficiently acidic to provide hydrogen bonding not only to the chloride ion, but also to solvent molecules (Mayer *et al.*, 1976). The "ionization" may be considered a replacement reaction by which the hydrogen-bonded salt (the contact ion pair) is converted into solvated cation and chloride ion (May *et al.*, 1976).

Thus the cation–solvent interaction is appreciably stronger than the ion–ion interaction in a solvent of high DN and AN due to strong ion–solvent interactions.

10.7. *A Simplified Reaction Scheme*

Based on phenomenological criteria the following terminology has been proposed (Gutmann, 1977*c*):

(1) The term (solvent-separated) ion pair is retained in the original sense of Bjerrum and as such it is phenomenologically characterized by dielectric separation into free solvated ions.

(2) The term "outer-sphere complex" is applied to ion–ion interactions leading to species that are not separated into the free ions in a

medium of high dielectric constant without appropriate donor–acceptor interactions. Thus this term is also applied to "solvent-shared ion pairs."

(3) "Contact ion pairs" are found for weakly solvated ions in media of low dielectric and low coordinating properties, in concentrated salt solutions, and in fused-salt media.

The term "association constant"—at present applied for different types of equilibria—may be replaced in the following ways:

(1) Formation of solvent-separated ion pairs from free solvated ions: $K_{assoc} = 1/K_{sep}$.

(2) Formation of outer-sphere complexes between ions of opposite charges from free solvated ions: K_{out}.

(3) Formation of un-ionized species from free solvated ions: $1/K_{form}$.

(4) Formation of weakly solvated ion pairs from free unsolvated ions (limiting case of the contact ion pair): $K_{CIP} = 1/K_{dissoc}$.

The following simplified reaction scheme (see Fig. 10.3) has been presented (Gutmann, 1977*c*):

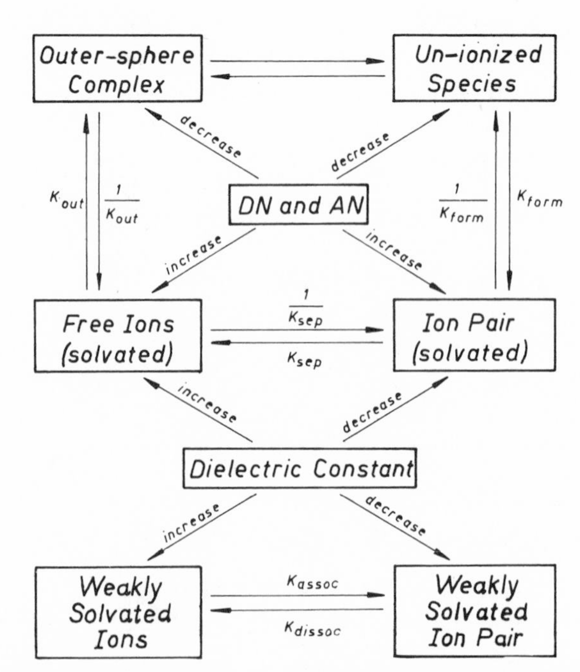

Fig. 10.3. Simplified reaction scheme for ionization equilibria (from Gutmann, 1977*c*, courtesy of Brunner Verlag, Zürich).

The stability of the solvated ion pair is enhanced both by increasing DN and AN of the solvent, by which both the un-ionized species and the outer-sphere complex between ions of opposite charge become less stable. In a solvent of moderate coordinating properties and at increasing solute concentration, mutual penetration of the solvation spheres and hence outer-sphere complexation is favored. K_{out} is also concentration dependent. Thus the simplified scheme does not represent the existence of the great number of different outer-sphere species in each particular system.

Chapter 11
Complex Stabilities in Solution

11.1. Qualitative Considerations

Among the great number of factors influencing and determining the stability constants of metal complexes (Beck, 1970), little attention has been paid to the consideration of specific solvent effects. Complex formation in solution can be regarded as the result of the competition of ligand and solvent molecules (S) for coordination at a metal in M^{n+}. When anions (X^-) are involved as ligands, the reaction can be considered as the inverse of an ionization reaction, for example:

$$MS_6^{2+} + 2X^- \rightleftharpoons MeX_2S_4 + 2S$$

For a given metal cation M^{n+} the equilibrium will be shifted to the right side (increasing complex stability) the greater the donor strength of X^- and the smaller the donor strength of S. The donor strength for a given free ligand varies with the acceptor properties of the solvent: The stronger the donor ligand X^- is solvated by solvent molecules acting as acceptors, the more its donor strength is lowered. Hence complex stability in solution is enhanced by decrease in both donor number and acceptor number of the solvent (Mayer and Gutmann, 1975).

The donor effect of the solvent has been studied by comparing the stabilities of a given complex in solvents of different donor number and of similar acceptor number. For example, the equilibrium constant for the replacement of a solvent molecule coordinated to antimony(V) chloride by a chloride ion to give the hexachloroantimonate ion,

$$D\cdot SbCl_5 + Ph_3CCl \rightleftharpoons Ph_3C^+SbCl_6^- + D \qquad (K_{Ph_3C^+SbCl_6^-})$$

is related to the donor number of the solvent (Fig. 11.1): The

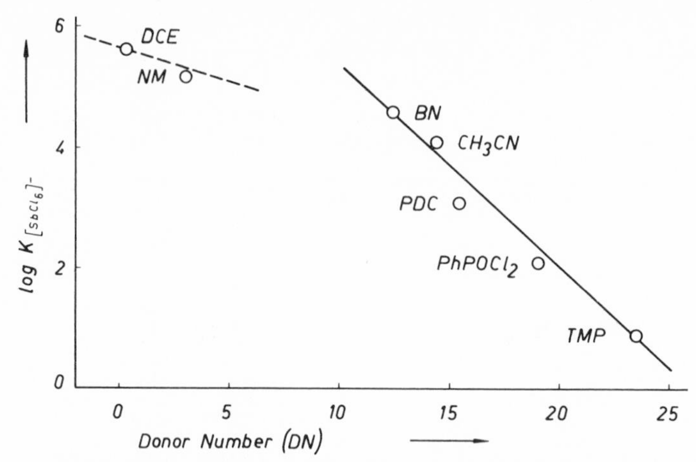

Fig. 11.1. Relationship between the $\log K_{Ph_3C^+SbCl_6}$ values for the reaction $D \cdot SbCl_5 + Ph_3Cl \rightleftharpoons Ph_3C^+SbCl_6^- + D$ in various solvents D and the donor numbers of these solvents.

$\log K_{Ph_3C^+SbCl_6^-}$ values are inversely proportional to the donor number (Gutmann and Schmid, 1971).

An analogous relationship holds for tetrahalocobaltate in different solvents of similar acceptor number except for water and DMA (Tschebull *et al.*, 1975). Addition of the stoichiometric amount of the halide ion to a solution of cobalt(II) perchlorate in nitromethane leads to nearly quantitative formation of tetrahalocobaltate. In a solvent of medium donor number, such as acetonitrile or acetone, an excess of ligand X^- is required, and a still higher excess is required in a solvent of high donor number, such as DMSO or HMPA. In these solvents complex stability increases in the sequence of ligand donor properties, $I^- < Br^- < Cl^- < NCS^- < N_3$, with CoI_4^{2-} and $CoBr_4^{2-}$ being unstable. The stability of a complex containing anionic ligands is greater the smaller the acceptor number of the solvent, as the donor properties of the anions are increased. Although the donor numbers for DMA and DMSO are similar, a given tetrahalocobaltate is more stable in DMA than in DMSO because of the smaller acceptor number of DMA. Stabilities of $Co(N_3)_4^{2-}$ and $CoCl_4^{2-}$ are similar in DMSO and HMPA since the former has a lower donor number, whereas the latter has a lower acceptor number. ES and PDC have similar donor numbers, but PDC has a lower acceptor number, and hence CoX_4^{2-} complex stability is greater in PDC. The stability (Beck, 1970) of $CoCl_4^{2-}$ in water is greatly increased in a strongly acidic solution since water molecules are removed from the hydration shell of the chloride ions in order to hydrate the hydrogen ions. CoX_4^{2-} ions are considerably more stable in acetone than in water since the former has weaker donor properties and a considerably smaller acceptor number.

Co^{2+} forms a hexasolvated species with most solvent molecules, but it gives the tetrasolvate with HMPA (Donoghue and Drago, 1962). There is a linear relationship between donor number and enthalpy of solvation of Co^{2+} for acetonitrile, DMF, and DMSO, but HMPA falls off the line (Gutmann *et al.*, 1969). The ^{1}H NMR spectra at different temperatures suggest that at room temperature coordination center and ligand shell are not at rest; the coordination center appears to be "rattling" within the rigid sphere of solvent molecules (Gutmann *et al.*, 1969) and the Co–HMPA bonds are weaker than expected, two HMPA molecules being readily replaceable by chloride ions (Gutmann and Weiss, 1969). Tetrahedral solvates of cobalt(II) are also known with other strong donor solvent molecules (Goodgame *et al.*, 1962; Cotton *et al.*, 1964). In TMP (Angerman and Jordan, 1969), ammonia (Glaeser *et al.*, 1965), DMA (Gutmann *et al.*, 1972), and even in water (Swift and Connick, 1962; Swift, 1964), equilibria are found between tetrahedral and oxtahedral species.

Replacement at an octahedral complex is predominantly S_N1 in character, involving an intermediate of lower coordination number. The equilibrium,

$$FeD_6{}^{3+} \rightleftharpoons FeD_5{}^{3+} + D$$

has been investigated in six donor solvents of similar acceptor number, namely, TMP, AA, DMF, DEF, DMA, and DMSO. The amount of the lower coordinated species was found greatest in DMF, and it is decreased as the donor number is either decreased or increased (Fig. 11.2) (Schmid *et al.*, 1976*b*). This has been explained by: (a) the displacement of one ligand from the inner-sphere into the outer sphere $FeD_6{}^{3+} \rightleftharpoons FeD_5{\cdot}D^{3+}$, and (b) dissociation of D from the outer sphere with rearrangement in the inner sphere $FeD_5{\cdot}D^{3+} \rightleftharpoons FeD_5{}^{3+} + D$. While the equilibrium constant of the first process appears to increase with increasing donor number (Gutmann and Schmid, 1974), this is not true for the second step: As the rearrangement energy is lowered with decreasing Fe—D bond strength, the reaction is favored by low donor number.

It has been stated that the donor number concept is not applicable for interactions between soft units (Section 2.2). Whereas tin(IV) iodide, as expected, is extensively ionized in DMSO (high donor number), mercury(II) iodide is not ionized in the same solvent since the soft mercury(II) binds iodide ions more strongly. Tin(IV) is preferentially coordinated in DMSO via the oxygen atom, whereas mercury(II) is linked via the sulfur atom of DMSO (Gaizer and Beck, 1967). The exclusive consideration of bond properties of the complex species is, however, not satisfying.

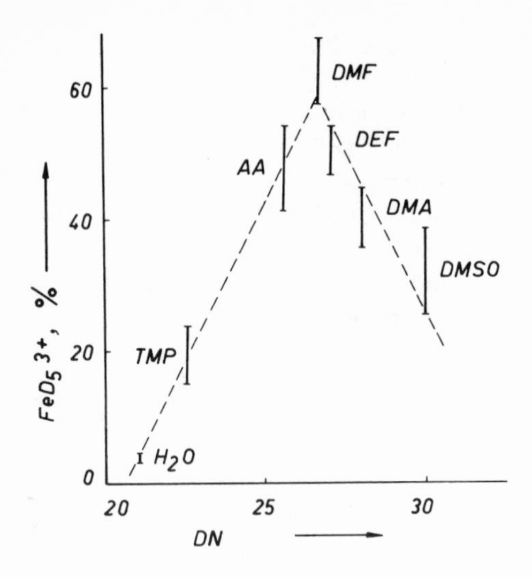

Fig. 11.2. Fraction of the activated species Fe_a^{3+} in 1×10^{-4} M solutions of FeL_6^{3+} in acetonitrile as a function of the donor number of the ligand L (from Schmid *et al.*, 1976*a*, courtesy of Verlag Chemie, Weinheim).

According to thermochemical calculations, the heterolytic dissociation enthalpies for the reaction

$$CH_3HgX_{(g)} \rightleftharpoons CH_3Hg^+_{(g)} + X^-_{(g)}$$

increase regularly in the series, $X = I^- < Br^- < Cl^-$ (Mayer and Gutmann, 1975). Consequently, ionization constants for the hydrogen halides in water should be expected to increase in the order $Cl^- < Br^- < I^-$. In fact the reverse stability order is found in water and this has been considered as a consequence of typical "soft acid" behavior of the CH_3Hg^+ ion (Schwarzenbach and Schellenberg, 1965). This behavior is explained in the following qualitative way: D_{het} values for CH_3HgX compounds are rather similar. The increase in bond energies in the series, $I < Br < Cl$ is therefore overcompensated by the strong increase (numerical values) in solvation energies of the halide ions. This would suggest that in solvents of poor solvating properties for halide ions, methylmercury halides should show normal stabilities, $K_{CH_3HgI} < K_{CH_3HgBr} < K_{CH_3HgCl} < K_{CH_3HgF}$,

$$K = \frac{[CH_3HgX]}{[CH_3Hg^+][X^-]}$$

a behavior actually found in HMPA due to results of preliminary conductometric studies (Mayer and Gutmann, 1975). Thus, soft behavior of methylmercury halides is not an inherent property of the CH_3HgX^+ ion, but it is influenced by the solvating properties of the solvent toward the anionic ligands.

11.2. Quantitative Considerations

General comments have been made on the reliability of stability constants (Rosseinski, 1971), as compiled for aqueous systems (Sillén and Martell, 1971). The favorite noncomplex forming anion, the perchlorate anion, is not as innocent in outer-sphere complexation as has been assumed for the evaluation of experimental data (Beck, 1970; Lederer and Mazzei, 1968; Gutmann and Schmidt, 1974). The situation is more complicated if simultaneous formation of inner- and outer-sphere complexes occurs. The usual equilibrium analysis cannot distinguish between the corresponding outer- and inner-sphere complexes, for instance, between $M(OH_2)_6L$ and $M(OH_2)_5L$, and consequently the stability constant of the first complex, derived in a usual way, is a composite constant. The drastic effects on stability changes due to outer-sphere effects may be visualized from the available evidence on solvated species in solvent mixtures (Erlich *et al.*, 1973).

Concerning the evaluation procedures, Rossotti *et al.* (1972) draw the following conclusions:

> Computerized methods of calculation can extract the maximum information from the data, but the data cannot be improved. Systematic errors cannot be treated by statistics designed only for random errors. The "best" set of constants, however obtained, is not necessarily the "right" one. Stability constants obtained by using sophisticated computer programs, like those obtained by any other method, can never rise above the status of values "compatible with" the available experimental data.

By means of thermochemical cycles, Mayer (1975) derived equations for complex formation equilibria involving substitution of a solvent molecule by an anion: The standard free enthalpy $\Delta\Delta G^0_{exp}$ of a given reaction in different solvents is related to the differences in donor (ΔDN) and acceptor ($\Delta\Delta G_{solv}[Cl^-]$) properties of the solvents and strength of the solvent–solvent interactions ($\Delta\Delta G_{vap}$ = differences in standard free enthalpies of vaporization). The symbol Δ indicates that all quantities are referred to a standard solvent:

$$\Delta\Delta G^0_{exp} = \Delta\Delta G^0_{(g)} + b\cdot\Delta\Delta G_{solv}[Cl^-] + c\Delta\Delta G_{vap} \quad (1)$$

$$\Delta\Delta G^0_{exp} = a\cdot\Delta DN + b\cdot\Delta\Delta G_{solv}[Cl^-] + c\Delta\Delta G_{vap} \quad (2)$$

For aprotic solvents a linear correlation has been found between $\Delta\Delta G_{solv}[Cl^-]$ and the acceptor number of the solvent (Mayer *et al.*, 1975) so that the latter quantity may be inserted instead of $\Delta\Delta G_{solv}[Cl^-]$ in Eqs. (1) and (2) (Mayer, 1977).

In Eq. (1), $\Delta\Delta G^0_{(g)}$ means the free energy of the corresponding gas-phase reaction

$$CoCl_3{\cdot}D^- + Cl^- \rightleftharpoons CoCl_4^{2-} + D$$

The $\Delta\Delta G^0_{(g)}$ values calculated from Eq. (1) are related linearly to the donor number of the solvent (Fig. 11.3). Similar correlations have been obtained for the rate constants of the forward and backward reaction and for the equilibrium constants of the iodine–triiodide equilibrium in different solvents (Fig. 11.4) (Mayer, 1975):

$$I_2 + I^- \equiv I_3^-$$

Conversely, Eq. (2) may be used to calculate equilibrium constants or rate constants of a given reaction in different solvents provided that the corresponding solvent properties (DN, AN, or $\Delta\Delta G_{solv}[Cl^-]$, $\Delta\Delta G_{vap}$) are available.

A relationship analogous to Eq. (2) but without the term $\Delta\Delta G_{vap}$, which represents the solvent–solvent interactions, has recently been proposed (Krygowski and Fawcett, 1975; Fawcett and Krygowski, 1975), where the Dimroth–Reichardt E_T values are used as a measure for the acceptor properties of the solvent.

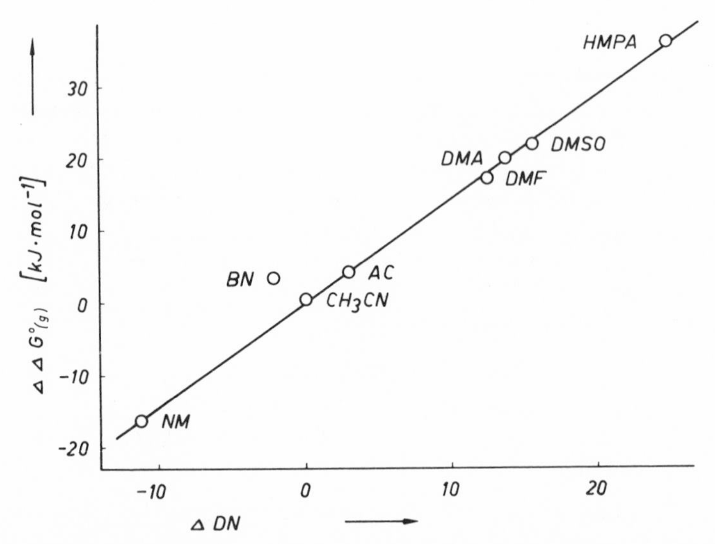

Fig. 11.3. Relationship between $\Delta\Delta G^0_{(g)}$ for the reaction $[CoCl_3D]^- + Cl^- \rightleftharpoons [CoCl_4]^{2-} +$ D and the donor number of D compared to that of CH_3CN (from Mayer, 1975, courtesy of Butterworth and Co., London).

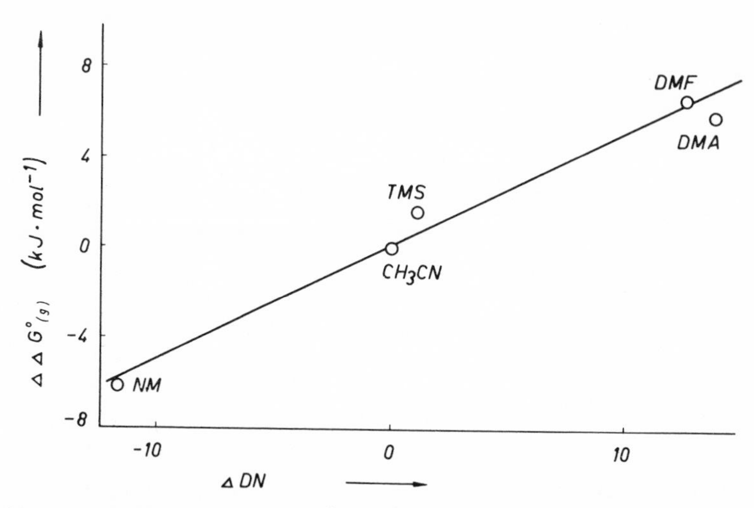

Fig. 11.4. Relationship between $\Delta\Delta G^0_{(g)}$ for the reaction $I_2 + I^- \rightleftharpoons I_3^-$ and the solvent donicity compared to the values in acetonitrile (from Mayer, 1975, courtesy of Butterworth and Co., London).

11.3. Autocomplex Formation

Ionization of trimethyltin iodide by donor molecules is an example of a heterolytic fission of a covalent bond. The situation is more complicated when the substrate more than one ionizable bond, and in particular, when anions are formed which compete with the donor molecules for coordination at the substrate. If the donor properties of the competing ligands are not vastly different, both complex cations and complex anions are formed at the same time; this process is known as "autocomplex formation" or "ligand disproportionation." The autocomplex formation reaction

$$nD + 2MX \rightleftharpoons D_nM^+ + MX_2^-$$

can be formally divided into the following steps:

(*i*) *Formation of a complex cation*:

$$nD + MX \rightleftharpoons D_nM^+ + X^-$$

(*ii*) *Formation of a complex anion*:

$$X^- + MX \rightleftharpoons MX_2^-$$

In a strong donor solvent a solute will undergo autocomplex formation if the M—X bond is not easily ionized, whereas in a weak donor solvent a compound which is easily ionized will give autocomplex ions.

Autocomplex formation frequently gives species each of which has only one kind of ligand, such as

$$2CoCl_2 + 6DMSO \rightleftharpoons Co(DMSO)_6^{2+} + CoCl_4^{2-}$$

The term "symbiosis" has been suggested for this phenomenon by Jørgensen (1964). The equilibria between adduct formation, ionization, and autocomplex formation are further influenced by the molar ratios of the donor and substrate applied, a factor frequently ignored in the interpretation of data. NMR measurements, spectrophotometric, kinetic, potentiometric, polarographic, and conductometric investigations are helpful in elucidating the various types of species present in solution.

Conductometric titration in a poorly coordinating medium of reasonable dielectric constant has proved very useful for obtaining indications about the superposition of autocomplex formation, adduct formation, and ionization. For example, the nonconducting solution of iron(III) chloride in dichloroethane turns lemon yellow after addition of a strong donor due to the formation of tetrachloroferrate (autocomplex formation). Its conductivity is a function of the donor number and the amount of the donor added. HMPA gives poorly conducting solutions. At a molar ratio HMPA: $FeCl_3 = 1:2$, $FeCl_4^-$ ions are present. Further addition of HMPA leads to a decrease in conductivity, which shows a minimum at a molar ratio 1 : 1, when tetrachloroferrate(III) has disappeared (Gutmann and Wegleitner, 1970).

$$[FeCl_4]^- + HMPA \rightleftharpoons HMPA{\cdot}FeCl_3 + Cl^-$$

$$[HMPA{\cdot}FeCl_2]^- + Cl^- \rightleftharpoons HMPA{\cdot}FeCl_3$$

Further addition of HMPA increases the conductivity due to the ionization to $(HMPA)_nFeCl_2^+$ and Cl^- ions: In the presence of excess HMPA the tetrachloroferrate ion is no longer stable, since the chloride ions cannot compete with the HMPA molecules for coordination at iron(III). For further examples, see Gutmann (1971*a*).

Complexes with metals in low oxidation number are even subject to autocomplex formation in the presence of strong donor molecules (Hieber *et al.*, 1957; Hieber and Werner, 1957; Hieber and Brendel, 1957; Hieber and Schubert, 1965; Chini, 1968), such as DMSO, ammonia, or the alcohols. Besides the cation which is coordinated by donor molecules, polynuclear anions, which can be degradated at higher temperatures, are formed. It may be noted that in this process of autocomplex formation, changes in the oxidation numbers are involved; for example:

$$3Co_2(CO)_{12} + 24D \rightleftharpoons 4\overset{+II}{Co}D_6(\overset{-I}{Co}(CO)_4)_2 + 4CO$$

11.4. Solubility Considerations

The solubility product K_s of a given compound A—B can be derived from thermodynamic quantities, according to the free enthalpy cycle:

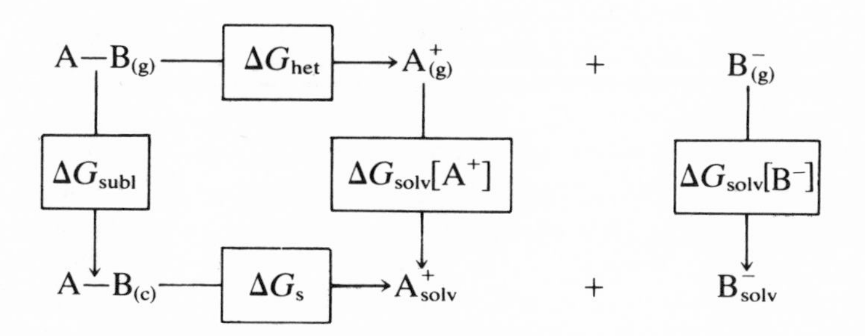

where $A{-}B_{(c)}$ represents the condensed phase of A—B.

$$\Delta G_s = -RT \ln K_s$$

$$K_s = [A^+_{solv}][B^-_{solv}]$$

$$\Delta G_s = \Delta G_{subl} + \Delta G_{het} + \Delta G_{solv}[A^+] + \Delta G_{solv}[B^-]$$

In comparing the solubilities of a given compound (e.g., constant values for ΔG_{subl} and ΔG_{het}) in different solvents, the solubility product is a function of the free enthalpies of solvation for the cation and the anion. Since the latter have been shown to be related to the donor number and acceptor number of the solvent, respectively, these empirical parameters may be used as a useful guide in accounting for solubility relationships.

Table 11.1. pK_s Values (K_s = Solubility Product) for NaCl and KCl in Solvents of Different Donor and Acceptor Numbers

			pK_s (molar scale)	
Solvent	DN	AN	KCl	NaCl
NM	2.7	20.5	—	10.0
CH_3CN	14.1	18.9	7.3	8.2
PDC	15.1	18.3	7.1	7.8
DMF	26.6	16.0	5.5	4.7
NMP	27.3	13.3	—	5.3
DMSO	29.8	19.3	3.4	2.7
CH_3OH	19.0	41.3	3.2	2.2
H_2O	18.0	54.8	−0.9	—

Accordingly, the solubilities of KCl and NaCl in solvents of nearly identical acceptor number increase in the order of increasing donor number, as long as their acceptor numbers are similar. The differences in solubilities in NM (AN = 20.5), CH_3CN (AN = 18.9), PDC (AN = 18.3), and DMSO (AN = 19.3) are related to the differences in the donor numbers of solvents (Table 11.1). Likewise, for solvents of similar donor number, such as DMF (DN = 26.6), NMP (DN = 27.3), and DMSO (DN = 29.8), the differences in solubility products can be accounted for by the different acceptor numbers. The pK_s value for NaCl in NMP would be expected to be smaller if the solvent had a greater acceptor number.

The role of the acceptor number in stabilizing the anions in solution can be seen by comparing the solubility products in DMSO and in methanol, which are in the same order of magnitude: The smaller donor number of methanol is overcompensated by its greater acceptor number; hence the solubility in water (higher acceptor number) is considerably higher.

By means of the above equations it is therefore possible to calculate solubility data for a given compound in various solvents, provided that either ΔG_{solv} values or donor and acceptor numbers are available.

Chapter 12

Kinetics of Substitution Reactions

12.1. *General Considerations*

We shall now discuss solvent effects on rate and mechanisms of substitution reactions. For the reaction

$$L_x^nM{-}L^l + L^e \longrightarrow L_x^nM{-}L^e + L^l$$

L^e is the entering, L^l the leaving, and L^n the nonleaving ligand or coordinated solvent molecule, solvent exchange being merely a special case of substitution; charges being omitted, rate and mechanism will depend on: (1) The nature of M; (2) the donor properties of the ligands L^e, L^l, and L^n; and (3) the donor and acceptor properties of the solvent. The effect of the particular ligands and of the solvent on the rate will differ for a dissociative (S_N1) and an associative (S_N2) mechanism. For an S_N1 reaction, bond breaking is rate determining:

$$L_x^nM{-}L^l \xrightarrow{k} L_x^nM + L^l$$

The rate coefficient k is a function of the free enthalpy for the formation of the activated complex and is increased by decreasing strength of the bond $M{-}L^l$. The bond strength of $M{-}L^l$ may be considered as the result of both the acceptor properties of L_x^nM toward L^l and the donor properties of the leaving ligand L^l toward L_x^nM. The acceptor strength of the center M is decreased by increasing donor properties of the ligands L^n, hence the bond $M{-}L^l$ is weakened by increasing donor number of L^n (Gutmann and Schmid, 1974), in agreement with the first bond-length variation rule:

$$\begin{array}{ccc} L^n & & L^n \\ & \searrow \quad \swarrow & \\ L^n \rightarrow & M & \leftharpoondown L^l \\ & \nearrow \quad \nwarrow & \\ L^n & & L^n \end{array}$$

Thus the rate of substitution is decreased by increasing donor number of the leaving ligand L^l and increased by increasing donor number of the nonleaving ligands L^n.

For S_N2 reactions the rate-determining step is that of bond making:

$$L_x^nM{-}L^l + L^e \longrightarrow L_x^nM\begin{matrix} \diagup L^l \\ \nwarrow L^e \end{matrix}$$

The higher the donor property of the entering ligand L^e, the greater the rate of the substitution reaction.

The situation is more complicated for substitution reactions which occur by a mechanism intermediate between S_N1 and S_N2 (Gold, 1956).

12.2. Substitution at Carbon

The rate of the S_N1 reaction,

$$R_3C{-}X \longrightarrow R_3C^+ + X^- \xrightarrow{Y-} R_3CY + X^-$$

will be slower the stronger the C—X bond. As long as the acceptor strength of the carbon atom is kept constant, the rate will be decreased as the donor strength of the leaving ligand X^- is increased. For example, the reactivity of the carbon–halogen bond for a given R decreases in all solvents in the order (Ingold, 1953)

$$RI > RBr > BCl > RF$$

The acceptor strength of the carbon atom depends on the nature of the R groups. R groups do not usually take part in substitution reactions and hence from the aspect of coordination chemistry they may be considered as nonleaving ligands. Their effects are known as substituent rather than ligand effects. The more the acceptor strength of the carbon atom is decreased by the R groups, the weaker the C—X bond becomes and the more readily substitution will take place. Table 12.1 shows the effect of the *para* substituents R on the rate of S_N1 solvolysis of 1-aryl-1-bromopropanes in 90% acetone–water at 75°C (Lloyd and Parker, 1970):

$$R{-}C_6H_4{-}\underset{\displaystyle Br}{\underset{|}{C}H}{-}CH_2CH_3 \rightleftharpoons R{-}C_6H_4{-}\overset{\oplus}{C}H{-}CH_2CH_3 + Br^-$$

Methyl and methoxyl groups are electron donating, which is also known as

Table 12.1. Rates of Solvolysis of 1-Aryl-1-bromopropanes with Changes in R Groups

R	Log *k*
NO_2	−7.3
H	−4.0
CH_3	−2.5
OCH_3	+0.1

the +I effect, whereas the nitro group or halogen atoms are electron withdrawing (−I effect).

The effect of the solvent on an S_N1 reaction is given by its ability to polarize the C—X bond and to stabilize the transition state produced by donor action D at the carbon atom and by acceptor action Ac at X:

$$D + R_3C{-}X + Ac \rightleftharpoons \underset{\underset{D}{\uparrow}\quad\underset{Ac}{\downarrow}}{R_3\overset{\delta+}{C}{-}\overset{\delta-}{X}} \rightleftharpoons R_3C^+_{solv} + X^-_{solv}$$

The ionization will be faster with increasing acceptor and donor number of the solvent. Neither the donor number nor the acceptor number alone will be sufficient to predict the solvent effect on rate. An example is presented by the S_N1 solvolysis of *trans*-4-*tert*-butylcyclohexyl tosylate (Table 12.2) (Beltrame *et al.*, 1972):

$$(CH_3)_2C{-}\langle\text{⬡}\rangle{-}OTs \xrightarrow{k} (CH_3)_3C{-}\langle\text{⬡}\rangle{-}\overset{\oplus}{C}{-}H + OTs^-$$

$$\downarrow$$

Solvolysis products

The relative influence on the rate of donor or acceptor properties of the solvent depends also on the acceptor and donor properties, respectively, of the ions being formed, because the transition state has much ionic character. In the above-mentioned reaction the developing tosylate ion is a moderately strong donor, but the large, developing carbonium ion is a weak acceptor.

For this reason, we should expect the influence of the acceptor number of the solvent to be greater than that of the donor number. This expectation is confirmed by the data given in Table 12.2. The solvolysis rate is smaller in acetone than in nitromethane, since the acceptor number of nitromethane is greater than that of acetone. The greater donor number of acetone cannot compensate for the effect of its much smaller acceptor number. Likewise, the rate coefficient is greater in acetonitrile than in acetone, since the acceptor number of acetonitrile is greater than that of acetone. The role of the donor number becomes, however, more apparent the greater the donor property of the solvent: In comparing the rates in acetonitrile, dimethylformamide, and dimethylsulfoxide, which are not vastly different in acceptor numbers, an increase is found as the donor number is increased in going from acetonitrile to dimethylsulfoxide.

In the solvolysis of *p*-methoxyneophyl tosylate (Smith *et al.*, 1961),

OMe OMe

$H_3C-C(CH_3)(C_6H_4OMe)-CH_2OTs \xrightarrow{k}$ $(H_3C)_2C \overset{\oplus}{-} CH_2$ (bridged by the *p*-methoxyphenyl ring) $+ OTs^-$

Solvolysis products

the carbonium ion is an even weaker acceptor, because of participation by the *p*-methoxyphenyl group (internal solvation). Thus it is mainly the acceptor property of the solvent toward the developing tosylate ion and hence the acceptor number of the solvent which determines the solvent effect on the rate, as shown in Fig. 12.1: The logarithms of the rate constants bear a roughly linear relationship with the acceptor numbers of

Table 12.2. Rates of Solvolysis of trans-4-Tert-butylcyclohexyl tosylate $(CH_3)_3C-C_6H_{10}-OTs$ at 75°C

Solvent	Log *k*	AN	DN
Acetone	−5.3	12.5	17
Nitromethane	−4.5	20.5	2.7
Acetonitrile	−4.5	18.9	14.1
Dimethylformamide	−3.8	16.0	26.6
Dimethylsulfoxide	−2.9	19.3	29.8

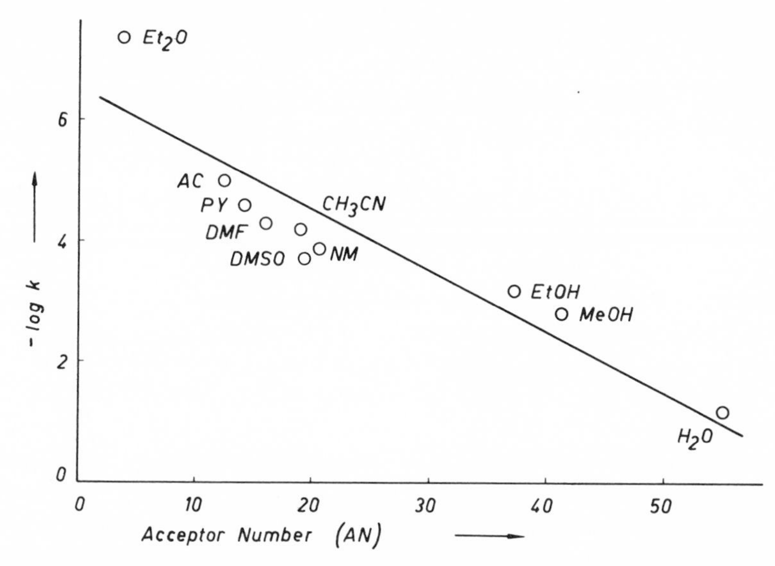

Fig. 12.1. Logarithm of the rate constants of solvolysis of *p*-methoxyneophyl tosylate vs. the acceptor numbers of various solvents.

the solvents, whereas no relationship exists with the donor numbers of the solvents. For an S_N2 reaction of the charge type,

$$Y^- + R_3C{-}X \rightleftharpoons [Y\cdots CR_3\cdots X]^- \longrightarrow Y{-}CR_3 + X^-$$

bond forming (C—Y) and bond breaking (C—X) are synchronous and the substitution rate depends on both the bond strength C—X and on the donor strength of the entering group. The weaker the leaving group is bound, the faster the reaction. For example, the rate of the reaction

$$CH_3X + SCN^- \longrightarrow CH_3SCN + X^-$$

increases in the order of $X = Cl < Br < I$ (Parker, 1969). If the entering group is an anion, the acceptor strength of the solvent is rather decisive. The donor strength of the anion will be decreased by solvation the greater the acceptor number of the solvent. Hence the reaction rate will be smaller the greater the acceptor number of the solvent. Since neither the reactants nor the transition states are strong acceptors, the donor number of the solvent is not important.

The solvent effect on the rate of the S_N2 reaction

$$MeI + Cl^- \xrightarrow{k} MeCl + I^-$$

has been reported by Parker (1969). Figure 12.2 shows the relationship that exists between the logarithm of the second-order rate constant for this reaction in different solvents and the acceptor number of the solvent. The situation is analogous for the reaction of methyl iodide with radioactively labeled iodide ions (Swart and Le Roux, 1957): The log k values are decreased by an increase in acceptor number of the solvent (Mayer *et al.*, 1975). A similar behavior is found for the interaction between *p*-nitrofluorobenzene and azide ions (Parker, 1969),

$$O_2N-C_6H_4-F + N_3^- \xrightarrow{k} O_2N-C_6H_4-N_3 + F^-$$

and the relationship between logarithm of the rate constant and the acceptor number of the solvent is shown in Fig. 12.3.

Rates in solvents with similar donor numbers and very different acceptor numbers [e.g., acetone (DN = 17; AN = 12.5) and methanol (DN = 19, AN = 41.3)] differ by six orders of magnitude, whereas nearly the same reaction rates are found in solvents with very different donor numbers, as long as their acceptor numbers are similar [e.g., DMSO (DN = 29.8; AN = 19.3) and NM (DN = 2.7; AN = 20.5]. A further example is the variation of the Arrhenius activation energies for the reaction of *n*-butyl bromide with azide ions, which also proceeds via an S_N2 mechanism, in different solvents (Delpuech, 1965): The greater the acceptor number of the solvent, the stronger the solvation of the azide ions and hence the higher the activation energy (Mayer *et al.*, 1975; Gutmann, 1976*b*).

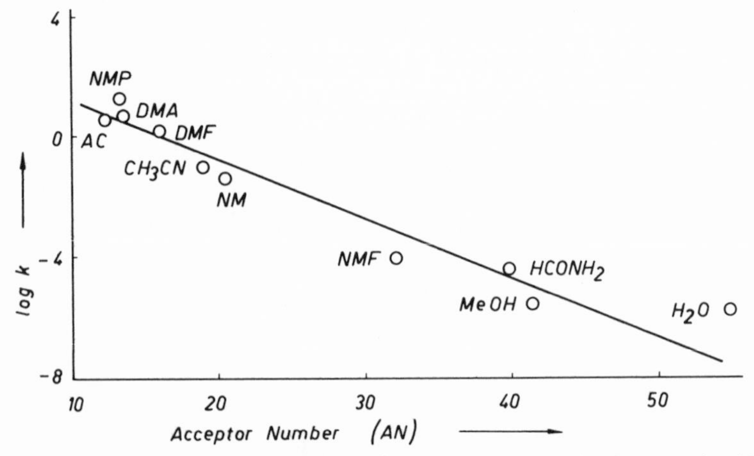

Fig. 12.2. Logarithm of the rate constants of the substitution reaction between the chloride ion and methyl iodide vs. the acceptor numbers of various solvents.

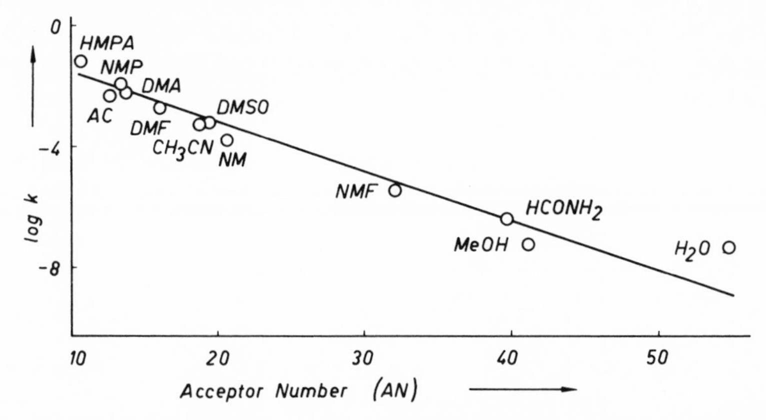

Fig. 12.3. Logarithm of the rate constants of the substitution reaction between the azide ion and *p*-nitrofluorobenzene vs. the acceptor numbers of various solvents.

It can therefore be stated that for the charge types discussed, S_N1 reactions will be faster as the donor *and* acceptor numbers of the solvent increase, while S_N2 reactions will be faster as the acceptor number of the solvent decreases.

On the other hand, the donor strength of the solvent is of importance if the entering group is a strong acceptor, as for the S_E2 reaction (Agami and Prévost, 1966):

$$R_2Zn + Cd^{2+} \rightleftharpoons [CdR_2Zn^*]^{2+} \longrightarrow R_2Cd + Zn^{2+}$$

The higher the donor number of the solvent, the more strongly Cd^{2+} will be solvated and hence the slower the substitution reaction.

12.3. Substitution at Metal Complexes

The rate of ligand substitution at metal complexes is related in part to the metal–ligand bond strength (Caldin, 1964; Steinhaus and Margerum, 1966; MacKellar and Rorabacher, 1971; Gutmann and Schmid, 1974). Much evidence has been accumulated in support of the hypothesis that the rate-determining step of a substitution reaction of octahedral complexes involves the loss of a ligand from the inner coordination sphere (Eigen and Wilkins, 1965; Pearson and Ellgen, 1967). This is thought to be preceded by the rapid formation of an outer-sphere complex in which the entering ligand is coordinated to the inner coordination sphere of the coordination center (Tobe, 1965). Thus the nature of the entering ligand is also important in determining the kinetic behavior.

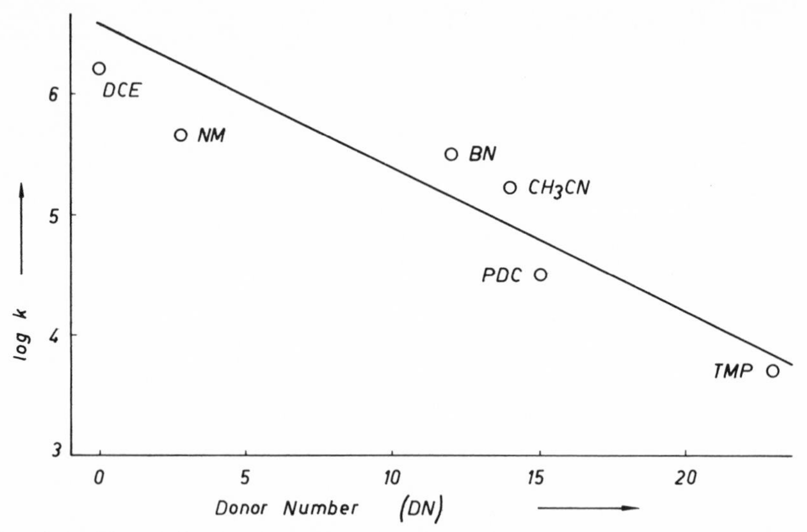

Fig. 12.4. Plot of the donor numbers of various solvents vs. the logarithm of the rate constants for the reaction of $SbCl_5L$ with Ph_3CCl.

Substitution at tetracoordinate metal complexes is usually associative and the existence of the five-coordinate intermediate is demonstrated in many cases.

We shall first discuss the situation where *the leaving ligand is varied.* The rate of replacing a solvent molecule L^l by Cl^- at antimony(V) chloride

$$Cl_5Sb{-}L^l + Ph_3CCl \longrightarrow Ph_3C^+SbCl_6^- + L^l$$

in six different solvents has been found roughly inversely proportional to the donor number of the solvent L^l (Fig. 12.4) (Gutmann and Schmid, 1971): The stronger L^l is bonded to Sb, the slower the substitution reaction. The observed deviations may be due to the different acceptor properties of the solvent toward trityl chloride, by which the donor properties of the latter may be varied. Likewise, the rate constant of the reaction

$$(H_2O)_5ML^{2+} + H_2O \longrightarrow M(OH_2)_6^{2+} + L$$

($L =$ methanol, water, or ammonia, and $M = Co^{2+}$ or Ni^{2+}) (MacKellar and Rorabacher, 1971) is decreased as the donor number of the solvent increases. Similarly, the rate of substitution at nickel(II) complexes with substituted pyridines by hydronium ions,

$$(H_2O)_5Ni{-}py^{2+} + H_3O^+ \longrightarrow Ni(OH_2)_6^{2+} + pyH^+$$

is decreased by increasing basicity of the pyridine bases (Moore and Wilkins, 1964).

The rate-determining step of the reaction

$$Cl_3CoL^- + Cl^- \xrightarrow{k} CoCl_4^{2-} + L^l$$

can be regarded as due to the formation of the pentacoordinate intermediate $Cl_4CoL^{l\,2-}$. The rate should be determined by both the donor and acceptor properties of the solvent. Increasing donor number of L^l causes a decrease in acceptor strength of the coordination center and hence a decrease in formation rate of the intermediate. In addition the reaction will be retarded by increasing the acceptor number of the solvent due to increasing solvation and hence to decreasing donor property of the entering chloride ion. Covering a range from 12.5 to 20.5 the acceptor numbers of the solvents recorded in Fig. 12.5 are not very different and thus the substitution rates correlate reasonably with donor numbers of the solvents (Tschebull, 1974).

We shall now discuss the situations where *the nonleaving ligands are varied.* The acceptor strength of L_5^nM is changed not only by varying the metal ion M but also by varying the nonleaving ligands L^n. The nature of the metal ion is the main factor determining the overall substitution rate. Unfortunately for metal ions no characteristic property is available for the acceptor strength. The rate constant for water exchange is not always inversely proportional to the ionic radius for metals of given charge type. Deviations have been regarded due to partially filled *d*-orbitals on the metal ion which cause considerable complications interpreted in terms of crystal-field stabilization and the Jahn–Teller effect (Eigen, 1963).

Within a homologous series of hydrated metal ions an increase in substitution rate is paralleled by decrease in hydration enthalpy. It might therefore be expected that in nonaqueous solutions the substitution rate in

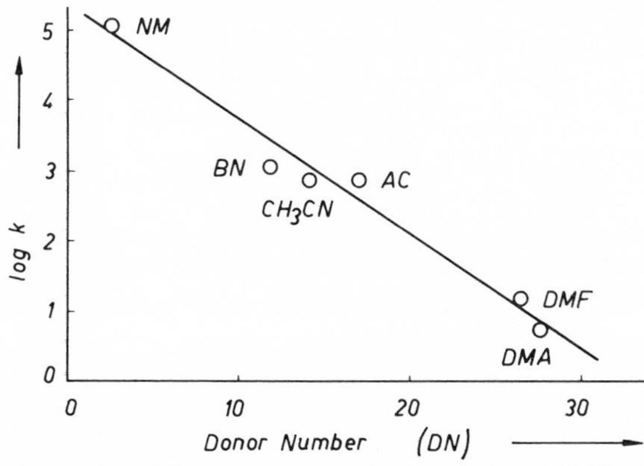

Fig. 12.5. Logarithm of the rate constants for the reaction of $CoCl_3L^-$ with Cl^-.

a solvate will be an inverse function of the solvation enthalpy. The weakening of the M—L^l bond by increasing donor properties of L^n will be manifested in an increase in M—L^l bond rupture rate and substitution reactions proceeding through a S_N1 mechanism will be faster (Gutmann and Schmid, 1974).

The rate of the reaction ($L^n = L^e = L$)

$$L_5Ni{-}NCS^+ + L \longrightarrow NiL_6{}^{2+} + SCN^-$$

is found to increase with increasing donor strength of L (Dickert *et al.*, 1970). The rate constants vary by eight orders of magnitude in going from methanol to ammonia (Dickert and Hoffmann, 1971; Hoffmann *et al.*, 1974). Surely an acceptor number effect on the leaving anion will also be relevant. Methanol and water act as acceptors toward an SCN group, so that the Ni—NCS bond is weakened:

$$L_5Ni{-}NCS{\rightarrow}H{-}O{<}^{H}$$

A relationship between substitution rate and the donor number of the solvent (Fig. 12.6) has been established for the reaction (Hoffmann *et al.*, 1974)

$$L_5Ni{-}TFA^+ + L \xrightarrow{k} NiL_6{}^{2+} + TFA^-$$

(TFA = trifluoroacetate). The water exchange rates in aquoammine-nickel(II) complexes are increased by an increase in the number of

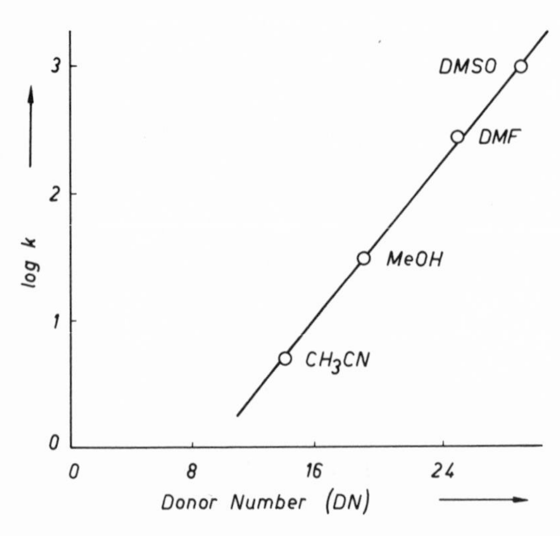

Fig. 12.6. Relationship between the logarithm of the substitution rate and the donor numbers of various solvents for the reaction of $L_5Ni{-}TFA^+$ and L (from Hoffmann *et al.*, 1974, courtesy of Verlag Chemie, Weinheim).

ammonia molecules in the inner coordination sphere of Ni(II) (Desai *et al.*, 1970), ammonia being a stronger donor than water:

$$H_3N \rightarrow \overset{\frown}{Ni - O}H_2$$

The incorporation of water molecules into the inner coordination sphere of nickel(II) methanol solvates enhances the rate of methanol exchange (Luz and Meiboom, 1964; Rogers *et al.*, 1965). $Ni(OH_2)_5(MeOH)^{2+}$ is the most labile species of solvated Ni(II) ions in methanol–water mixtures. Alcohols are weaker donors than water and hence the acceptor strength of all possible $NiL_5^{n\,2+}$ ions is smallest for $Ni(OH)_2)_5{}^{2+}$. The rate of dissociation of MeOH from $Ni(OH_2)_5(MeOH)^{2+}$ is therefore greater than for any other composition of the water–methanol solvate (MacKellar and Rorabacher, 1971). The rate constant for the nickel–methanol bond rupture in this species is approximately 800 times larger than the rate constant established for breaking the nickel–methanol bond in $Ni(MeOH)_6{}^{2+}$, and it is 20 times larger than the rate constant for nickel–water bond ruptures in $Ni(OH_2)_6{}^{2+}$.

The same interpretation—the variation of the acceptor strength of a metal ion by variation of the nonleaving ligands—is applicable to ligand exchange in tetrahedral complexes involving a bimolecular mechanism. Pignolet and Horrocks (1968) found that the lability of ligands L^l in complexes of the type $L_2^nML_2^l$ (M = Co or Ni) increases in the order $L^n = I^- < Br^- < Cl^-$.

The acceptor strength of L_x^nM is not only varied by variation of nonleaving ligands, but also by variation of the coordination number. Rare-earth metal ions react much more rapidly than expected on the basis of the charge–radius rule (Caldin, 1964). The comparatively high water-exchange rates for these trivalent ions have been attributed to their coordination numbers, which are greater than six (Geier, 1965); this is in agreement with the third bond-length variation rule (Section 1.5).

We have seen that the rupture rate of $M—L^l$ in $L_5^nM—L^l$ is increased with increasing donor number of the nonleaving ligands L^n and with decreasing donor number of the leaving ligand L^l. Hence for substitution reactions at octahedral unmixed solvates, i.e., where all six ligands are identical solvent molecules [e.g., $Co(DMSO)_6{}^{2+}$ in DMSO], the effects of increasing the donor number of the ligand are opposite to each other. This means that the effect of the ligand's donor number on substitution rates at unmixed complexes might be leveled. In fact, relatively small differences are observed for the exchange rates between solvent molecules coordinated to a metal ion and the same bulk solvent molecules in different solvents (Langford and Sastri, 1972; Bennetto and Caldin, 1971).

Leaving and nonleaving donor effects become more obvious, however, when solvents of considerably different donor numbers are compared.

MacKellar and Rorabacher (1971) have found that solvent exchange proceeds, relatively to aqueous solutions, more rapidly in the strong donor solvent ammonia than in the much weaker donor solvent methanol (Table 12.3). Since five coordinated solvent molecules contribute to the overall acceptor strength of the intermediate S_5M^{n+}, but only one ligand leaves for exchange of ligands at MS_6^{n+}, it is reasonable that an increase in the donor number of the solvent should cause the relative decrease in acceptor strength of MS_5^{n+} to have a greater effect than the relative increase in donor number of the single solvent molecule which is substituted. Hence one ligand is more readily dissociated from the hexacoordinated complex MS_6^{n+} as the donor number of the solvent is increased (Gutmann and Schmid, 1974). An increase in bond strength of the existing bonds in the species MS_5^{n+} will weaken the sixth bond being formed:

$$S_5M + S \longrightarrow S_5M \leftarrow S$$

The equivalence of all six metal-ion–ligand bonds in solution is only statistically valid. In other words, the solvated ion is not a perfectly rigid unit, but rather involves continuous rapid interchanges in bond energies and distances within certain limits.

12.4. Outer-Sphere Effects

A further effect that requires consideration is that produced by outer-sphere coordination; this may appear in inorganic as well as in organic chemistry. The outer-sphere effect for donor ligands causes an increase in coordinate bond strength in the inner sphere and hence a decrease in dissociation rate of the inner-sphere ligand (Gutmann and Schmid, 1974). Shu and Rorabacher (1972) studied the kinetics of the solvent–ammonia

Table 12.3. Solvent Exchange Rates for Ni(II) Solvates in Various Solvents

Complex	Solvent	Log k
$Ni(MeOH)_6^{2+}$	MeOH	3
$Ni(OH_2)_6^{2+}$	H_2O	4.5
$Ni(NH_3)_6^{2+}$	NH_3	5

exchange on quinquedentated Ni(II) as a function of methanol–water solvent composition:

$$\text{ZNi—NH}_3 + \text{S} \underset{k_{21}}{\overset{k_{12}}{\rightleftharpoons}} \text{ZNi—S} + \text{NH}_3$$

Both the dissociation rate k_{12} and the formation rate k_{21} are increased with increasing methanol content of the solvent mixture due to the different outer-sphere effects by water and methanol, respectively: The nickel–nitrogen bond is strengthened by outer-sphere solvation, the effect being greater for H_2O than for MeOH:

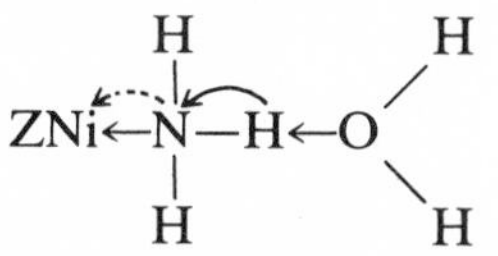

$$\text{ZNi} \leftarrow \underset{\text{H}}{\overset{\text{H}}{\text{N}}} \text{—H} \leftarrow \text{O} \begin{matrix} \diagup \text{H} \\ \diagdown \text{H} \end{matrix}$$

If water is substituted in the outer sphere by methanol the Ni—N bond will become weaker and hence the dissociation rate is increased.

The rate for the reaction of the Ni(II) ion and acetylacetone yielding monoacetylacetonate nickel(II) is increased by addition of propanol to the aqueous solution (Casazza and Cefola, 1967). The change in the dielectric constant of the solvent mixture within a range between 24 and 26 was thought to be responsible for the change in rate. However, within this small range of dielectric constant a difference in the rate constant of nearly 100% is observed. A linear relationship between rate of water substitution and inner-sphere composition in aquoamminenickel(II) complexes exists in water without noticeable change in dielectric constant of the medium (Desai *et al.*, 1970). Since the dielectric constant in water–methanol mixtures is also a linear function of solvent composition (Amphlett *et al.*, 1948), there is a linear relationship not only between rate and dielectric constant, but also between rate and solvent composition. Since the inner sphere remains intact, there should be a continuous change in outer-sphere composition with change of medium composition. The very rapidly increasing rates in propanol–water mixtures of small water content indicate a continuous change from outer-sphere to inner-sphere coordination of propanol at the metal ion, the highest lability for the metal–propanol bond being expected in $(H_2O)_5Ni(ROH)^{2+}$.

The rate constant of the solvolysis of $Cr(NH_3)_2(NCS)_4^-$, the ion of Reinecke's salt, with replacement of thiocyanate, is greater in solvents of the ROH type than in non-hydrogen-bonding solvents such as nitromethane or dioxane (Adamson, 1958). Likewise, a decrease in rate has been found for the aquation of $Cr(en)(NCS)_4^-$ in water–methanol mixtures with increasing methanol content (Thomas and Holba, 1969). These phenomena may be regarded as due to the outer-sphere effects of the acceptor ligands, by which the metal–inner-sphere ligand bond is weakened:

$$Cr{-}N{=}C{=}S \rightarrow H{-}O{-}H$$

Decrease in bond strength

The same explanation may be valid for the slight retardation of the replacement reaction,

$$Cr(NH_3)_5Cl^{2+} + H_2O \longrightarrow Cr(NH_3)_5(OH_2)^{3+} + Cl^-$$

by admixing of organic solvents which possess weaker acceptor abilities than water. As a consequence the M—Cl bond is weaker the stronger the outer-sphere interactions with an acceptor (first bond-length variation rule):

$$M{-}Cl \rightarrow H{-}O{-}H$$

In addition the structural features in solvent mixtures should not be disregarded (Burgess, 1973; Watkins and Jones, 1964). Outer-sphere effects by acceptor ligands labilize the metal–ligand bond and hence they are responsible for the acid catalysis of certain aquation reactions (Basolo *et al.*, 1956; Fehrmann and Garner, 1961; Matts and Moore, 1969).

Metal ions may also be capable of outer-sphere coordination with labilization of the leaving group; for example, the rate of loss of a chloride ion from the first coordination sphere of the complex *cis*-$(NH_3)_4(H_2O)Ru_4Cl^+$ is 30 times faster in the presence of chromium(III) ions (Movius and Linck, 1970):

$$\overset{+II}{Ru}{-}Cl \rightarrow Cr^{3+}$$

The rate of aquation of chloropentaaquochromium(III) is drastically increased in the presence of mercury(II) (Espenson and Birk, 1965) and the release of chloride ions is also accelerated by silver(I) (Elving and

Zemel, 1957). Likewise the aquation of salicylatopentaamminecobalt(III) is catalyzed by Fe(III) or Al(III) (Dash and Nanda, 1973):

$$\left[(NH_3)_5Co{-}O{-}C(=O)\text{-}C_6H_4\text{-}O{\rightarrow}M \right]^{4+}$$

Electrophilic catalysis of S_N1 reactions is also well known in carbon chemistry (Ingold, 1953). For example, the hydrolysis of benzyl chloride (Roberts and Hammett, 1937) and of several simple primary and secondary alkyl bromides is accelerated by mercuric salts:

$$R{-}Cl{\rightarrow}Hg^{2+} \xrightarrow{\text{slow}} R^+ + HgCl^+$$

$$R^+ + H_2O \xrightarrow{\text{fast}} ROH + H^+$$

A variety of the outer-sphere solvent effects is suggested in the "solvent assisted dissociation" mechanism (SAD) (Jones *et al.*, 1961). It is applied to solvent effects which occur when the solvent molecules slip a bond in between the metal ion and the leaving ligand that is stretched beyond a critical length, so that the dissociation of the ligand is completed, an effect which is in agreement with the first bond-length variation rule. It has been observed recently that the exchange of vanadyl acetylacetonate with labeled acetylacetone in dichloroethane is decreased by addition of a donor solvent. The exchange rate has been found to decrease as the donor number of the solvent is increased (Saito and Nishizawa, 1977). The reaction is thought to be initiated by the attack of Hacac at the vacant apical site of square pyramidal $VO(acac)_2$ followed by an intramolecular interchange. The effect of the donor solvent is interpreted by its competition with free Hacac for the vacant site. This may be considered as a "solvent suppressed" effect.

Chapter 13
Kinetics of Redox Reactions of Metal Complexes

13.1. General Considerations

Rates of reduction of metal complexes demonstrate a tremendous sensitivity to a change in ligands. On the other hand, no data are available on the oxidation rates of complexes with different ligands (Schmid *et al.*, 1976*a*). Likewise, few data are available for solvent effects on electron-transfer reactions between different metal ions (Matthews and Watts, 1974).

This chapter deals mainly with the effect of ligands on the kinetics of redox reactions between different metal complexes, and refers in particular to the papers published on the kinetics in nonaqueous solvents, whereby the distinction which is usually made between outer-sphere and inner-sphere mechanism will be followed.

13.2. Outer-Sphere Mechanism

In the outer-sphere mechanism, the inner coordination spheres of the reactants remain intact and no ligand is common to the inner coordination spheres of reducing and oxidizing agents. The actual electron transfer takes place through an outer-sphere complex. As a rule, rate measurements yield second-order rate constants k_{obs} which cannot be broken down into their elementary components, namely, the equilibrium constant K of the outer-sphere complex formation and the rate constant k of the electron-transfer step:

$$k_{obs} = K \cdot k$$

The hitherto unique exception, where the outer-sphere complex could be established, demonstrates the importance of the coordination chemical properties of the coordination spheres for outer-sphere complex formation (Gasvick and Haim, 1971):

$$Fe(CN)_6^{4-} + Co(NH_3)_5OH_2^{3+} \overset{K}{\rightleftharpoons} (NC)_5FeCN{\rightarrow}H{-}\overset{\overset{\displaystyle H}{|}}{O}{-}Co(NH_3)_5^{-} \longrightarrow \text{products}$$

The hydrogen atoms of coordinated water molecules offer excellent acceptor sites, whereas the nitrogen atom of the cyano group acts as a fairly good donor (see Section 8.1). The important contributions of the inner coordination spheres to outer-sphere complex stability have been emphasized recently (Gutmann and Schmid, 1974; Haim and Sutin, 1976).

The rate of the reduction

$$L_xM^{n+} + e \longrightarrow L_xM^{(n-1)+}$$

is determined by the ability of L_xM^{n+} to accept an electron (electron-acceptor strength). This depends on the nature of the metal M and the donor strength of the ligands L. For hard-metal ions an increase in the donor number of the ligands leads to a greater decrease in the free enthalpy of the (more positively charged) oxidized form as opposed to the reduced form. The transition state for the above-mentioned reaction is expected to have some of the properties of the reduced species and thus is expected to be less susceptible to the stabilizing effect of the ligands than the oxidized form. Thus the rate of reduction is decreased with increasing donor number of the ligands. The rate of electron transfer between two metal complexes

$$\underset{\text{Oxidant}}{L_xA^{n+}} + \underset{\text{Reductant}}{L'_yB^{m+}} \longrightarrow L_xA^{(n-1)+} + L'_yB^{(m+1)+}$$

is expected to be determined both by the ability of the reductant to donate an electron (electron-donor strength) and the ability of the oxidant to accept the electron. If the reductant is kept constant, the rate of electron transfer will decrease with increasing donor number of the ligands coordinated to the oxidant. Conversely, the rate will increase with increasing donor number of the ligands coordinated to the reductant, the oxidant being kept constant. An example is provided by the reduction of FeL_6^{3+}

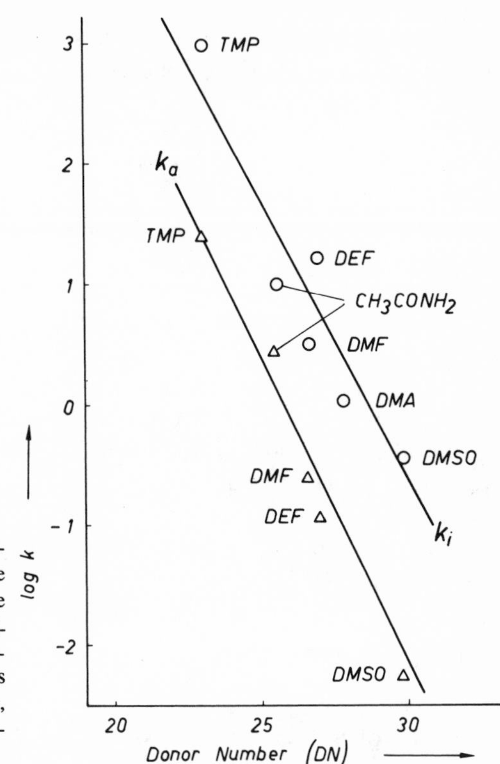

Fig. 13.1. Logarithm of the outer-sphere reduction rate k_a and the inner-sphere reduction rate k_i for the reaction of FeL_6^{3+} with *tris*-(3,4,7,8-tetramethyl-1,10-phenanthroline)-iron(II) vs. donor numbers of various solvents (from Schmid *et al.*, 1976*a*, courtesy of Verlag Chemie, Weinheim).

with different ligands by *tris*-(3,4,7,8-tetramethyl-1,10-phenanthroline)iron(II) in acetonitrile (Schmid *et al.*, 1976*b*). The outer-sphere reduction rate k_i decreases with increasing donor number of the ligand. The oxidation of $V(OH_2)_6^{2+}$ is more readily accomplished by I_2 than by I_3^- (Malin and Swinehart, 1969): This may be related to the greater electron-acceptor property of iodine as compared to that of the triiodide ion. The rate of the reduction of hard-metal ions by hydrated electrons in water increases as the donor properties of the ligands decrease, e.g., within the series of ligands (Anbar and Hart, 1965), $CN^- < F^- < Cl^- < I^- < NH_3 < H_2O$. Likewise, the outer-sphere reduction rate of $Co(NH_3)_5X^{2+}$ by $Cr(bipy)_3^{2+}$ or by $Ru(NH_3)_6^{2+}$ is increased by variation of the ligands X^- following the same order (Candlin *et al.*, 1964; Endicott and Taube, 1964).

The effect of the solvents PDC, CH_3CN, and NM on the reduction of $Fe(OH_2)_6^{3+}$ by methyl-substituted $Fe(phen)_3^{2+}$ has been interpreted in terms of an outer-sphere solvent effect through the hydrogen atoms of water molecules coordinated in the inner sphere. The reduction rate is decreased by increasing solvate bond strength with the water ligands caused by an increase in the donor number of the solvent (Schmid *et al.*, 1976*b*):

$$\mathrm{Fe{-}O} \begin{matrix} \diagup \mathrm{H} \leftarrow \mathrm{Solvent} \\ \diagdown \mathrm{H} \leftarrow \mathrm{Solvent} \end{matrix}$$

13.3. Inner-Sphere Mechanism

In the inner-sphere mechanism, the electron transfer is preceded by the movement of a ligand from one reactant into the coordination sphere of the other reactant to form a bridged complex. In this complex the two metal ions are connected through a bridging ligand common to both inner coordination spheres, through which the actual electron transfer takes place. An example of the inner-sphere mechanism is the following reaction (Taube *et al.*, 1953):

$$(H_2O)_6Cr^{2+} + ClCo(NH_3)_5^{2+} \rightleftharpoons (H_2O)_5\overset{e^-}{Cr \leftarrow ClCo}(NH_3)_5^{4+} + H_2O$$

This type of mechanism is possible if at least one of the ligands of one reactant demonstrates donor capacity toward the reaction partner acting as an acceptor. Examples of such ligands are halides and pseudo-halide ions, oxide and sulfide ions, or organic ligands which dispose of functional groups, thus remaining free to coordinate a second metal ion (Cannon and Gardiner, 1970; Taube, 1965). Measurements in acetonitrile have shown that the phenanthroline group of $Fe(phen)_3^{3+}$ is also capable of binding a pentacoordinated iron(II) solvate. The formation of this intermediate is obviously assisted by participation of solvent molecules (Schmid *et al.*, 1976*a*). Moreover, multidentate ligands can undergo one-ended dissociation, the free end of the ligand attaching to the reaction partner (Sapunov and Schmid, 1977). An example is the oxidation by acetylacetonates as will be discussed later. The inner-sphere mechanism can be considered as occurring in at least four steps, namely, (a) loss of one or more ligands from the labile partner, (b) formation of the bridged complex, (c) electron transfer, and (d) subsequent dissociation into the products. The bridged complex is formally called "precursor complex" before the electron transfer takes place and "postcursor complex" after the electron has been transferred (Sutin, 1968). The postcursor complex dissociates according to the relative stabilities of the possible products. The stabilities depend mainly on the relative acceptor properties of the metal ions connected by the bridging ligand. The acceptor strength increases with increasing oxidation number. The atoms or groups forming the bridge will generally prefer the metal ion of higher oxidizing properties and hence a ligand transfer from the reduced to the oxidized form is frequently observed. For instance, the oxidation of $Cr(OH_2)_6{}^{2+}$ by $Co(NH_3)_5Cl^{2+}$ has been shown to involve chloride transfer from the oxidizing to the reducing agent (Taube and Myers, 1954). The postcursor may be formulated as

$$(H_2O)_5\overset{+III}{Cr} \leftarrow Cl - \overset{+II}{Co}(NH_3)_5{}^{4+}$$

in which both $(H_2O)_5Cr^{3+}$ and $Co(NH_3)_5{}^{2+}$ ions compete for the chloride ion. Since the former is the stronger acceptor, $(H_2O)_5CrCl^{2+}$ is produced by accepting a chloride ion from $ClCo(NH_3)_5{}^{+}$.

The effect of the ligands on inner-sphere redox reactions may vary. Distinctions are made between the effects of bridging and nonbridging ligands, as well as between electron-transfer-controlled or substitution-controlled redox processes. For the interaction of the reducing agent L_mA^{2+} with the oxidizing agent XBL'^{2+}_n (A and B representing the different metal ions, L and L′ neutral ligands, and X^- a univalent bridging ligand), the following reaction schemes are possible (Linck, 1972):

Scheme 1. The formation of the precursor complex is rate determining, e.g., the rate of electron transfer occurs rapidly as soon as the precursor complex is formed:

$$L_m\overset{+II}{A}{}^{2+} + X - \overset{+III}{B}L'^{2+}_n \longrightarrow L_{m-p}\overset{+II}{A} \leftarrow X - \overset{+III}{B}L'^{4+}_n + pL$$

Assuming a pure S_N1 mechanism for the substitution at the A center, the reaction rate will increase with increasing donor number of L as the acceptor strength of A decreases, whereas the reaction rate will be independent from the oxidizing agent. Examples are vanadium(II) redox reactions, for instance, reactions between Co(III) complexes of the type $Co(CN)_5X^{3-}$ and $V(OH_2)_6{}^{2+}$ (Davies and Espenson, 1969). For different complexes, similar values of k have been found for X^- ($X^- = Cl^-$, Br^-, SCN^-, $N_3{}^-$, and H_2O) (Linck, 1972). Assuming an S_N2 mechanism for precursor formation, the rate will be increased with increasing donor strength of the binding ligand X^-, and is also increased with increasing donor number of L′ (Gutmann and Schmid, 1974). The oxidation of iron(II) by various acetylacetonates in acetonitrile is preceded by one-ended dissociation (k_1) and the formation of a binuclear bridged complex (k_2) (Sapunov and Schmid, 1977):

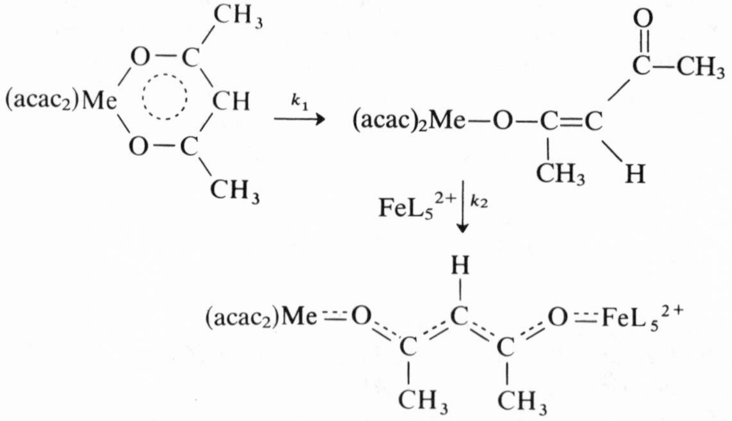

$$(acac_2)Me\!\!\begin{array}{c}O-C(CH_3)\\ \quad CH\\ O-C(CH_3)\end{array} \xrightarrow{k_1} (acac)_2Me-O-C(CH_3){=}CH-C(=O)-CH_3$$

$$\downarrow k_2\ \ FeL_5^{2+}$$

$$(acac_2)Me \cdots O{\cdots}C(CH_3){\cdots}C(H){\cdots}C(CH_3){\cdots}O \cdots FeL_5^{2+}$$

After the electron transfer has taken place through the bridging ligand, the complex dissociates into the products $Me(acac)_2$ and $Fe(acac)L_4^{2+}$. The higher the acceptor strength of the metal ion the smaller the k_1 and the smaller the donor strength of the oxygen at the free end of the bidentate ligand, resulting in a decrease in the second-order rate constant k_2 of the formation of the bridged complex. Thus the values of k_1 follow roughly the same trend as k_2.

Scheme 2. The precursor complex is readily formed and the electron transfer is rate determining:

$$L_{m-p}\overset{+II}{A}\overset{e}{\leftarrow}X-\overset{+III}{B}L'^{4+}_n \longrightarrow L_{m-p}\overset{+III}{A}\leftarrow X-\overset{+II}{B}L'^{4+}_n$$

In this case the ligands L and L′ will have the same effect on the rate as in outer-sphere reactions. The rate is increased with increasing donor number of L and decreasing donor number of L′. For example, the logarithm of the rate constant k_a of the electron transfer within the intermediates,

$$L_5\overset{+III}{Fe}-phen-\overset{+II}{Fe}(phen)_2(CH_3CN)_x^{5+} \quad (e \text{ transferred from Fe(+II) to Fe(+III)})$$

decreases linearly with increasing donor number of the ligands L (Schmid *et al.*, 1976*b*) (Fig. 13.1). The bridging ligand will contribute in two different respects to the overall rate constant. It will determine the stability constant of the bridged complex and act as conductor of the electrons. The ability to channel electrons should increase with increasing polarizability, e.g., $F^- < Cl^- < Br^- < I^-$, whereas the stability of the bridged complex is increased by increased donor property, e.g., for hard-metal ions in aqueous

Table 13.1. Kinetic Parameters for the Reduction of $[Co(NH_3)_5X]^{2+}$ by Fe^{2+} in DMF

Oxidizing agent	K (liter·mol^{-1})	k (s^{-1})
$[Co(NH_3)_5F]^{2+}$	3200	10.6
$[Co(NH_3)_5Cl]^{2+}$	1.61	19.1
$[Co(NH_3)_5Br]^{2+}$	0.34	12.4

solution: $I^- < Br^- < Cl^- < F^-$. The bridging ligands accelerate the Cr^{2+} reductions of $Co(NH_3)_5X^{2+}$ in the order of the polarizabilities (Candlin and Halpern, 1965), while for Fe^{2+} and Eu^{2+} reductions, the inverse order is followed (Diebler and Taube, 1965), indicating that in the latter case the effect of the bridging ligand on the stability constant is of major importance.

The oxidation of $Fe(DMF)_6{}^{2+}$ by $Co(NH_3)_5X^{2+}$ in DMF shows relationships that are more complicated. Matthews *et al.* (1976) broke down the overall rate constants into their elementary components, the stability constant of the bridged complex, K, and the rate constant of the electron transfer, k (Table 13.1). The rate constants of electron transfer are not significantly different for $X = F^-, Cl^-, Br^-$, whereas the stability constant of the bridged complex for $X = F^-$ is considerably greater than for $X = Cl^-$ or Br^-. This indicates that in DMF, fluoride is particularly capable of binding two metal ions simultaneously. This ability is suppressed by water, a solvent which has a very high acceptor number (AN = 54.8) and which therefore solvates fluoride particularly well (Table 13.2). The overall rate constant is increased, the smaller the acceptor number of the solvent; e.g., it is greater in DMF (AN = 16) than in DMSO (AN = 19.3). The overall rate constants for the chloro and bromo compounds decrease in the same

Table 13.2. Overall Rate Constants for the Reduction of $(NH_3)_5CoX^{2+}$ by Iron(II)

	k_2 (liter·mol^{-1}·s^{-1})		
Solvent	$(NH_3)_5CoF^{2+}$	$(NH_3)_5CoCl^{2+}$	$(NH_3)_5CoBr^{2+}$
H_2O[a]	7.6×10^{-3}	1.6×10^{-3}	0.9×10^{-3}
DMSO[b]	4.7	9.7×10^{-3}	2.5×10^{-3}
DMF[c]	34	3.0×10^{-2}	4.2×10^{-3}

[a]From Diebler and Taube, 1965. [b]From Matthews and Watts, 1974.
[c]Calculated from data in Table 13.2.

Table 13.3. Reductions by Cr^{2+} via Scheme 3

Oxidant	k (s^{-1})
$Ru(OH_2)_5Cl^{2+}$	$1.2+0.36/a_{H^+}$
cis-$Ru(NH_3)_4(OH_2)Cl^{2+}$	1.2×10^2
$Ru(NH_3)_5Cl^{2+}$	4.6×10^2

order, e.g., $H_2O < DMSO < DMF$, although the differences are less pronounced than for the fluoro compound. This is because the donor properties decrease in the order $F^- > Cl^- > Br^-$, and hence the interactions with acceptor solvents decrease in the same order. The donor number of the solvent cannot be expected to have an effect in this case, since the donor numbers of the three solvents under consideration are similar.

Scheme 3. The overall reaction rate depends on the rate of bond rupture in the postcursor complex, for which the usual pathway may be assumed; i.e., the bridging ligand remains at the oxidized metal complex:

$$L_{m-p}\overset{+III}{A}—X—\overset{+II}{B}L_n'^{4+} \longrightarrow L_{m-p}\overset{+III}{A}—X^{2+} + \overset{+II}{B}L_n'^{2+}$$

The rate of bond rupture is related to the bond strength. The bond strength X—B is decreased by increasing donor number L′, which decreases the acceptor strength of $BL_n'^{2+}$, whereas the donor properties toward A are increased (Gutmann and Schmid, 1974):

$$\begin{array}{ccc} & L' \quad L' & \\ & \searrow \; \swarrow & \\ A\leftarrow X & — B & \leftarrow L' \\ & \nearrow \; \nwarrow & \\ & L' \quad L' & \end{array}$$

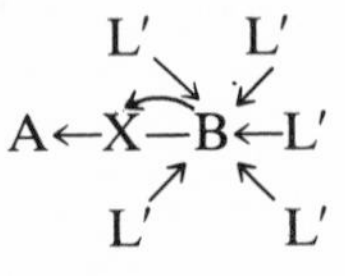

Hence the rate of bond rupture X—B is increased by increasing donor number of L′ as well as by decreasing donor properties of X^-. The donor function of X^- is decreased by decreasing donor property of X^- and by increasing acceptor strength of $L_{m-p}A^{3+}$. The latter is increased with decreasing donor number of L.

Movius and Linck (1970) found that reduction of $Ru(NH_3)_5Cl^{2+}$ by Cr^{2+} proceeds via the mechanism of Scheme 3. In this case the rate coefficient is increased by increasing donor number of L′ (Seewald *et al.*, 1969) (Table 13.3). The ligand effects have been summarized (Gutmann and Schmid, 1974), as given in Table 13.4.

Table 13.4. The Effect of Increasing Donor Number of the Ligands L and L′ on the Rate of a Redox Reaction, with AL_m^{2+} as the Reductant and $XBL_n'^{2+}$ as the Oxidant

	Change in reaction rate	
Mechanism	L	L′
Outer-sphere	increase	decrease
Inner-sphere	increase	decrease
Scheme 1	increase	no effect[a]
	increase	increase[b]
Scheme 2	increase	decrease
Scheme 3	decrease[c]	increase[c]
	increase[d]	decrease[d]

[a] Assuming an S_N1 mechanism for precursor formation.
[b] S_N2 mechanism for precursor formation.
[c] X^- remains at the oxidized complex.
[d] X^- remains at the reduced complex.

Chapter 14

Solvent and Reactant Effects on the Course of Substitution and Insertion Reactions

14.1. General Considerations

Solvent effects on the rates of various reactions in carbon chemistry (Cram, 1965; Parker, 1969) have been presented from the point of view of the donor–acceptor concept in Section 12.2. Solvent effects on the reactivities of ions and their relation to thermodynamic properties have been dealt with in Sections 7.2 and 10.1. We shall now turn our attention to interactions in organometallic and covalent inorganic chemistry, where an ionic equilibrium state is not reached, and hence to the solvent effects on the formation of *anionoid* and *cationoid* structures, respectively.

For example, the highly polarized complexes of various acceptor molecules are more reactive with DMF or DMSO than in the absence of the said solvents (Hall, 1956), and they can lead to compounds formed by decomposition of the solvent (Lappert and Smith, 1961). Acid halides are known to react rigorously with DMSO to give monohalodimethyl sulfides (Heininger and Dazzi, 1951) and phosgene, just as thionyl halides in DMF convert carboxylic acid into the acid halides at room temperature or below. Even trichloroacetic acid gives its acid chloride when allowed to react with thionyl chloride in the presence of DMF (Boshard and Zollinger, 1959).

It has been shown in Sections 1.2 and 10.1 that by donor attack of the substrate A—B at A the donor property at B is enhanced, whereas by acceptor attack at B the acceptor property at A is increased. The bond polarity is increased and the heterolytic bond energy (Section 10.3) is decreased by either of these interactions. All of these properties can be varied as the donor and acceptor properties of the molecular environment, such as solvent or catalyst, are changed.

For an anionoid structure of high donor property a solvent or a catalyst must have: (i) a "suitable" donor number, and (ii) an acceptor number as low as possible. The suitable donor number depends on the ease of heterolysis of the bond A—B. The donor number must be high enough to induce the required donor property at B, whereas it should not be high enough to give rise to the formation of anions B^-. It has been shown in some of the preceding sections that the donor properties at a donor site are decreased by increasing interaction with an acceptor; hence increasing acceptor properties of solvent or catalyst will decrease the donor property at B. In order to maintain the donor property at B induced by attack of a "suitable" donor at A, the acceptor number of the molecular environment of A—B should be low as possible. Both of these requirements are best fulfilled by etheral solvents or by mixtures of ethers with HMPA (Gutmann, 1976*a*, *b*).

For a cationoid structure of high acceptor property a solvent or a catalyst must have: (i) a high acceptor number, and (ii) a donor number as low as possible since a donor solvent would decrease the acceptor property at A. Such requirements are provided by nitrobenzene, chloroform, or dichloroethane.

We shall now examine the applicability of these rules to various types of interactions.

14.2. Reactions Involving Organo–Alkali Metal Compounds

The high reactivities of organolithium compounds are usually regarded as due to the high polarities of the lithium–carbon bonds. The polarities of the corresponding sodium or potassium compounds are considered even higher but their reactivities in solution are usually lower than those of the lithium compounds.

The differences are readily accounted for by considering the different acceptor properties of the alkali metal ions. The unsolvated lithium ion is a stronger acceptor than any other unsolvated alkali metal ion and hence it is more strongly solvated by a given donor solvent. The interaction of the solvent at the lithium atom of the lithium–carbon bond leads to an increase in bond polarity greater than that of a potassium–carbon bond under analogous conditions. In a solvent of low acceptor number the extent of bond polarity in R_3C—Li is mainly due to donor interactions with the metals, which are related to the solvent donor properties as represented by the donor number, and to the acceptor properties of the metal ions.

In HMPA, lithium iodide gives a higher conductivity and hence it is more strongly ionized than a tetraalkylammonium iodide in the same solvent (Mayer *et al.*, 1973). In comparing these solutes we have to consider the different behavior of Li^+ and R_4N^+ ions in this solvent. The lithium ion is a stronger electron-pair acceptor than the tetraalkylammonium ion (see Section 2.3). Hence the radius of the strongly solvated $Li(HMPA)_n^+$ ion is greater and the cation–anion interaction weaker than in the case of the tetraalkylammonium salt. This means that in HMPA the Li salt is more strongly ionized than the tetraalkylammonium salt.

The tetraphenylborate ion is weakly solvated in most solvents. The cesium salt shows different behavior in THF and in dimethoxyethane (DME), the dissociation constant being 20 times greater in DME than in THF, although the solvent dielectric constants are nearly identical (Carvajal *et al.*, 1965). In THF the cesium ion is weakly solvated, while it is more strongly solvated in DME (Carvajal *et al.*, 1965). These effects may be rationalized by the donor number concept: The higher DN of DME stabilizes the cesium cation, enhances the dissociation, and hence the concentration of the anions which are weakly solvated in this system. To account for the conductance behavior of $Na^+BPh_4^-$ it is necessary to assume that the salt is ionized to give solvent-coordinated Na^+ ions in both THF and DME, since Na^+ ion is a stronger acceptor than the Cs^+ ion (see the series in Section 2.3).

Methyl lithium is tetrameric in the solid state: Four lithium atoms form a tetrahedron with methyl groups at each of its faces (Weiss and Lucken, 1964), so that each lithium atom is coordinated to three methyl groups. The tetrameric structure prevails also in benzene solution. Addition of donor molecules leads to coordination to each of the lithium positions (Lewis and Brown, 1969) with subsequent increase in polarity of the lithium–carbon bonds.

The extent of metal–carbon bond polarity and of ionization in solution depends on the donor strength of the solvent (Normant, 1967; Waack *et al.*, 1969; Schlosser, 1973) and follows the order of the donor numbers as long as their acceptor properties are comparable:

$$\text{benzene} < \text{diethyl ether} < \text{THF} < Me_3N < \text{PY} < \text{HMPA}$$

The different reactivities of organolithium compounds in solvents of different donor properties are seen from the following example: Aliphatic nitriles (O'Sullivan *et al.*, 1961) and aliphatic ketones (Puterbaugh and Hauser, 1959; Hauser and Dunnavant, 1960; Hauser and Puterbaugh, 1963) are deprotonated by organolithium compounds in diethyl ether, while no reaction is observed in petrol ether. Coordination at the lithium by diethyl ether molecules induces a strong polarity of the Li--C bond, and thus carbanionoid properties, possibly yielding carbanions which are separated from the cations by ether molecules coordinated to the lithium ions. Both the carbanionoid and the carbanion remain virtually unattacked by diethyl ether and hence they are readily available for electrophilically attacking the methyl groups of a nitrile or of a ketone with deprotonation:

$$n(Et_2O) \rightarrow \underset{\text{Hetero-lysis}}{\overset{\frown}{Li-R}} \rightarrow \underset{\text{Hetero-lysis}}{\overset{\frown}{H-CH_2CN}} \rightleftharpoons RH + (Et_2O)_nLi^+ + CH_2CN^-$$

In petrol ether (low donor and low acceptor properties) the Li—R species remains un-ionized, and in this state the organic group bound to the lithium is not sufficiently reactive to permit the occurrence of an analogous reaction.

The different behavior of diphenylmethyl lithium and diphenylmethyl potassium in liquid ammonia is also readily explained by the different acceptor properties of the lithium and potassium ions: Diphenylmethyl potassium is not decomposed by liquid ammonia (Wooster and Ryan, 1932), while the lithium compound is ammonolyzed to give lithium amide and diphenylmethane (Wittig, 1958; Schlosser, 1973). These remarkable differences have been regarded as due to the differences in charge densities at the 1,1-diphenylmethyl carbanion and at the amide ion (Schlosser, 1973), but they may be interpreted in the following way (Gutmann, 1976*a*): The lithium compound is subject to considerable solvation at the metallic center with formation of carbanions, which interact with coordinated ammonia molecules with hydrogen-bond formation to give an outer-sphere complex. The charge transfer transmitted toward the metallic center (Waack *et al.*, 1969) leads to deprotonation of a coordinated ammonia molecule:

$$\begin{array}{ccccccccc} & NH_3 & & H & & & Ph & & \\ & \downarrow & & | & & & | & & \\ NH_3 \longrightarrow & Li^{\oplus} & \longleftarrow & N & \longrightarrow & H \longleftarrow & C^{\ominus} & — H & \rightleftharpoons \\ & \uparrow & & | & & & | & & \\ & NH_3 & & H & & & Ph & & \end{array}$$

Coordination | Induced effect (heterolysis) | Outer-sphere coordination

$$Li^{+}NH_2^{-} + H—\underset{Ph}{\overset{Ph}{C}}—H$$

If the free carbanions were formed, the outer-sphere complexation would be expected to be weaker in the potassium compound, since the hydrogen atoms of ammonia molecules are less acidic when coordinated to a potassium ion, as compared to those coordinated to a lithium ion:

$$\begin{array}{cccc} & NH_3 & & Ph \\ & \downarrow & & | \\ NH_3 \longrightarrow & K & ---- & C — H \\ & \uparrow & & | \\ & NH_3 & & Ph \end{array}$$

Increase in polarity without bond cleavage

It should be expected that the relative reactivities of a series of organo–alkali metal compounds will be changed considerably in the presence of a crown ether, which is capable of specific and strong solvation of a particular alkali metal ion, e.g., K^+.

Potassium-*t*-butanolate acts as a stronger base in DMSO (Cram, 1965) than it does in *t*-butanol. This behavior indicates that the K—C bond is more polar in DMSO than in *t*-butanol. This is due to the greater donor strength of DMSO, which has a lower acceptor number than *t*-butanol. Thus in the latter solvent the increase in anionoid donor strength, induced by the solvent donor attack at potassium, is partly decreased by the stronger acceptor attack of *t*-butanol at the anionoid constituent. The same explanation holds for the observed increase in kinetic activity of alkoxide anions in DMSO as compared to that in hydroxylic solvents: The rate of isotopic exchange catalyzed by potassium-*t*-butoxide–0.9 M *t*-butyl alcohol in DMSO is about 1×10^{13} times that observed in methanol–potassium methoxide.

14.3. Grignard Reactions

The excellent solvent properties of diethyl ether for Grignard reactions are well known. The elucidation of the crystal structure of the monomeric phenylmagnesium bromide dietherate (Guggenberger and Rundle, 1964) has established that two ether molecules are coordinated to a single magnesium atom. Dimethylmagnesium forms high-molecular-weight chains of alternating magnesium atoms and methyl groups (Weiss, 1964), and in concentrated solutions oligomeric species are present. In the solvent, diethyl ether, these chains are coordinated by the solvent molecules (Schlosser, 1973).

The composition of the Grignard compounds RMgX in solutions depends on the nature of the solvent and the nature of the R and X groups, as well as on the concentration. At concentrations below 0.1 M in diethyl ether, monomeric species appear to be dominant, and the mechanism of the addition of Grignard compounds to ketones has been presented in terms of an attack of monomeric or dimeric RMgX species (Ashby and Smith, 1964). Various Grignard reactions have been interpreted in terms of attack of the dimeric species (Dessy *et al.*, 1957; Miller *et al.*, 1961; Bikales and Becker, 1962; Cowan and Mosher, 1962), but for the present discussion we shall assume a mechanism involving monomeric alkylmagnesium halide species and a four-membered ring as a transition center. We shall assume further that the same mechanism should prevail in different solvents.

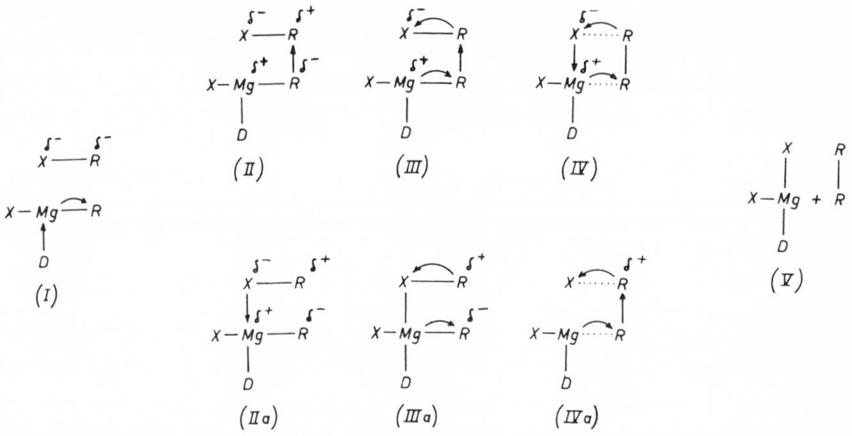

Fig. 14.1

The coordination of the donor molecules of RMgX species increases the polarities of metal–carbon bonds (I) (Fig. 14.1). The negative fractional charges, and hence the donor properties at the carbon atoms, are increased, so that interaction of this carbanionoid with the alkyl groups of RX readily takes place (II). This coordinating interaction induces increased polarities of both the Mg—R and the R—X bonds (III). The induced increase in acceptor properties at the magnesium atom and the induced increase in nucleophilic properties at X in the alkyl halide allow the electrophilic attack at the carbon atom in R—X (IV). In this way, further increase in polarity for both the Mg—R and the X—R bonds is induced to the extent that heterolysis (V) takes place. Alternatively, the reaction may be initiated by nucleophilic attack of the X in RX at the magnesium (IIa) followed by the steps represented as (IIIa) and (IVa) to give finally (V).

For this mechanism a certain range in donor properties of the solvent is required to provide carbanionoids of suitable nucleophilic properties. A weak donor solvent cannot initiate the reaction due to insufficient polarity of the Mg—R bond. In a very strong donor solvent such as HMPA (DN = 38.8), the Mg—R bonds are heterolyzed to give carbanions which are so reactive that undesired reactions take place (see below); the solution of benzylmagnesium chloride in HMPA shows the red color, characteristic of the benzyl carbanion (Ebel and Schneider, 1965):

$$C_6H_5CH_2MgCl + n\text{HMPA} \rightleftharpoons C_6H_5CH_2^- + (\text{HMPA})_nMgCl^+$$

Furthermore a strong donor solvent may decrease the acceptor properties of the cations to the extent that they cannot be active in step (IV), as may be seen from the retardation of the addition of di-*n*-butylmagnesium to acetone if diethyl ether is replaced by the stronger solvating tetrahydrofuran (Holm, 1966). The decisive role of the donor properties of the solvent toward magnesium is also illustrated by the reaction of diisopropylketone with ethylmagnesium bromide in HMPA, which gives the enolate (Fauwarque and Fauwarque, 1966): The free ethyl carbanion is formed in HMPA; this deprotonates the ketone, with formation of a diisopropylketone carbanion, which displaces HMPA from the coordination sphere of the $MgBr^+$ ion. On the other hand, in diethyl ether the tertiary carbinolate is formed (Ritchie and Unschold, 1967). Ethyl ether polarizes the Mg—C bond to a considerably lesser extent than does HMPA, and coordination of magnesium to the carbonyl group induces a suitable fractional positive charge to the C atom of the C=O group, so that coordination by the ethyl group takes place followed by heterolysis of the Mg—C bond.

$$(Me_2CH)_2C{=}O \xrightarrow[\text{HMPA}]{MgBr^+CH_3CH_2^-} (Me_2CH)(Me_2C{\leftarrow}MgBr)C{=}O + CH_3CH_3$$

$$\xrightarrow{H_2O} \underset{\text{Enolate}}{(Me_2CH)(Me_2C{=})C{-}OH} \rightleftharpoons (Me_2CH)_2C{=}O$$

$$(Me_2CH)(Me_2Ch)C{=}O \rightarrow Mg(Et)(Br)(OEt_2)_2 \xrightarrow{\text{ether}} (Me_2CH)_2C{=}O \cdots Mg(Et)(Br)(OEt_2)_2 \xrightarrow{H_2O} (Me_2CH)_2C(Et)OH$$

Protonic solvents such as water or ethyl alcohol (DN ≈ 19) cannot be used for such reactions, since they lead to deprotonation (Zerevitinov reaction).

The unique usefulness of etheral solvents for Grignard reactions is due to: (i) their moderate donor numbers, (ii) their very weak acceptor properties, and (iii) their very weak ability to protonate the Grignards. Ether molecules in the etherates virtually retain their molecular structure after coordination to magnesium and remain practically nonacidic.

The most appropriate bond polarization for a particular Grignard reaction may be achieved by applying a mixture of an ether and HMPA, both of which have poor acceptor properties. Thus a solvent of any desired DN between 19 and 38.8 with very low acceptor number can be obtained by appropriate mixtures of these solvents (see Section 6.5). For example, mixtures of 80% THF and 20% HMPA have been used for the lithium–organic interactions with organic molecules containing C=O or C=S groups in order to achieve the "Umpolungs" reactions (Seebach and Enders, 1975). These spectacular reactions are examples of decisive changes in chemical reactivities due to suitable charge-density rearrangements within the organic molecules.

14.4. *Reactions of Carbon Dioxide*

Although CO_2 shows no dipole moment, it does have highly polar carbon–oxygen bonds: The carbon atom provides a center for nucleophilic attack, while the oxygen atoms have donor properties. In a typical donor solvent of low acceptor but moderate donor properties such as diethyl ether, the coordination number of the carbon atom is increased, and the polarities of the C=O bonds are enhanced:

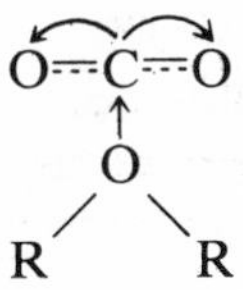

$$O \overset{\curvearrowleft\curvearrowright}{=\!=} C =\!= O \quad (\uparrow\ O R_2)$$

```
O==C==O
   ↑
   O
  / \
 R   R
```

Alternatively, an acceptor solvent of low donor properties, such as $CHCl_3$, will increase the fractional positive charge at the carbon atom of CO_2 and hence increase the acceptor properties of that carbon atom:

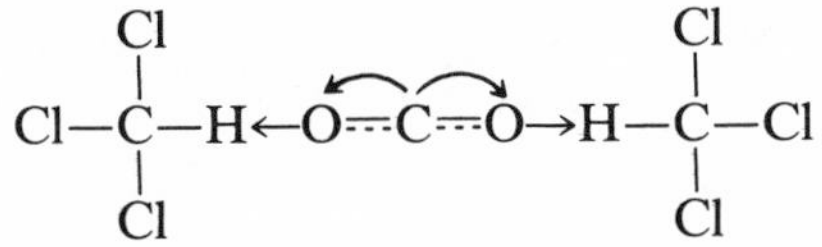

```
    Cl                      Cl
    |                       |
Cl—C—H←O==C==O→H—C—Cl
    |                       |
    Cl                      Cl
```

The reactions of carbon dioxide with organometallic compounds are used in organic synthesis and in the study of transmetallation and other organometallic reactions (Gilman and Harris, 1931; Volpin and Kolomnikov, 1975). The reactions involve CO_2 insertion into the metal–carbon bond according to the acceptor properties of the carbon atom and the donor properties of the oxygen atoms. For example, alkyl lithium reacts with carbon dioxide with subsequent heterolytic cleavage of the lithium–carbon bonds:

$$\begin{matrix} R-Li \\ O=C=O \\ Li-R \end{matrix} \longrightarrow \begin{matrix} R & Li \\ | & | \\ O-C- & O \\ | & | \\ Li & R \end{matrix} \xrightarrow{H_2O} R_2CO + 2LiOH$$

The reactions are usually carried out by pouring the solution of the organolithium compound in a noncoordinating solvent, such as hexane or toluene, on solid carbon dioxide. Under these conditions the organopotassium compounds are even more reactive than the corresponding lithium compounds, while in the presence of a donor solvent the reactivities may be reversed. The insertion of CO_2 at the M—H bond normally gives formate complexes. Depending on the polarity of the transition metal–hydrogen bond, the insertion may proceed in different ways (Volpin and Kolomnikov, 1975):

Normal:

$$\begin{matrix} O=C=O \\ \uparrow \quad \downarrow \\ {}^{\delta-}H-M^{\delta+} \end{matrix} \longrightarrow \begin{matrix} O=C-O-M \\ | \\ H \end{matrix}$$

Abnormal:

$$\begin{matrix} O=C=O \\ \uparrow \quad \downarrow \\ {}^{\delta-}M-H^{\delta+} \end{matrix} \longrightarrow \begin{matrix} O=C-O-H \\ | \\ M \end{matrix}$$

14.5. *Reactions of Molecular Nitrogen*

For coordination compounds containing molecular nitrogen (Allen, 1971), such as $Ru(NH_3)_5N_2X_2$, a strong sharp band in the infrared spectrum is found somewhere between 1900 and 2200 cm^{-1}. This is regarded as good, an almost sufficient indication of the presence of an N_2 molecule linearly coordinated —M—N≡N.* There is some connection between the

*One example is known for M---(N≡N, side-on) bonding.

position of this band and the strength of the interaction between the N_2 and the metal. The N_2 stretching frequency (Raman) of the free nitrogen molecule is at 2331 cm^{-1} and the extent to which the frequency is lowered in the complex has been taken as an indication of the extent of π-bonding from the metal to the ligand:

$$-M \leftrightarrows N \equiv N$$

The induced lengthening of the N≡N bond is highly influenced by the other ligands coordinated to the metal. In osmium(II) complexes the effect appears to be increased by increasing donor properties of the ligands present. This bonding concept is further supported by the observed increase in metal-to-ligand π-electron transfer in going from the higher to the lower oxidation state.

There is, no doubt, an increased reactivity of coordinated nitrogen. Due to the increase in negative fractional charge at the terminal nitrogen atom, the formation of dimeric N_2-bridged compounds is easily accomplished (Harrison *et al.*, 1968):

$$Ru(NH_3)_5N_2^{2+} + (H_2O)Ru(NH_3)_5^{2+} \rightleftharpoons (NH_3)_5Ru-N_2-Ru(NH_3)_5^{4+} + H_2O$$

In this complex the metal—N—N—metal group is essentially linear and the N—N bond distance (112 pm) is only slightly longer than that in free N_2 (109 pm) (Treitel *et al.*, 1969). For this reason there is little tendency for further reaction of bridged nitrogen molecules. The N_2 stretching frequencies are considerably lower when bridging occurs between two different metals, such as in the deeply colored N_2-bridged compounds obtained from $ReCl(N_2)(PMe_2Ph)_4$ with early transition-metal halides (Chatt *et al.*, 1969). In the same respect, since there is an increase in C═O bond lengths in carbonyl compounds by acceptor solvents (Gutmann, 1977*b*), the N—N polarities are increased by strong acceptors (Chatt *et al.*, 1970, 1971).

14.6. *Reactions of Silicon Compounds*

Silyl cations are formed by interactions between bromides or iodides of silicon and strong bases. The compound $SiL_4(PY)_4$ is formulated as $[(PY)_4SiI_2]^{2+} \cdot 2I^-$ (Wannagat and Schwarz, 1954; Wannagat and Vielberg, 1957; Schnell, 1961; Beattie *et al.*, 1964). Based on spectral results the pyridine complexes of silyl bromides and silyl iodides are assigned ionic structures with pentacoordinated silicon (Campbell-Ferguson and Ebsworth, 1965), e.g., $SiH_3(PY)_2^+Br^-$ and $SiH_3(PY)_2^+I^-$, which give highly conducting solutions in acetonitrile, in contrast to the behavior of the compounds of the corresponding chlorides or fluorides. Even trichlorosilane undergoes ionization by amines in solution of acetonitrile (Benkeser *et al.*, 1970; Bernstein, 1970). $Si(CH_3)_3I$ is completely ionized in HMPA, DMF, DMA, PY, and tetramethylurea, with formation of tetracoordinated siliconium ions $DSi(CH_3)_3^+$ (Mayer *et al.*, 1977*c*), which have been sought for some time (Aylett *et al.*, 1955; Corey and West, 1963; Beattie and Parrett, 1966; Campbell-Ferguson and Ebsworth, 1967).

Silyl cations are formed by ionization of silyltetracarbonylcobalt $H_3SiCo(CO)_4$ or silylpentacarbonylmanganese $H_3SiMn(CO)_5$ under the influence of trimethylamine or pyridine (Aylett and Campbell, 1969).

While NH_3 and BF_3 form a stable adduct, a cleavage reaction takes place when $(H_3Si)_3N$ and BF_3 are reacted (Burg and Kuljian, 1950; Sujishi and Witz, 1954, 1957; Aylett, 1968). The formation of the coordinate N→B bond induces an increase in fractional negative charge at the fluorine atoms with appropriate increase in donor properties, as well as increase in fractional positive charge at the Si atoms with corresponding increase in acceptor properties (the π-bonding between N and Si is reduced). These changes lead to formation of F→Si bonds, with increased polarity of both the B—F and the Si→N bonds and shortening of the original N→B bond. Subsequent repetitions of the electronic cycle lead finally to heterolysis of one B—F and one Si—N bond:

$$\begin{matrix} H_3Si & F \\ H_3Si-N\rightarrow & B-F \\ H_3Si & F \end{matrix} \quad (\text{heterolysis}) \longrightarrow H_3Si-\underset{}{\overset{H_3Si}{\overset{|}{N}}}-BF_2 + H_3SiF$$

The analogous interactions between other Si—F positions lead to further cleavage reactions with simultaneous p_π–d_π strengthening and shortening of the N—B bonds to give the trimer $(H_3SiNBF)_3$. Polymers composed of $(SiH_2NR)_x$ have been prepared according to the following equation:

$$SiH_2I_2 + 3RNH_2 \longrightarrow \frac{1}{x}(SiH_2NR)_x + 2RNH_3I$$

This condensation reaction may be explained by analogous electronic cycles: Formation of the coordinate N→Si bond leads to increasing Si—I and N—H bond distances, thus increasing both the donor properties of the iodine atoms and the acceptor properties of the hydrogen atoms attached to the nitrogen atom. The formation of a coordinate I—H bond leads to further lengthening of the N—H and Si—I bonds, while the N→Si bond as well as the Si—H bonds are further shortened. The reaction proceeds to heterolysis of the Si—I and N—H bonds and to further analogous condensation with other neighboring molecules:

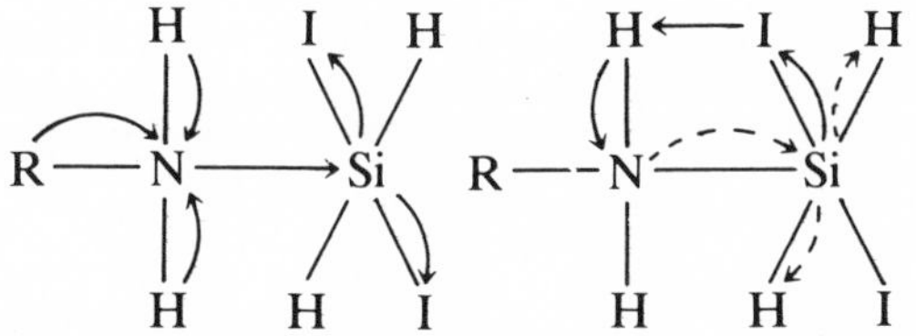

Noll (1963) has pointed out that acidolysis of siloxanes is highly dependent on the inductive effects caused by the substituents. If an acid-activated Fuller's earth is allowed to act as a catalyst on hot octamethylcyclotetrasiloxane, the ring is readily split and polymerization to long-chain molecules takes place. With a mixture of octamethylcyclotetrasiloxane and octaphenylcyclotetrasiloxane, only the former is attacked by the catalyst, while the phenyl compound remains unaltered. Similarly, the reactivities of siloxanes toward acceptors depend on the nature of the substituents (Noll, 1963).

Hexamethylsiloxane is cleaved by BF_3 after intermediate formation of a weakly bonded adduct (Wiberg and Krüerke, 1953), while no reaction is observed with hexachlorodisiloxane (Kriegsmann, 1959) and with siloxene (Hengge and Pretzer, 1963). In the latter compounds the donor property of the bridging oxygen atom, as well as the Si—O bond distances, are smaller than in hexamethyldisiloxane (+I effect by the alkyl groups and −I effect by chlorine):

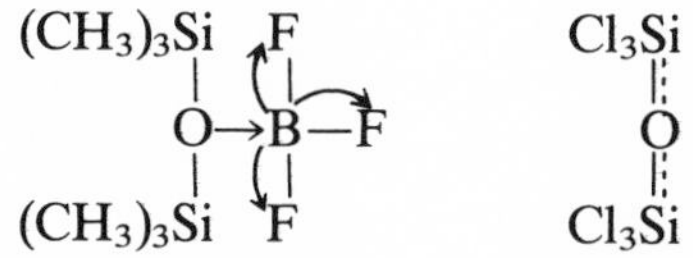

Hexamethyl-disiloxane

Hexadichloro-disoloxane

An aminomethyl group acts as a strong electron donor toward the Si—O—Si group, so that the polarities of the Si—O bonds are increased. $NH_2CH_2(CH_3)_2Si—O—Si(CH_3)_3$ undergoes molecular rearrangement reactions due to the high reactivity of the Si—O—Si bonds.

For the Si—Si bond cleavage in hexaalkylsilane by sodiummethylate, the anions must be present in a highly reactive form and hence the reaction is best carried out in HMPA (Sakurai *et al.*, 1971). This solvent, a strong donor, solvates Na^+ ions, whereas it interacts weakly with carbanions due to its weak acceptor properties.

14.7. Reactions of Boron Compounds

While boron trifluoride is un-ionized in donor solvents, boron triiodide is easily ionized by pyridine (Muetterties, 1960):

$$2PY + BI_3 \rightleftharpoons [(PY)_2BI_2]^+ + I^-$$

The adduct $(CH_3)_3NBH_2I$ is ionized even by weaker donor solvents, such as acetonitrile (Ryschkewitsch and Zutshi, 1970), but the still weaker donor solvent benzonitrile gives an un-ionized adduct.

Boron hydrides undergo ionization in the presence of strong bases (Pitochelli *et al.*, 1962; Achran and Shore, 1965; Ryschkewitsch *et al.*, 1970):

$$B_{10}H_{12}(CH_3CN)_2 + 2Et_3N \rightleftharpoons 2Et_3NH^+ + B_{10}H_{10}{}^{2-} + 2CH_3CN$$

$$B_2H_6 + 2DMSO \rightleftharpoons BH_2(DMSO)_2{}^+BH_4{}^-$$

$$(CH_3)_3NBHBr_2 + 3D \rightleftharpoons D_3BH^{2+} + 2Br^- + (CH_3)_3N$$

The adduct $R_3N{\cdot}BH_3$ undergoes symmetrical cleavage (Long, 1972). The exchange of BH_3 as an entity in diborane has been established by a study of isotopic exchange (Mayburg and Koski, 1953). In the H_3BCO complex, the B—H bond distances are slightly longer (119.4 pm) than the terminal B—H bond distances in diborane (119.2 pm) (Kuchitsu, 1968). The B—C bond is somewhat shorter than the normal B—C distance in organoboranes, while the C—O bond distance of 113.1 pm is slightly longer than in free carbon monoxide (112.9 pm).

Anions as well as uncharged donors may also act as donor ligands toward boron (Parshall, 1964), for example,

$$Mn(CO)_5{}^- + \tfrac{1}{2}B_2H_6 \rightleftharpoons H_3BMn(CO)_5{}^-$$

Normally the reaction between diborane and donor D proceeds further: A second molecule of donor is coordinated with unsymmetrical cleavage of B—H bonds (Parry and Shore, 1958); for example, with DMSO (Achran and Shore, 1965) or diethyl ether as donor solvents, (Long, 1972) the reaction is

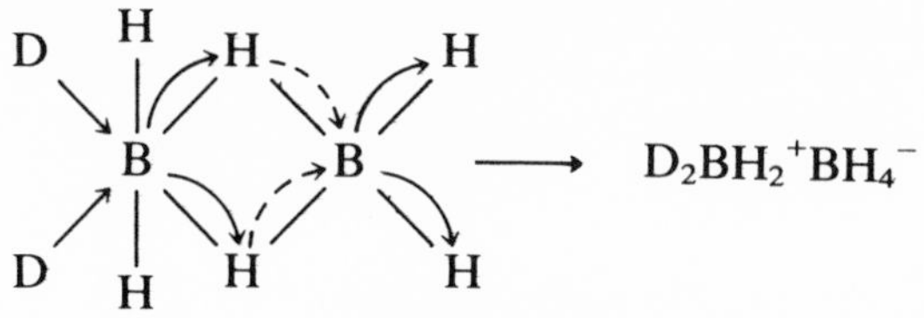

$$\ce{D2BH2H2BH2} \longrightarrow D_2BH_2^+BH_4^-$$

Ammonia attacks a BH_2 group of tetraborane in the gaseous state or in the presence of ether at −78°C, with heterolysis of the adjacent B—H (bridge) bonds. There are changes in B—B distances from 171 pm in compound (I) to 180 pm in compound (II):

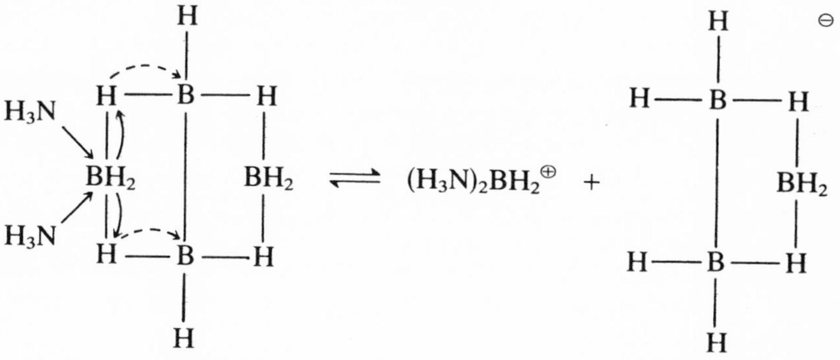

$$H_3N,\ H_3N + B_4H_{10}\ \text{(I)} \rightleftharpoons (H_3N)_2BH_2^{\oplus} + B_3H_8^{\ominus}\ \text{(II)}$$

(I) (II)

The reaction of NaB_3H_8 with NH_4Cl in ether leads to the substitution of a terminal H atom by an ammonia molecule and to the following bond-length variations (Nordman, 1957; Nordman and Reimann, 1959; Peters and Nordman, 1960): The bond distances of $B_{(1)}$—$B_{(2)}$ and $B_{(2)}$—$H_{(4)}$ are increased from 180 to 182 pm, and from 120 to 123 pm, respectively (see page 216). The internuclear distance of $H_{(4)}$—$B_{(3)}$ is shortened from 150 to 139 pm and that of $B_{(3)}$—$H_{(7)}$ considerably lengthened, namely, from 150 to 175 pm. The $B_3H_7NH_3$ unit is less symmetrical than $B_3H_8^-$; this is reflected in the long $B_{(1)}$—$B_{(3)}$ distance of 180 pm and the short $B_{(1)}$—$H_{(7)}$ bond distance of 112 pm, which would be expected to be longer according to the first bond-length variation rule. This might be due to a *trans* effect of the ammonia molecule toward the $B_{(1)}$—$B_{(2)}$ bond, which is continued through the cyclic arrangement with subsequent shortening of the $H_{(7)}$—$B_{(1)}$ bond. The NMR spectra of such adducts depend critically on the relative donor strength of the ligand and the solvent (Eaton and Lipscomb, 1969).

A bridge proton from decaborane is removed from the longest B—H—B bridge. It results in dramatic shortening of the $B_{(5)}$—$B_{(6)}$ bond distance to 165 pm (Sneddon *et al.*, 1972). $B_{(7)}$—$B_{(8)}$ is slightly lengthened

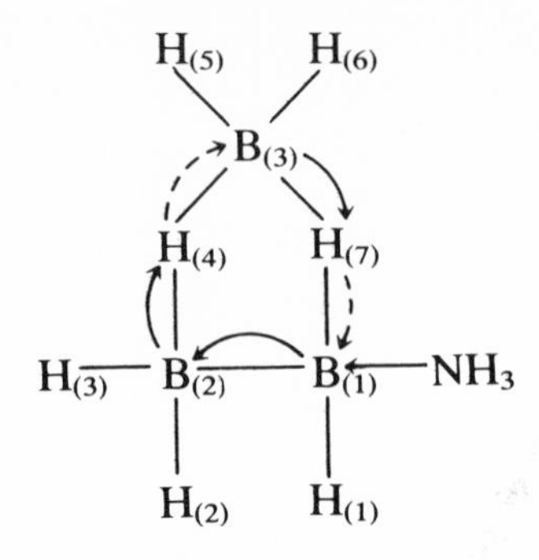

to 204 pm, while the $B_{(5)}—B_{(10)}$ bond distance is shortened from 197.3 pm to 186 pm.

Convenient, large-scale synthesis of salts of boranate anions have been developed using ammonia as the deprotonating agent (Brubaker *et al.*, 1970):

$$NH_3 + Bu_4NBr + B_4H_{10} \xrightarrow{CH_2Cl_2} Bu_4N^+B_4H_9^- + NH_4Br$$

From proton competition reactions such as the following example, the relative Brønsted acidities of several boron hydrides have been established (Remmel *et al.*, 1975):

$$B_4H_{10} + B_5H_8^- \rightleftharpoons B_4H_9^- + B_5H_9$$

An alkyl-substituted B_5H_9 or B_6H_{10} is less acidic than the parent hydride (+I effect of the alkyl group), while a halo substituted molecule is more acidic than the parent hydride (−I effect of the halogen atom) (Brice and Shore, 1973; Remmel and Shore, 1977).

Removal of a bridging proton from a boron hydride creates a B—B bond susceptible to insertion of an electrophilic agent; by inserting BH_3 into this site a series of previously unknown boron hydride anions was generated (Remmel *et al.*, 1975), for example:

$$B_4H_9^- + \tfrac{1}{2}B_2H_6 \longrightarrow B_5H_{12}^-$$

Conversion of the anion to the neutral boron hydride allows for the preparation of boron hydrides containing one more boron than the starting compound. For example, starting from $CH_3B_5H_8$ the compound $CH_3B_6H_{11}$ has been prepared in 80% yield (Long, 1973; Jaworiwsky, 1975).

The anions have proved to be susceptible to insertion of a variety of electrophilic agents in addition to BH_3 insertion. Thus group (IV) elements (Gaines and Iorns, 1967; Gaines, 1973) and a number of metals have successfully been inserted into the boron–boron bond (Brice and Shore, 1970, 1975):

$$K^+B_6H_9^- + Cu(PPh_3)_3Cl \longrightarrow (PPh_3)_2CuB_6H_9 + KCl + PPh_3$$

Addition of a proton to B_6H_{10} yields $B_6H_{11}^+$, which has been isolated as $B_6H_{11}^+BCl_4^-$, and addition of transition-metal carbonyls is represented by the following equation (Davison *et al.*, 1972, 1974; Brennan *et al.*, 1973):

$$B_6H_{10} + Fe_2(CO)_9 \rightleftharpoons Fe(CO)_4B_6H_{10} + Fe(CO)_5$$

The reaction between decaborane and diethyl cadmium (Greenwood and Travers, 1966, 1967) proceeds only in the presence of an electron-pair donor, such as diethyl ether or bipyridyl, and yields a dimeric product of composition $(D_2CdB_{10}H_{12})_2$ and ethane. According to X-ray information (Greenwood *et al.*, 1972), the attack must occur at the $B_{(5)}$—H—$B_{(6)}$ bridge, which is the weakest in the $B_{10}H_{14}$ molecule. The hydrogen atom is abstracted by the ethyl group, suggesting the following mechanism: Coordination of D at the cadmium atom enhances the donor properties of the ethyl group toward the weakest hydrogen bond. The deprotonation leads to an increase in electron density at the boron atoms, which act as donor sites toward the partially solvent-coordinated cadmium ion:

$$\begin{array}{ccccc} B_{(5)} & - & H^{\delta+} & \leftarrow & Et^{\delta-} \\ | & & & & \| \uparrow \\ B_{(6)} & & \longrightarrow & & Cd — Et \\ & & & & \uparrow \\ & & & & D \end{array}$$

The exchange reaction,

$$BF_3 + BCl_3 \rightleftharpoons BF_2Cl + BCl_2F$$

proceeds via a bridge structure with subsequent cleavage of a B—F and of a B—Cl bond (Long and Dollimore, 1954; Muetterties, 1967), in accordance with the first bond-length variation rule:

$$\begin{array}{ccccc} F & & \curvearrowright F & & Cl \\ & \diagdown B & & \searrow B \diagup & \\ F \diagup & & \nwarrow Cl \swarrow & & \diagdown Cl \end{array}$$

Chapter 15
Miscellaneous Applications

15.1 Outer-Sphere Effects on Oxo Anions

In the dichromate ion, the bridging oxygen atom may be considered as an ionic coordination center coordinated by two chromium(VI) oxide molecules. Likewise, the chromate ion may be regarded as resulting from the interaction between an oxide ion and a CrO_3 molecule. The donor–acceptor interaction induces increasing Cr—O bond lengths, which are 159.9 pm in the CrO_3 molecule and 165.9 pm in the chromate ion (Stephens and Cruickshank, 1970). The basicity originally present at the oxide ion has been spread over to the other oxygen atoms in the chromate ion, all of which show weakly basic properties (Gutmann, 1971*a*):

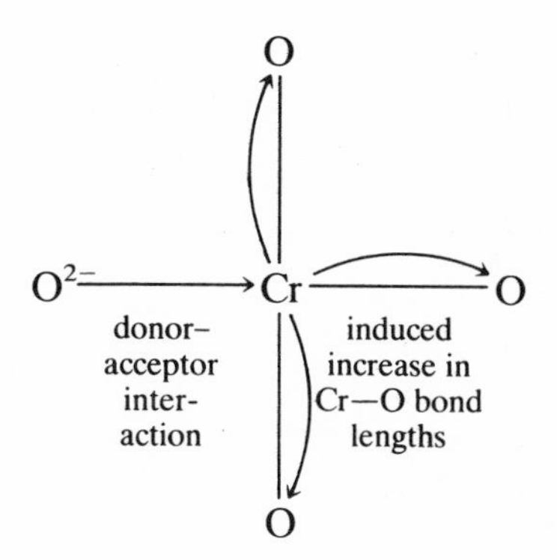

The conversion of CrO_4^{2-} into $Cr_2O_7^{2-}$ may be considered as due to the formation of a coordinate link between one terminal oxygen atom of the CrO_4^{2-} ion and one chromium atom of CrO_3. This leads to an increase of both the Cr—O (bridging) bond length and the Cr—O bond lengths originating from CrO_3 (primary effects). As a secondary effect, the terminal Cr—O bonds originating from the chromate ion are shortened:

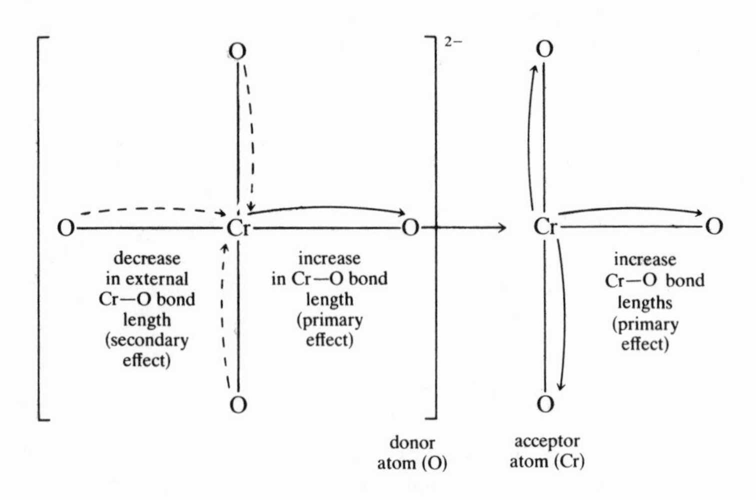

In the dichromate ion, the terminal Cr—O bonds (Brandon and Brown, 1968) are longer (162.9 pm) than in CrO_3 (159.9 pm) and shorter than in the chromate ion (165.9 pm), and hence the basicities of the terminal oxygen atoms of dichromate ions are weaker than in the chromate ion. For this reason CrO_4^{2-} ions are more strongly hydrated in water than $Cr_2O_7^{2-}$ ions (Lederer and Mazzei, 1968). The chromate ion may alternatively be coordinated either by sulfur trioxide to give the sulfatochromate ion (Haight *et al.*, 1964) or by two BF_3 molecules (Gutmann *et al.*, 1974*b*). K_2CrO_4, insoluble in acetonitrile, is readily dissolved in the presence of acceptor cations, such as Al^{3+}, to give a dark brown solution.

By coordination of acceptors to the oxo anion, both the acidic and oxidizing properties are enhanced, just as in the presence of mineral acids: The acidity of the hydrogen-phosphate ion is enhanced by coordination with boron(III) fluoride due to stronger stabilization of the ion with higher negative charge. The more acidic the medium, the more electronic charge is withdrawn from the coordination center and the greater the oxidizing properties. The rate of oxidation by chromic acid depends also on the identity of the mineral acid.

The intermolecular action between NO_3^- and BF_3 is strong enough to heterolyze the adjacent N—O bond (Gutmann *et al.*, 1974*b*), according to the first bond-length variation rule, while the terminal O—N bonds are

$$\left[\begin{array}{c} \mathrm{O} \\ | \\ \mathrm{H—O—P—O \rightarrow BF_3} \\ | \\ \mathrm{O} \end{array}\right]^{2-} \rightleftharpoons \mathrm{H^+ + [PO_4BF_3]^{3-}}$$

increase in bond length (heterolysis) | decrease in bond length | increase in bond length | increase in B—F bond length

donor atom (O) | acceptor atom (B)

shortened (second bond-length variation rule):

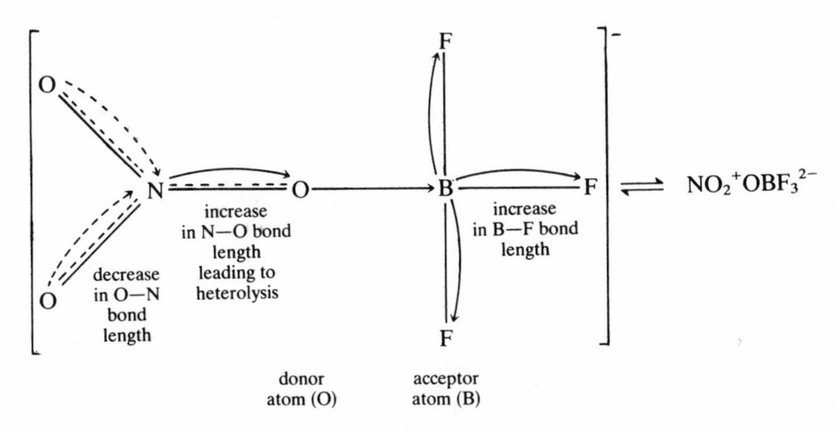

This is in agreement with the formation of nitrobenzene when tetraethylammonium nitrate and boron(III) fluoride are added to benzene (Gutmann *et al.*, 1974*b*). Thus the nitrate ion in the presence of a strong acceptor behaves as it does toward sulfur trioxide when nitronium ions are produced:

$$\left[\mathrm{O_2N—O \rightarrow SO_3}\right]^{-} \rightleftharpoons \mathrm{NO_2^+SO_4^{2-}}$$

decrease in N—O bond length | increase in N—O bond length leading to heterolysis | increase in S—O bond length

donor atom (O) | acceptor atom (S)

15.2. Molten Salts

The donor–acceptor approach has been applied to oxide melts by Lux (1939, 1942, 1948, 1949), who defined a base as a donor of oxide ions and an acid as an oxide-ion acceptor:

$$\underset{\text{donor 1}}{CaO} + \underset{\text{acceptor 2}}{CO_2} \rightleftharpoons \underset{\text{acceptor 1}}{Ca^{2+}} + \underset{\text{donor 2}}{CO_3^{2-}}$$

This concept has been successfully applied (Flood and Förland, 1947; Flood *et al.*, 1947; Flood and Muan, 1950).

By analogy, chloride-ion transfer reactions—in accordance with the formulation given by Gutmann and Lindqvist (1954)—have been ascribed to reactions in sodium tetrachloroaluminate:

$$2AlCl_4^- \rightleftharpoons Al_2Cl_7^- + Cl^-$$

This equilibrium is displaced to the right-hand side with increasing temperature and also with substitution of lithium for sodium ions (Torsi and Mamantov, 1972). Chloride ions are set free by the addition of oxide ions, which are stronger donors, e.g.,

$$2AlCl_4^- + O^{2-} \rightleftharpoons Al_2OCl_5^- + 3Cl^-$$

whereas polymeric anions are produced upon the addition of acceptors, such as hydrogen ions or $AlCl_3$ (Cyvin *et al.*, 1970; Fannin *et al.*, 1972):

$$2AlCl_4^- + H^+ \rightleftharpoons Al_2Cl_7^- + HCl$$

$$AlCl_4^- + 2AlCl_3 \rightleftharpoons Al_3Cl_{10}^-$$

Displacement reactions between liquid steel and a slag have been generally formulated as follows (Herasymenko, 1938, 1941; Herasymenko and Speight, 1950):

$$S_{(alloy)} + O^{2-}_{(slag)} \rightleftharpoons O_{(alloy)} + S^{2-}_{(slag)}$$

A quantitative measure for the basicity of a slag has been defined by Wagner (1975) in order to correlate equilibrium data for various reactions.

Solubilities (Janz, 1967) and chemical reactions proceeding in molten salts can be described readily by the donor–acceptor approach. For example, the solution of transition-metal cations in chloride and nitrate melts leads to coordination as is clearly shown from their spectra (Kerridge, 1975). Many ligand-exchange reactions have been established (Sundermeyer, 1961, 1963; Jackl and Sundermeyer, 1973), for example:

$$Me_3SiCl + KCN \longrightarrow Me_3SiCN \qquad \text{in LiCl–KCl}$$

$$COCl_2 + KSCH \longrightarrow CO(SCN)_2 \quad \text{in LiCl–KCl}$$

$$Me_2SiCl_2 + NaN_3 \longrightarrow Me_2Si(N_3)_2 \quad \text{in KCl–ZnCl}_2$$

Such exchange reactions have been extended to prepare fluorocarbon compounds (Kerridge, 1975), for example:

$$CCl_4 + F^- \longrightarrow CCl_3F, \quad CCl_3F \text{ in KF–KCl at } 675°$$

Many other reactions of organic substances can also be carried out in molten salts (Sundermeyer, 1965).

Various metals are soluble to a certain extent in molten halides and their solutions contain the metal ions in low oxidation states. These compounds are stronger electron donors (reducing agents) than the corresponding metal ions in a higher oxidation state. Consequently they are stabilized in the presence of nonoxidizing acceptors, such as $AlCl_3$. They may be obtained pure by crystallization from alkali metal tetrachloroaluminate melts. For example, bismuth has been found to form Bi^+, Bi_5^{3+}, and Bi_8^{2+} cations in molten NaCl—$AlCl_3$. Bi_5^{3+} is isoelectronic with the pentaboranate ion $B_5H_5^{2-}$ (Corbett, 1968). Other examples are zirconium(II) (Larsen *et al.*, 1976), hafnium(III) (Struss and Corbett, 1970), vanadium(II) (Oye and Gruen, 1964), and tantalum(III) (Gut, 1960). On the other hand, in oxidizing salt media the highest oxidation state can frequently be obtained, for example, iron(VI) in fused KNO_3 or neptunium(VII) in $NaNO_3$–KNO_3.

15.3. Liquid–Liquid Extraction

Liquid–liquid extraction is one of the most important and versatile of all analytical techniques, and is even more important in mineral processing, for example, of uranium, copper, or nickel. The rapid development of practical applications was, however, not due to a systematic approach, nor was it accompanied by a similar progress in the theory of extraction processes. It is still impossible to predict values of distribution coefficients or the changes provoked by altering an extraction parameter (Marcus and Kertes, 1969). The absence of a quantitative theory is a consequence of the fact that the total change in free enthalpy is the sum of partial changes in free enthalpies of various processes, such as dehydration, complex formation, or solvation in the organic phase (Markl, 1972). The situation is further complicated due to the high concentrations of several solutes and the miscibilities of aqueous and nonaqueous phases, the extent of which may vary with the amounts of species extracted. A few rather qualitative considerations may be made by applying the donor–acceptor concept.

The chemical equilibrium in a multiphase system depends on the activities of the components in the different phases. The system is in equilibrium when the chemical potential of each component is equal in all phases in which the component is present. Consequently, the factors influencing the activity of a certain component in the different phases will determine the distribution of this component between the equilibrium phases, so that the analysis of a multiphase system will require a description of all of the donor–acceptor interactions in each phase. It was shown in Chapter 11 that there is a relationship between the complex stability on the one side, and the donor properties, as well as the acceptor properties, of the solvent on the other side (Mayer, 1976).

The solvent extraction system H_3PO_4–H_2O–dibutyl ether shows the following characteristics: At 5°C the distribution of H_3PO_4 is almost completely on the H_2O side, up to a concentration of 75% H_3PO_4 in the aqueous phase (Baniel and Blumberg, 1968). At that concentration a third phase appears containing about 25% H_3PO_4 in ether with a water content below 1%. By further increase of the H_3PO_4 concentration, the phase concentrations remain constant and the phase ratios will change until the whole amount of ether is saturated by H_3PO_4, e.g., the third acid-free ether phase disappears. With further increase in H_3PO_4 concentration there is a parallel increase in the concentration of both the aqueous and the etheral phases, and at about 90% H_3PO_4 only one homogeneous phase remains.

Using the donor–acceptor approach the following interpretation is offered: At low H_3PO_4 concentration water provides the strongest interactions; its bulk donor number is greater than that of ether, so that solvation of the protons is favored by water. The solvation of phosphate ions is accomplished exclusively by water molecules with hydrogen-bond formation because ether has extremely weak acceptor properties. At 75% H_3PO_4 only two molecules of water are available for each H_3PO_4 molecule: Two protons are present as H_3O^+ ions, while for the third proton present in unhydrated HPO_4^{2-} no water molecule is available, so it cannot be hydrated. At this stage the donor properties of ether come into action; H_3PO_4 added to this system is carried into a new phase.

$$
\begin{array}{c}
\mathrm{O} \\
| \\
\mathrm{Bu_2O \rightarrow H{-}O{-}P{-}O{-}H \leftarrow OBu_2} \\
| \\
\mathrm{O} \\
| \\
\mathrm{H} \\
\uparrow \\
\mathrm{OBu_2}
\end{array}
$$

When all of the ether molecules have been coordinated, a single homogeneous phase is formed, which does not contain any solvent since it is composed of H_3O^+ and HPO_4^- ions as well as of $H_3PO_4{\cdot}3Bu_2O$ molecules.

For the extraction of metal ions by amines the organic phase has considerably higher donor properties than the aqueous phase, the differences being even more pronounced in the highly acidic range. The extracted species is a quaternary ammonium salt of an anionic metal complex, e.g., in the presence of hydrochloric acid a chlorometallate ion such as $CoCl_4^{2-}$ or $FeCl_4^-$. These ions are hardly formed in water even in the presence of excess chloride ions, unless fairly concentrated hydrochloric acid solutions are added (Bobtelsky and Spiegler, 1949). This is because hydration of both chloride ions and of cobalt ions is drastically decreased in strongly acidic solutions since most of the water molecules have been used for the formation of the strongly hydrated $H(OH_2)_n^+$ ions. Increasing the HCl content of the aqueous phase favors both the stability of the chlorometallate ions and its extractability into the organic phase: The distribution coefficient for various tetrahedral chlorometallates is increased by increasing the HCl content of the aqueous phase. However, for a certain amount of acid added, a maximum in distribution coefficient is reached with a corresponding decrease by further increasing the acidity. Although the solubility of the organic phase in water is increased in this manner and hence so is that of the extracted species, the distribution coefficient is decreased due to the high concentration of free acid, which is increased by increasing the hydronium-ion concentration.

15.4. Stabilization of Colloids in Liquid Dispersion Media

The state of affairs in colloid chemistry has been well characterized by Overbeck (1972):

> In the earlier days of colloid science the emphasis was mainly on aqueous and "electrocratic" systems. But with the arrival of non-ionic surfactants we are getting into a much better position to understand and control the stability of suspensions and emulsions in non-aqueous and, particularly, in non-polar media.

A colloidal system differs from a true solution by the greater size of the dispersed particles. In an electrolyte solution, equilibria are established between solvated un-ionized molecules and solvated ions (Section 10.1). By analogy the colloidal system may be considered to contain both solvated neutral and solvated ionic particles in equilibrium

$$\begin{matrix}\text{Neutral} \\ \text{colloid} \\ \text{(lyophilic)}\end{matrix} \rightleftharpoons \begin{matrix}\text{Charged} \\ \text{colloid} \\ \text{(lyophobic)}\end{matrix} + \text{Counter ion}$$

and hence the donor–acceptor concept should be applicable.

Lyophilic colloidal particles are solvated by molecules of the dispersion medium (Weiser, 1950). They are relatively insensitive to the addition of electrolytes, but they are coagulated by great amounts of ions since these are more strongly solvated than the neutral colloidal particles.

Lyophobic colloids are stable when charged and hence they are identical with the "charged" colloid in the above equation. In the neutral state, solvation is too weak to stabilize the colloid. Dilute solutions of this type of colloid are stable (at most up to 1%), and their precipitation occurs readily upon the addition of electrolytes, which compete for the solvent molecules due to their stronger solvation.

The terms "lyophobic" or "hydrophobic" are unfortunate, since the neutral colloids are actually not repelling the dispersion medium. In fact, both lyophilic and lyophobic colloids owe their stabilities to solvation: Lyophilic colloids are solvated as neutral species, while lyophobic colloids* are sufficiently solvated only in the charged state. Equilibria established between these two different kinds of colloids are decisively influenced by the relative solvate stabilities. If the stability of the solvation sphere is lowered, for example, by replacing water by a weaker solvating medium, such as alcohol, the stability of the colloid is decreased correspondingly.

Since there is no principal difference between normal ionic equilibria in solution and ionic equilibria in a colloidal system (the main differences being the sizes of the particles), the equilibria will depend on the nature and size of the dispersed particles, on the donor and acceptor properties of the dispersion medium, on the concentration, and on the presence of other solutes.

$Si(OH)_4$ gives a neutral colloid in water, stabilized by hydration. The silicic acid hydrosol obtained by the reaction

$$n\mathrm{Si(OCH_3)_4} + 4n\mathrm{H_2O} \rightleftharpoons 4n\mathrm{CH_3OH} + (\mathrm{H_4SiO_4})_n$$

is relatively stable upon the addition of electrolytes. The surface of silica is ordinarily covered by a monolayer of OH groups generally termed "bound water," and when silica is heated to 500–600°C, this layer is partly removed without sintering the silica (Shapiro and Weiss, 1953; Iler, 1955). Part of the surface can be slowly rehydrated upon exposure to water, but stronger donor molecules, for example, amines, are more strongly adsorbed from the gas phase on silica gel (Iler, 1955).

The neutral colloids are converted into negatively charged colloidal particles in alkaline solution as hydroxide ions are attached to the colloid, which is increased in donor property and hence in stability due to increased

*An emulsion of benzene in water is hydrophobic in the sense of water-repelling; this is unstable, since the particles of the same kind tend to flock together.

coordination by the acceptor property of the solvent. The charged colloid does not precipitate gelatin and albumen, in contrast to the action of the colloidal silica solution of $pH < 6$. In the alkaline range the charged colloid hitherto called "hydrophobic" is therefore more stable than the neutral colloid. Thus colloid stability is increased as the donor properties of the donor toward the OH groups on the surface of the silica particles $-Si-O-H \leftarrow D$ increase: $ROH < H_2O < OH^-$.

Likewise the influence of nonelectrolytes on the coagulation of negative selenium dioxide sol in the presence of barium chloride (Chatterji and Tewari, 1955) cannot be explained by considering the differences in dielectric constants, but rather by considering the donor properties, which increase in the order

$$\text{dioxane} \approx \text{ethyl alcohol} < \text{water} < \text{urea}$$

Colloidal solutions of silver iodide in the presence of excess silver ions contain positively charged colloidal particles. In neutral solutions a number of colloidal anions are present since the concentration of free iodide ions ($c_{I^-} \approx 10^{-10}$) is considerably smaller than that of silver ions ($c_{Ag^+} \approx 10^{-6}$). This is usually ascribed to the preferred adsorption of iodide ions at the surface of the colloidal silver-iodide particles. In accordance with the donor–acceptor concept the equilibrium may alternatively be reached by ionization of the neutral colloid with formation of silver ions. Silver ions are more strongly hydrated than the iodide ions: The acceptor properties of silver ions are more strongly decreased by hydration than are the donor properties of the iodide ions; hence the former remain in solution to a greater extent than the latter:

$$(AgI)_x + mH_2O \rightleftharpoons Ag(OH_2)_m^+ + Ag_{x-1}I_x^-$$

15.5. *Solvent Effects on Electronic Spectra*

Solvent effects on electronic spectra are well known (see, for example, Reichhardt, 1969; Matage and Kubota, 1970). They have frequently been interpreted by the polarity or the dielectric constant of the solvent (Dimroth, 1953). Effects by nonpolar solvents have been attributed to intermolecular dispersion forces (Ferguson, 1956).

In view of the relationships that exist between spectroscopic data, such as Kosower's Z values or Dimroth and Reichhard's E_T values and the acceptor number of the solvent (Section 3.2), it seems reasonable to relate solvent effects on spectroscopic properties to the empirical parameters of the solvent according to the donor–acceptor approach.

Table 15.1. Transition Energy ΔE for the π–π^* Transition of Mesityl Oxide in Various Solvents

Solvent	AN	ΔE (kJ·mol^{-1})
Isoçetane	low	519.1
CH_3CN	19.3	511.7
DCE	20.4	507.8
i-Propanol	33.5	506.7
EtOH	37.1	506.2
CH_3OH	41.3	505.4
H_2O	54.8	493.4

For example, a rough correlation exists between the π–π^* transition energy of mesityl oxide in various solvents (Kosower, 1958) and the acceptor number of the solvent (Mayer *et al.*, 1975) (Table 15.1).

A relationship exists also between the transition energies for the visible absorption bands of the hexabromoiridate anion (Jørgensen, 1962) and the acceptor number of the solvent (Mayer *et al.*, 1975) (Table 15.2).

Table 15.2. Transition Energies ΔE for Visible Absorption Bands of $IrBr_6^{2-}$ in Various Solvents

		ΔE (kJ·mol^{-1})	
Solvent	AN	Band III	Band IV$_b$
AC	12.5	197.2	220.4
DCE	20.4	198.1	222.8
NM	20.5	200.1	224.5
$CHCl_3$	23.1	198.8	224.1
CH_3OH	41.3	200.8	228.3
H_2O	54.8	205.3	235.1

Chapter 16

Homogeneous and Heterogeneous Catalysis

16.1. Catalytic Ability

Catalytic ability is the property of a given chemical material, called a catalyst, to affect the rate of reaction of another chemical material, called the substrate. Specific relations between catalyst and substrate exist. The catalytic ability is expressed by the rate of the catalyzed reaction, in moles of reacting substrate per second relative to the rate of the uncatalyzed reaction.

In studies of catalysis a distinction is usually made between heterogeneous and homogeneous catalysis. Entirely different approaches are used in the description of the electronic properties of a solid catalyst for heterogeneous catalysis as compared to those of catalytically active molecules or ions dissolved in a homogeneous reaction medium. Links between homogeneous and heterogeneous catalysis have been sought for years (Yates, 1974).

The extended donor–acceptor approach to catalysis appears to provide a framework for a unified treatment of both heterogeneous and homogeneous catalysis, since this approach is not concerned with the interpretation of the binding forces between substrate and catalyst. It is, however, difficult to establish relationshps between sets of available data, since most of the experiments have been carried out on actual catalysts, which have not been precisely characterized structurally. It may be appropriate, however, to indicate some of the requirements for more systematic research based on the extended donor–acceptor concept.

A catalytic system may be characterized by the following requirements (Gutmann and Noller, 1971):

(i) The electronic properties of the catalyst and substrate must allow an intermolecular interaction in a highly specific way: Electron transfer and (or) polarization effects must provide suitable bond polarities in the intermediate molecular adduct. A catalyst functioning as a donor toward the substrate will increase the donor properties of the substrate in the adduct, e.g., the base or (and) the reducing property. If the catalyst acts as an acceptor toward the substrate, it will increase the acceptor property, e.g., the Lewis acid or (and) the oxidizing property of the substrate in the adduct.

(ii) The induced changes in structural and in chemical properties in the substrate–catalyst adduct must be suitable for the reaction to proceed.

(iii) The catalyst must be brought into the position to regain its original state by an appropriate electron rearrangement. Thus electronic charge appears to oscillate between catalyst and substrate; hence the catalytic system may be considered as an "electron pump."

16.2. Catalytic Selectivity in Homogeneous Catalysis

The optimal catalytic activity for a given type of catalyst can be adjusted by varying its electron density (Strohmeier, 1968).

For homogeneous catalysis this adjustment can be effectuated (a) by substituent or (b) by solvent effects.

The variation of the catalytic ability of the homogeneous so-called Vaska-hydrogenation catalyst $Ir(CO)Cl(PR_3)_2$ was studied by changing the substituents R (Strohmeier and Onada, 1968). It was found that the electron density at the catalytically active site of the catalyst, e.g., the iridium atom, increased in the order $R = C_6H_{11} < i\text{-}C_3H_7 < C_4H_9 < CH_2C_6H_5 < p\text{-tolyl} < C_6H_5 < OC_6H_5$, whereas the catalytic ability of the various substituted catalyst molecules, as measured by the rate of hydrogen consumption per millimole ethylene per unit time, showed a marked maximum for $R = p$-tolyl (Fig. 16.1). In this series of catalysts the activation enthalpies remained similar, namely, between 45 and 50 kJ, while the rate coefficients were found to differ by three orders of magnitude. Thus the changes in catalytic ability of $Ir(CO)Cl(PR_3)_2$, with change of R, are due mainly to changes in the entropy of activation.

No systematic work has been done to investigate specific solvent effects on the activity of a given homogeneous catalyst. However, the situation is complicated because not only the catalyst, but also the reactants, will be affected by a change in solvent.

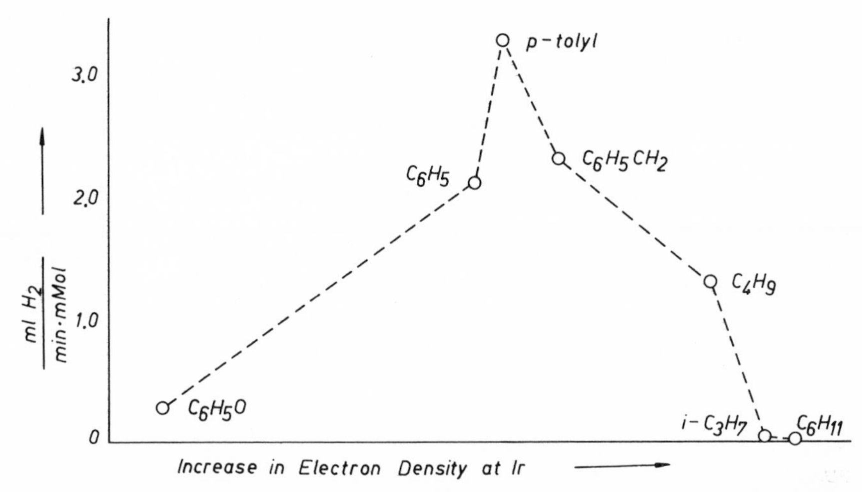

Fig. 16.1. Relationship between H_2 consumption for the reaction $Ir(CO)Cl(PR_3)_2 + H_2 \rightleftharpoons Ir(CO)Cl(PR_3)_2H_2$ as a function of the electron density at Ir due to different R groups (from Strohmeier and Onada, 1968, courtesy of Verlag der Zeitschrift für Naturforschung, Tübingen).

16.3. *Catalytic Selectivity in Heterogeneous Catalysis*

For heterogeneous catalysis at a solid surface, the electronic properties at the surface area will be decisive in determining the catalytic ability of the solid catalyst. For a given material the surface structure has been shown to be influenced by its morphology, by crystal lattice defects, and by the nature of any support for the catalyst.

The support often plays a significant role in catalytic reactions. For example, the activation energy for the decomposition of formic acid on a nickel surface is reduced and thus the rate is enhanced by changing the supporting material for the nickel in the following order: $Ni < Al_2O_3 < NiO < NiO$ doped with Li_2O (Schwab *et al.*, 1957; Szabó and Solymosi, 1961).

In the absence of systematic structural investigations as related to catalytic activity, I may postulate that the surface reactivity of a given material will depend on: (i) the thickness of the supported surface layer (according to Section 5.3 and 5.4 the surface energy will be smaller the thinner the surface layer), and (ii) the interaction of the surface layer with the supporting material (the surface energy will be smaller the stronger the intermolecular interaction).

For a particular material the optimal catalytic ability may be adjusted by a suitable combination of supporting material and by the thickness of the catalyst.

Important investigations on atomically clean single-crystal surfaces mark a new area in research in heterogeneous catalysis. Thus, Yates and King (1972) have shown that the so-called α-CO state of carbon monoxide adsorbed on a clean tungsten surface is similar to the "carbonyl-type" CO ligands as found in hexacarbonyl tungsten(0). Here the C≡O bonds are considerably polarized due to back donation from tungsten toward the C≡O ligands (see Section 3.4). As the CO coverage of the tungsten surface increases, a slight strengthening of the C≡O bonds is found, which has been explained by the back donation model of Yates and King (1972): The availability of tungsten *d*-electrons for back donation toward adsorbed carbon-monoxide molecules decreases with increasing CO coverage since both the primary bonded tungsten atoms (those involved directly in the linear W=C=O species) and the neighboring tungsten atoms contribute to the π-bonding. This concept is in agreement with the finding that the photoelectron C (1*s*) and O (1*s*) bonding energies for the adsorbed carbon monoxide are lower than for the free gaseous carbon-monoxide molecules. There is a systematic decrease in photoelectron binding energy as the adsorption of CO is increased (Yates *et al.*, 1974).

At the same time, theoretical developments on the electronic properties of surfaces and adsorbed layers are taking place. The correct prediction of energetically stable, long-range ordered adsorbate structures is a major achievement (Grimley, 1967).

We may now proceed to discuss a few catalytic systems to which the donor–acceptor approach is easily applicable, namely, elimination and addition reactions.

16.4. Elimination Reactions on Solid Catalysts

Elimination reactions are known to proceed not only in the gas phase or in the liquid phase, but also over solid, usually polar, catalysts. An important difference of eliminations in the liquid phase is that in the liquid phase the whole substrate is in contact with the reaction medium, while in an elimination on a plane surface only the leaving groups may be in immediate contact with it. For this reason eliminations in the liquid phase are faster and occur at lower temperatures than on a solid catalyst.

The reaction on a polar catalyst may be considered analogous to that in a solvent in that the polar catalyst causes heterolysis of the bonds due to coordination (Pines and Manassen, 1966). Thus elimination reactions can be related to ionization processes in solution (Noller and Kladnig, 1976):

$$\text{Donor}\rightarrow\overset{\delta+}{\text{H}}-\overset{\delta-}{\text{C}}-\overset{\delta+}{\text{C}}-\overset{\delta-}{\text{X}}\rightarrow\text{Acceptor}$$

Thus the action of a polar catalyst parallels that of an amphoteric solvent, in that its negative sites act as donors toward the hydrogen atom of the substrate, while its positive sites will act as acceptors toward the halogen atom of the substrate. These interactions lead to loosening of both the H—C and C—X bonds and hence to lowering of the activation energy for heterolysis; the effects increase with increasing donor and acceptor strengths of the active sites of the catalyst. This applies primarily to reactions proceeding by an $E2$ mechanism, e.g., H and X are abstracted in one step. However, the movement of the atoms need not be fully synchronous. In the $E1$ mechanism, the reaction begins with the abstraction of X. Since X is generally more electronegative than H, the first step, which is rate determining, gives a carbonium ion, which loses a proton in the second step:

$$\text{X—C—C—H} \longrightarrow \text{X}^- + {}^+\text{C—C—H} \longrightarrow \text{C=C} + \text{HX}$$

Thus the interaction with the polar catalyst is that of an acceptor at X and that of a donor at the carbon atom bound to X:

$$\text{Acceptor}\leftarrow\text{X}-\underset{\underset{\text{Donor}}{\uparrow}}{\text{C}}-\text{C}-\text{H}$$

In the second step, removal of the proton is assisted by donation at hydrogen:

$${}^+\text{C—C—}\underset{\underset{\text{Donor}}{\uparrow}}{\text{H}} \longrightarrow \text{C=C} + \text{DH}^+$$

It has been found generally for catalyzed reactions that cationic centers in the catalyst play a more important role than anionic centers, so that acceptance from X is the major determinant of the reaction (Noller and Kladnig, 1976). According to Table 16.1 the activation energies for catalyzed eliminations increase with decreasing acceptor strength of the cations in the catalyst: $Li^+ > Na^+ > K^+$ and $Ca^{2+} > Sr^{2+} > Ba^{2+}$. A similar order of metal-ion reactivity has been found for the dehydration of ethanol on metal-sulfate catalysts (Mochida *et al.*, 1971). The study of catalysis by zeolites offers the possibility of varying the cation in essentially the same structural framework. In zeolites a given cation is usually more active than in simple salts. This is probably due to different coordination numbers. In simple salts the usual coordination number at the surface may be assumed

Table 16.1. Activation Energies E_a of the Dehydrochlorination on Alkali-Metal and Alkaline-Earth-Metal Chlorides and Sulfates

Reactant	Catalyst	E_a (kJ·mole^{-1})	Temperature (°C)
2-Chloropropane[a]	LiCl	83	242–271
	NaCl	105	362–393
	KCl	192	295–315
2-Chloropropane[b]	$CaCl_2$	75	116–189
	$SrCl_2$	83	128–171
	$BaCl_2$	117	82–142
2-Chlorobutane[c]	$MgSO_4$	54	150–300
	$CaSO_4$	67	180–320
	$BaSO_4$	92	250–400

[a] From Noller and Ostermeier (1956).
[b] From Noller and Wolff, 1959.
[c] From Correa *et al.* (1968).

to be 5, whereas it is 3 in the zeolites. Thus the acceptor strength of cations is higher in zeolites than in salts and oxides; this is in agreement with the third bond-length variation rule. For zeolites of type X the reaction rates for dehydration of 1-butanol follow the sequence LiNaX > NaX > KNaX > RbNaX (Galich *et al.*, 1965), in order of decreasing metal-ion strength.

As already mentioned (Section 12.3), for metal ions no characteristic property is available for the acceptor strength. Within a homologous series of metal ions a relative measure for the acceptor strength are the ionic radii, which are inversely proportional to the hydration enthalpies.

In a recent study with LiBr, Li_2SO_4, Li_2CO_3, Li_3PO_4, $BaCrO_4$, $BaCO_3$, and $Ba_3(PO_4)_2$, Andréŭ and Madrid (1973) found that the activation energy of the dehydrochlorination of 1- and 2-chlorobutane increases roughly with the donor property of the anion. This may be interpreted in terms of the effect of the anionic sites on the acceptor properties of the cationic sites. An increase in donor strength of the anion will decrease the acceptor strength of the cation (Noller and Kladnig, 1976). For example, although Mg^{2+} is a weaker acceptor than Al^{3+}, the specific activity of $MgSO_4$ for the dehydrochlorination of ethyl chloride is up to three orders of magnitude greater than that of Al_2O_3, the oxide ion being a stronger donor than the sulfate ion (Heinzelmann *et al.*, 1971).

Thus the activity of an elimination catalyst is determined mainly by the acceptor property of the cation and, to a lesser extent, by the donor property of the anion. Furthermore the relative donor and acceptor strength of the active sites toward the substrate will be decisive for the mechanism observed.

The *E*1 mechanism is favored by high acceptor strength of the catalyst and decreasing C—X bond strength. Increase in charge of the cationic center in the catalyst shifts the mechanism more toward *E*1. A C—Cl bond usually is more easily broken than a C—O, C—N, or C—C bond. Hence catalyzed dehydrochlorination has more *E*1 character than dehydration, deamination, and dealkylation. Elimination of HX from haloalkanes over $Ca_3(PO_4)_2$ follows a pure *E*1 mechanism (Noller *et al.*, 1966), whereas butan-1-ol dehydrates mainly by *E*2 on the same catalyst (Thomke and Noller, 1972).

On the other hand the *E*2 mechanism is favored by low acceptor strength and high donor strength at the catalyst sites and by low donor strength of X and high fractional charge of the hydrogen atom. Dehydrohalogenation of 1-chlorobutane proceeds by an *E*1 mechanism on γ-Al_2O_3, but by an *E*2 mechanism on the spinel $MgAl_2O_4$. Both catalysts show the same crystal-structure type, but in $MgAl_2O_4$ the acceptors properties of Al^{3+} centers are not as great as those in γ-Al_2O_3, and the donor properties at the terminal oxygen atoms are greater in $MgAl_2O_4$ than in γ-Al_2O_3, as derived from ESCA measurements. This is exactly the variation needed to shift the mechanism toward *E*2 (Vinek and Noller, 1976).

16.5. Addition of Hydrogen Halides to Alkenes on Solid Catalysts

Catalysis of the addition of hydrogen halides to alkenes is known to proceed either in homogeneous media or on saltlike catalysts, such as alkaline-earth halides (Noller and Wolff, 1959; Letterer and Noller, 1969; Andréŭ *et al.*, 1969):

(i) Addition of hydrogen halides at a surface of a salt may be initiated by interaction of the hydrogen-halide molecule with the metal ion of the salt. The latter functions as an acceptor toward the halogen atom of the hydrogen-halide molecule. According to the first bond-length variation rule the H–Hal bond length is increased. This interaction may lead to deprotonation of the halogen-halide molecule bonded to the metal ion. The addition follows Markovnikov's rule and this is an indication of the involvement of ions. The proton or the hydrogen atom of the bonded hydrogen-halide molecule interacts with the electron pair of the donor alkene and bonds to carbon. Consequently the second carbon atom of the carbon–carbon double bond acquires a positive charge and acts as a stronger acceptor for the halide atom attached to the metal ion of the cayalyst, from which it is removed. In this way the catalyst is regenerated. This addition mechanism may be considered as the reverse of the $E1$ elimination mechanism on a saltlike catalyst.

(ii) Alternatively, the hydrogen-halide molecule may act as an acceptor at the hydrogen atom toward the anion donor of the saltlike catalyst. This interaction induces an increased polarity of the H–Hal bond and an increased fractional negative charge at the terminal halogen atom. The alkene will interact with the proton, either before or after heterolysis of the H–Hal bond.

(iii) A third possibility is that of double adsorption of the hydrogen halide at the catalyst surface, i.e., so that its hydrogen atom unites with an anion of the catalyst and the halide atom with a cation of the catalyst. This corresponds to the "push–pull effect" in an ionizing solvent (Section 10.1): The H–Hal bond is ionized due to both the donor attack at the hydrogen and the acceptor attack at the halogen atom of the hydrogen-halide molecule. The alkene interacts with the adsorbed proton, which bonds to carbon, whereas the second carbon atom of the original double bond binds to the halide ion. In this way both proton and halide ion are removed from the catalyst, which is recovered.

Chapter 17
Biochemical Applications

17.1. General Considerations

Biochemicals can be structurally characterized only after they have been taken out of the functioning system. Efforts have been made to gain an understanding of more complicated molecules by studying "subunits," and numerous model compounds are being investigated. However, the conclusions drawn from the properties of such systems are unlikely to be valid for the more complicated biosystem. Furthermore the influence of the environment on the structure and on the properties of the isolated crystal have not yet been investigated systematically: The crystal which has been investigated may undergo substantial structural changes when dissolved in the actual biosystem.

Several transition-metal ions constitute an integral part of the protein structure, and they influence the electronic and structural arrangement of the protein and hence its chemical properties (Malmström and Neilands, 1964). Despite many efforts, mainly by coordination chemists, no significant progress has been made, since transition-metal stereochemistry is rather complex even for simple inorganic coordination compounds. No satisfying model or theoretical approach is at present available to describe in detail all aspects of structure and reactivity of metal complexes. For many of the metal ions the unrealistic ionic model is complicated by the nonspherical nature of the crystal-field effects and by the very considerable deviations brought about by the environment (Buckingham, 1973). The attempts to assign and to discuss oxidation numbers for the metallic nuclei in the complicated macromolecules are of no value, and the applicability to biological solutions of stability constants, which are valid for highly dilute aqueous solutions of defined ionic strength, is not justified (Sigel and McCormick, 1970).

Although I am not in a position to overcome any of these difficulties, I shall attempt to illustrate briefly the usefulness of the donor–acceptor approach by applying it to a few selected points. In emphasizing the structural variability of biomolecules the study of their functionality in different environments should be encouraged. The importance of cooperative effects was appreciated in biochemistry long before this concept had been accepted by solution chemists; hence the extended donor–acceptor concept—in particular the application of the rules for the structural variability—may be acceptable to a number of biochemists.

Various biochemical aspects of the approach presented in this book have been mentioned in previous sections, namely, weak complexes (Section 3.1), the structure of an oxygen carrier (Section 3.2), interface phenomena (Chapter 5), hydrogen bonding (Section 6.1), the decisive role of solutes in determining the structures of aqueous solutions (Sections 6.2 and 9.2), the relationship between redox properties and coordination (Chapter 8), the effects induced in imidazoles by coordination (Section 10.1), the reactions of carbon dioxide (Section 14.5) and of molecular nitrogen (Section 14.6), polyelectrolytes (Section 15.5), and catalytic phenomena (Chapter 16). Points of interest to biochemists may also be found in other parts of preceding sections.

17.2. Remarks on the Role of Water in Biological Systems

Water is a *conditio sine qua non* for biological processes. By removal of water, irreversible denaturation of biosystems occurs. For example, DNA requires 30% water in its native conformation in the crystalline state, but even partial dehydration leads to denaturation. Single crystals of all proteins are intrinsically hydrated (Drabkin, 1950). The important role of water in biological systems extends to the stabilization of the ternary and quaternary structures and further to the large free enthalpy changes associated with chemical reactions.

Unfortunately, the hydration structures of biopolymers are essentially unknown, and one of the serious obstacles for the understanding of the properties of proteins and other biological materials is that adequate means for assessing the role of hydration are lacking (Conway, 1972). A large body of proteins and enzyme proteins are polyelectrolytes, and treatment of such ionic solutions is based on unrealistic models in which the solvent is treated as a nearly unperturbed medium (see Section 6.2). A distinction is made between hydrophobic and hydrophilic solutes in water. Hydrophobic bonds play a role in association and in conformational changes, for example, in stabilizing the helical structure by incorporation between two neighboring residues of the α-helix. The helical form of DNA is stabilized by the addition of electrolytes (Eagland, 1975), and the effects are greatly influenced by the concentration.

The donor–acceptor concept is useful in describing the influence of ions upon the behavior of concentrated solutions: The stronger the acceptor properties of the cation and the donor properties of the anion, the more strongly these ions will be hydrated and the smaller the number of water molecules which remain available for the other solutes. Hence the solubilities depend not only on the solvent properties (Section 11.4) but also on the nature and on the concentration of other solutes present. For example, the solubilities of thymine and adenine are increased as the added ions are less strongly solvated (Luck, 1976*a*). The more water is taken up by the ions, the smaller the solubility of a given hydrated polymer (Hippel and Wong, 1965). Likewise, the melting temperature T_M of the helix-coil transition of DNA increases as the relative acceptor strength of the metal ion decreases, e.g., in the series $Li^+ \geqq Na^+ > K^+ > Rb^+ > Cs^+$, and as the relative donor strength of the anion decreases, as expressed by the so-called Hofmeister series, e.g., $HCOO^- > Cl^- > Br^- > ClO_4^- > CCl_3COO^-$ (Robinson and Grant, 1966). The capacity for reducing the optical rotation of poly-L-proline decreases as the hydration enthalpy and the concentration of the ions present increases (Schleich and Hippel, 1969). Hence the hydration properties of added solutes change the structure and the properties of the whole liquid system (see Sections 6.2 and 9.2).

The dependence of the structural order of aqueous solutions throughout the liquid phase on the composition has been discussed in Section 6.2. Such "supermolecular" interpretation was proposed for systems of biochemical significance a few years ago (Cope, 1973), providing an explanation for the influence of salt addition to the aqueous phase on the properties of the organic phase. Addition of salts to water have remarkable effects on interfaces. For example, properties of a monomolecular film of $C_{18}H_{37}$—$(O$—CH_2—$CH_2)_nOH$, with $n = 1$, 2, or 3, are changed (Luck, 1976*a*), as the composition of the aqueous solution is altered.

The specific structures at interfaces, such as at cell membranes, were pointed out more than 20 years ago (Jacobson, 1955), but the relationships to composition and structure of the solution have not yet been investigated systematically.

17.3. Oxidation of Hemoglobin

The transport and the storage of oxygen is an extremely important physiological function. Hemoglobin is one of the oxygen carriers. The structure of this macromolecule has been well established by high resolution X-ray diffraction (Kendrew *et al.*, 1960; Nobbs *et al.*, 1966; Perutz *et al.*, 1968; Bolton and Perutz, 1970). The tetrameric molecule has a molecular weight of 64,500 and consists of two identical pairs of units, the α- and β-units, roughly arranged in a tetrahedron. Each unit contains a heme group with a protein and is capable of binding oxygen reversibly by coordination to the iron atom incorporated in a porphyrine molecule. The latter is essentially planar, but two pyrrole rings are tilted up and two are tilted down, so that the nitrogen atoms are slightly out of plane. In the oxidized form the iron nuclei are found nearly in plane (the protein X-ray studies can determine the plane of the porphyrin, but not the plane of the pyrrole nitrogens). The molecule is not completely rigid, since the metal–nitrogen bond length in metalloporphyrins varies slightly. The length of the Fe—N bond varies between 206 pm in high-spin ferric and 199 pm in low-spin ferric compounds, while in metal-free porphyrin the distance from the center of the ring of the nitrogen atoms is 201 pm (Perutz, 1970). This is smaller than in the high-spin and larger than in the low-spin compound (the consideration of the ionic radii of the metal ions would require an Fe—N distance of 218 pm in the high-spin compound).

It has been shown for iron–porphyrin dioxygen complexes, which bind oxygen reversibly in solution or in the solid state, that in the crystal the oxygen molecule is bound "end-on" to the iron atom (Collman *et al.*, 1975), with an Fe—O—O bond angle of 136°. In the adduct the oxygen–oxygen distance is 124 pm; hence it is greater than in the free oxygen molecule (120.7 pm). The Fe—O distance of 175 pm is 10 pm shorter than expected from summation of the covalent radii, thus indicating multiple bonding. In the five-coordinated deoxy form, the iron atoms lie slightly out of plane of the four nitrogen atoms, but they lie in plane in the oxy and carbonmonoxy forms, which are six-coordinated (Perutz, 1970). By oxygen addition the Fe—N distances are increased from 201 to 207 pm. They are 17 pm lower than expected from the sum of the covalent radii. The electronic changes resulting from the addition of an oxygen molecule change the metal–ligand bond lengths and they are transmitted and amplified through the ligand structure in such a way that the reactivities of the second heme group toward oxygen, some 2500 pm away, are changed considerably.

The observed structural data are in accordance with the bond-length variation rules (Sections 1.3 and 1.5) as the bonds adjacent to the intermolecular Fe—O interaction are lengthened. The nature of the bond between iron and dioxygen has not been defined at the atomic level. According to MO theory, O→Fe σ-bonding would, however, lead to a decrease in O—O bond length. The following interpretation may be offered: Iron(II) is strongly bonded to five nitrogen atoms and hence its fractional positive charge will be low enough to allow π-back-bonding toward an appropriate ligand. When it acts as a π-donor toward the oxygen molecule a net transfer of electron charge will occur from the iron atom toward the ligated oxygen molecule (Section 3.4). In this way the O—O distance is increased and so are the donor properties at the terminal oxygen atom. The concept of back donation is also consistent with expectations from crystal-field theory and it accounts for the other observed structural changes. The increase in fractional positive charge at the iron nucleus leads to an increase in oxidizing property. Since the actual charge transfer toward the oxygen molecule is considerably lower than would be expected for a one-electron-transfer process to give an O_2^- ion (in which an O—O bond distance of 128 pm would be expected), it appears of little value to formulate the changes in redox property for changes of oxidation number. The actual change in fractional positive charge is further affected by the ligand structure, which is expected to compensate partly for the loss of negative charge due to back donation. The back-donation interpretation is also in accordance with the strength of the Fe—O bond and the lengthening of the N—Fe bonds from 199 pm in the deoxygenated molecule to 207 pm in the oxygenated form. This is in agreement with the first bond-length variation rule and with the expected increase of the Fe—N bond polarity by increase in fractional positive charge at the iron atom:

$$\overset{\delta+}{\mathrm{Fe}}-\overset{\delta-}{\mathrm{N}}$$

The movement of the metal into the porphyrin plane may be due to the *trans* effect exerted by the oxygen ligand, which comes into axial position with regard to the porphyrin plane:

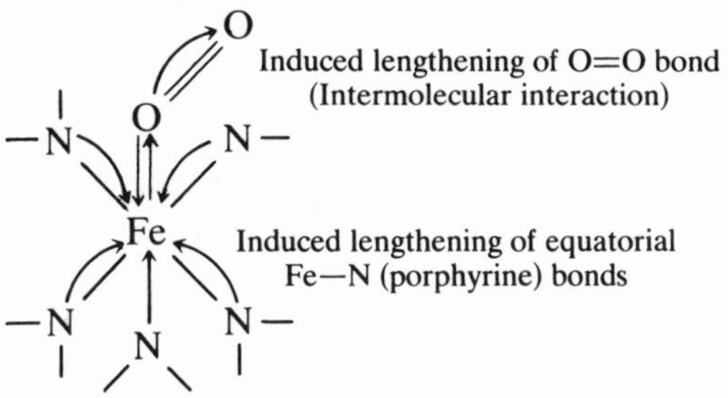

Induced lengthening of O=O bond
(Intermolecular interaction)

Induced lengthening of equatorial
Fe—N (porphyrine) bonds

The fifth N-ligand to iron, which is *trans* to oxygen in HbO_2, is the imidazole group of histidine, which will influence the electron distribution: It should strengthen the Fe—O bond and hence weaken the equatorial Fe—N bonds (see Section 4.4).

However, too much strengthening of the iron–oxygen bond would prevent the biological function of the molecule. Histidine has both σ- and π-bonding properties, as well as donor and acceptor sites. Its donor properties are increased when it is coordinated to donors through its hydrogen atoms, and the acceptor property of the hydrogen atom is increased the stronger the bonding of a nitrogen atom to a metal ion (Section 10.1). For example, the pK_a value of 8 for the uncomplexed imidazolium group ($\gtrless NH \rightleftharpoons \gtrless N^- + H^+$) is decreased to 2.8 when chelated to iron(II) (Hanania and Irvine, 1962). Likewise the acidity of chelated 2-(2-pyridyl) benzimidazole is increased by increased complex stability (Harkins and Freiser, 1956).

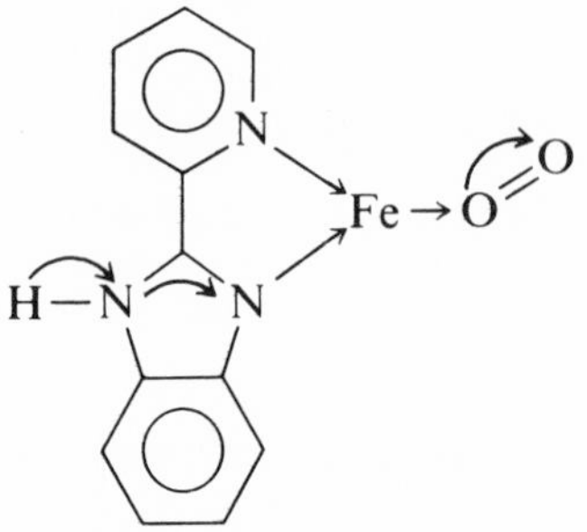

The histidine group in *trans* position to the oxygen molecule is part of the polypeptide chain. If its imidazole group is replaced by another group, such as a less basic tyrosine residue, the system no longer binds oxygen reversibly (Perutz and Lehmann, 1968; Morimoto *et al.*, 1971).

Perutz (1970) gives a stereochemical explanation for the cooperative effects occurring in hemoglobin due to oxygen binding. The last amino-acid residue in the α-chain is arginine, and in the β-chain, histidine. The last residues in both the α- and the β-chains are tyrosine units. In oxyhemoglobin the C-terminal residues of all chains have complete freedom of rotation and the penultimate tyrosines have partial freedom. In deoxyhemoglobin, on the other hand, each of the terminal residues is doubly anchored by salt bridges. All penultimate tyrosines are firmly anchored in pockets between two helices by hydrogen bonds between their OH groups and the carbonyls of valines. The constrained molecule in the deoxy form has a narrowing of the pocket containing the tyrosine residue near the end of the chain. Addition of oxygen to the iron atom causes cleavage of salt links between the terminal acid and other groups. Thus the effect due to Fe—N bond lengthening is transmitted throughout the structure leading to both decreasing acceptor properties of the hydrogen atoms and to decreasing donor strengths of the donor centers.

A very important aspect of the structure of hemoglobin is the fact derived from X-ray results, that most of the polar groups of the protein are on the outside of the molecule, thus producing a hydrophobic interior. Recent X-ray work on oxyhemoglobin revealed that each heme group of this molecule is in weak contact with about 60 atoms of the protein. Almost all of these atoms belong to amino-acid residues, all of which appear necessary for function. Stabilization of the heme–globin bonding appears to be provided by H-bonding interactions of acceptor carboxyl groups of the porphyrin with the donor group of the protein.

This means that, as has been mentioned, while certain H-atoms are weakened, others will be strengthened in acceptor property. Likewise there will be donor groups which are weakened, whereas others are enhanced in donor properties. It is important to investigate in more detail the functions of units capable of acting as regulating centers for electron distribution.

17.4. Carbonic Anhydrase

For enzymes, an old established analogy is that of the lock and the key. This analogy, however, does not indicate the electronic or conformational effects which are taking place (Hughes, 1972).

Carbonic anhydrase is essential in catalyzing the dehydration of carbonic acid under physiological conditions:

$$HCO_3^- \rightleftharpoons CO_2 + OH^-$$

The ellipsoidically shaped molecule has a molecular weight of 30,000 and it contains one atom of zinc per molecule and some 260 amino-acid residues. The structure determination revealed a narrow slit (Fridborg *et al.*, 1967). The zinc atom has a tetrahedral environment with three of the bonds to protein ligands, probably all imidazole groups or histidine residues. The fourth site is available for other groups, such as a hydroxide ion (Strandberg and Liljas, 1971).

In the hydration reaction of carbon dioxide a hydroxide ion appears to be added to a negative center (Dennard and Williams, 1966*a*, *b*). The rate of reaction between carbon dioxide and water has been studied as a function of pH, together with the effect of a number of inhibitor anions. No relationship was found between the pK_a of the conjugate acid of the anion and the inhibiting power. Infrared studies on the carbon-dioxide molecule bound in the hydrophobic slit at the active site of the enzyme indicated that it is not coordinated to the zinc atom (Riepe and Wang, 1967). Evidence is available for the implication of an imidazole residue in the activity of the enzyme, possibly in proton transfer (Hughes, 1972). From the point of view of the extended donor–acceptor concept, the following mechanism may be possible: The $H_{(1)}$ atom of the OH group is increased in acceptor strength by coordination of the oxygen atom $O_{(1)}$ to the zinc atom, as the $H_{(1)}—O_{(1)}$ bond is lengthened (first bond-length variation rule). Hence it may form a hydrogen bond with the nitrogen atom of the imidazole group, by which the $N—H_{(2)}$ bond is lengthened and the $H_{(2)}$ atom increased in acceptor property (see Section 10.1). Thus $H_{(2)}$ can form a hydrogen bond with the donor atom $O_{(2)}$ of the carbon dioxide molecule, which leads to lengthening of the bond $C—O_{(2)}$ and hence to increase in acceptor property at the carbon atom (I). Thus a donor–acceptor interaction $O_{(1)}\rightarrow C$ takes place, by which all of the said effects are increased (II). The $Zn—O_{(1)}$ bond and the $O_{(1)}—H_{(1)}$ bonds are weakened, whereas in accordance with the second bond-length variation rule the bonds $N—H_{(1)}$ and $O_{(2)}—H_{(2)}$ are strengthened (II). Eventually heterolysis takes place of the bonds $N—H_{(2)}$, $H_{(1)}—O_{(1)}$, and $O_{(1)}—Zn$, with formation of the free hydrogen carbonate ion.

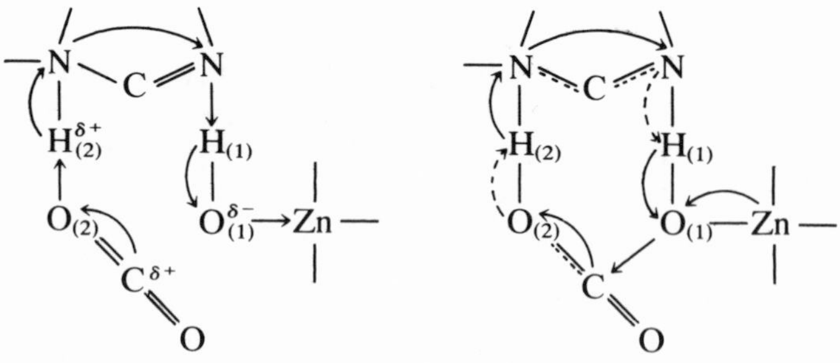

(I) (II)

In the course of the dehydration process the hydrogen carbonate ion will be attached to the zinc atom through one of its oxygen atoms (III). In this way negative charge will be piled up at the coordinated oxygen atom $O_{(1)}$ (see Section 3.2) as the bond $C—O_{(1)}$ is lengthened, the bond $O_{(2)}—C$ is shortened (second bond-length variation rule), and the bond $H_{(2)}—O_{(2)}$ is lengthened (III). This leads to the donor–acceptor interaction $N\rightarrow H_{(2)}$, which induces lengthening of the bond $N—H_{(1)}$ and hence increase in acceptor property at $H_{(1)}$ (III). The donor–acceptor interaction $O_{(1)}—H_{(1)}$ follows in due course, by which all of the said effects are magnified (IV), e.g., further lengthening of $H_{(1)}—N$, $H_{(2)}—O_{(2)}$, and $C—O_{(1)}$, whereas the bonds $N—H_{(2)}$ and $O_{(2)}—C$ are shortened. At the same time the bond $Zn—O_{(1)}$ is lengthened. Eventually heterolysis of the $H_{(2)}—O_{(2)}$, $C—O_{(1)}$, and $Zn—O_{(1)}$ bonds takes place with formation of carbon dioxide.

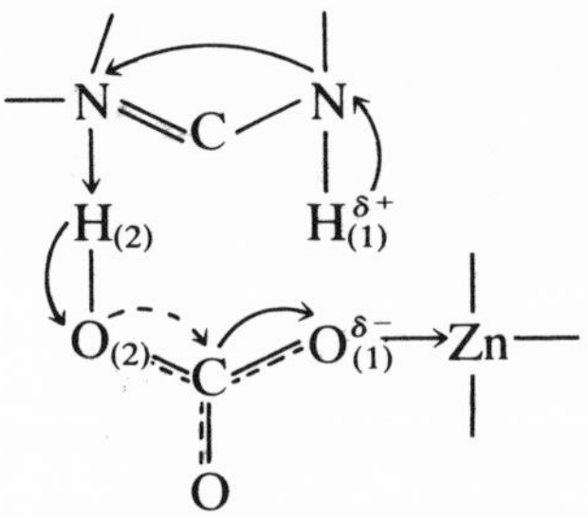

(III) 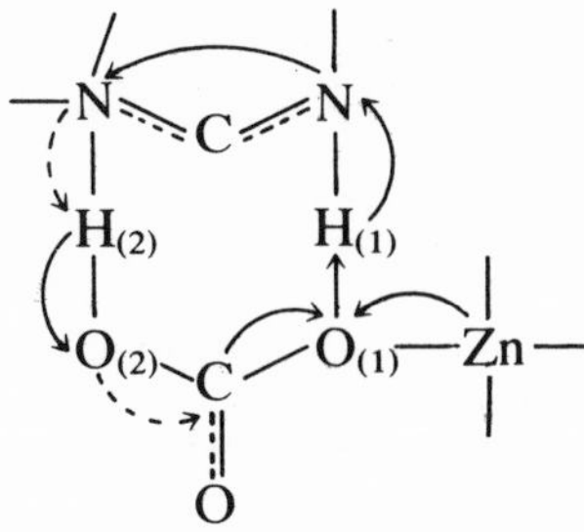 (IV)

The mechanism involves oxygen coordination at the zinc atom, analogous to cobalt anhydrase, where the binding of the bicarbonate ion to the metal has been established by ^{13}C NMR measurements (Yeagle *et al.*, 1975). Two binding sites have been identified, the same as has been found for coordination of the acetate ion (Bertini *et al.*, 1976). The function of carbonic anhydrase is inhibited by anions such as azide, cyanide, or cyanate ions, which bind strongly to the zinc atom. Likewise, sulfonamides (Coleman, 1973) and acetate ions (Lanir and Navon, 1974) are known to act as inhibitors and become bonded to the metal atom.

17.5. Corrinoids

The study of vitamin B_{12} chemistry is the quintessence of the interaction between inorganic chemistry and biochemistry. It demonstrates the intimate relationships between coordinating and redox interactions.

According to the results of structure determinations on corrinoids, the cobalt atom lies approximately in the plane of the four nitrogen atoms of the corrin ligands, which are macrocyclic monoanionic ligands (Hill, 1973). The unique structural feature is the presence of a cobalt–carbon bond with the carbon atom in an axial position in the distorted octahedron (Lenhert, 1968).

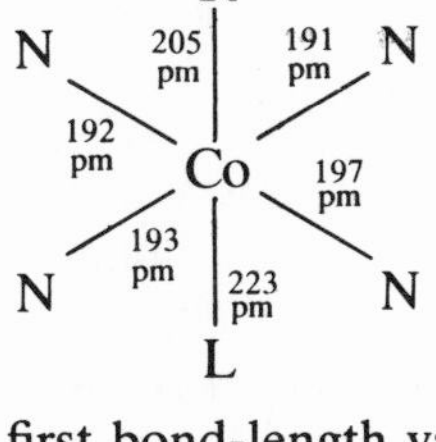

In agreement with the first bond-length variation rule, the nature of the ligand L has an influence on the length and hence on the properties of the other Co–ligand bonds. While the equatorial bonds are hardly affected, the Co—C bond is lengthened as the strength of L is increased. Hence the R group becomes more carbanionoid in character, the stronger the ligand L is bonded to the cobalt. The formation constants of cobalt(III) corrinoids with polarizable ligands, such as carbanions, are high. The ligands L, which are *trans* to the C positions, are weakly bonded and hence they are readily exchanged.

The strength and hence the properties of the cobalt–carbon bond are greatly varied by varying either the ligands or the oxidation number. According to the first bond-length variation rule the Co—C bond will be shortened as the Co—L bond is weakened by decreasing donor strength of L. By reduction of cobalt(III) to cobalt(II), all of the bond lengths originating from the coordination center are expected to become longer as the σ-acceptor properties of Co are lowered, after having accepted an electron. Indeed the additional electron in cobalt(II) as compared to cobalt(III) has an effect on the properties not unlike that of a coordinated group (Hill *et al.*, 1969; Firth *et al.*, 1967). The effects are well pronounced for the axial bonds, which are considerably lengthened by reduction of cobalt(III), whereas the equatorial bonds are hardly affected. This may be due to a certain amount of π-back-donation from Co(II) toward the equatorial N atoms: the decrease in σ-acceptor properties of cobalt being compensated for by increase in π-donor properties. The small effects due to the nature of the equatorial ligands can be seen from the comparison of the cobalt(III)–cobalt(II) redox potentials of complexes having Schiff bases as equatorial ligands. The redox potentials are slightly shifted to more negative values as the donor properties of the said ligands are increased (Costa *et al.*, 1972).

For cobalt(II) corrinoids, disproportionation occurs more readily than homolytic Co—C bond fission:

$$2\text{Co(II)—R} \rightleftharpoons \text{Co(III)—R} + \text{Co(I)—R}$$

Cobalt(I) corrinoids are square planar, with the axial ligands very weakly bonded, if at all. The cobalt(II)–cobalt(I) redox potentials are only slightly affected by the properties of the equatorial ligands, which may be considered to be bonded by π-back-donation. The most striking property of Co(I) is that of a powerful donor. The donor properties of the various cobalt(I) complexes follow the order of the Co(II)—Co(I) redox potential (Schrauzer and Deutsch, 1969; Costa *et al.*, 1969). Hence cobalt(I) complexes are capable of accepting carbonium ions rather than carbanions.

The different properties of cobalt–carbon bonds in the various oxidation states of cobalt are indeed decisive for the reactions of vitamin B_{12}. These reactions involve a rearrangement in which an organic group R and hydrogen exchange place. The transfer of an alkyl group utilizes cobalt(III), which accepts this group readily. Its transfer from the cobalt easily takes place after reduction of cobalt into the +1 oxidation state (Costa, 1971). The reactivities, which are further influenced by the ligands, have been summarized as follows (Costa, 1972): Co(III)—R: carbanionoid = donor, attacked by acceptors; Co(II)—R: radical, susceptible to homolytic bond cleavage, and capable of disproportionation; Co(I)—R: carbonium ionoid = acceptor, attacked by donors.

17.6. Activation of Phosphate Transfer

One of the key roles of phosphates in biology is that they are involved in the transfer of energy. In higher organisms, energy is produced through the reduction of oxygen, stored in adenosine triphosphate (ATP), through direct synthesis from adenosine diphosphate (ADP) and orthophosphate. Energy utilization then proceeds by the transfer of ATP phosphoryl groups to the acceptors responsible for muscle contraction, for ion transport across membranes against a concentration gradient, and for various other biosynthetic pathways (Lehninger, 1965; Racker, 1965). Finally, the transferred phosphoryl group is hydrolyzed to orthophosphate and made available for resynthesis of ATP (Spiro, 1973).

The role of metal ions in phosphate transfer has been considered from the point of view of donor–acceptor interactions. The metal ion acts as an acceptor toward one of the oxygen atoms of an orthophosphate group, so that the adjacent P—O bond is weakened and the fractional positive charge at the phosphorus atom is increased. Yet, in the absence of enzymes, the metal ions are unimpressive in this role (Bender, 1963). An impressive proliferation of iron–sulfur centers is one of the new features in biological electron transport (Buchanan and Arnon, 1970; Malkin and Bearden, 1971; Hall *et al.*, 1975). All of the iron–sulfur proteins contain iron atoms bonded to sulfur atoms—both cysteine sulfurs from the protein, and (except in rubredoxin) sulfur-atom bridges between iron centers. The 2Fe ferredoxin has two iron centers:

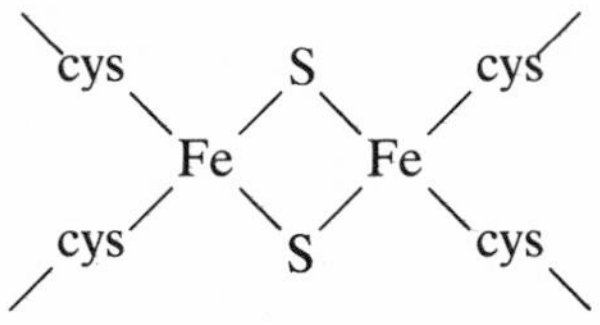

```
 \cys       S       cys/
     \     / \     /
      Fe      Fe
     /  \    /  \
  cys    S      cys
 /                 \
```

Unfortunately no accurate X-ray data are available. The indications are that the Fe—S bond lengths are not identical.

We may try to answer the following questions: (i) Why is the orthophosphate unit attacked by a given acceptor preferentially to the diphosphate? (ii) In what ways are iron–sulfur centers more powerful than other metal-ion proteins?

(i) The P—O bonds in orthophosphate are longer than the terminal bonds in diphosphate, an effect which is analogous to the structural differences between chromate and dichromate ions, discussed in Section 15.1. This means that the donor properties of the oxygen atoms in orthophosphate are greater than in diphosphate and hence the donor–acceptor interaction is stronger in the first case. The greater bond length in orthophosphate implies that the bond is weaker and more readily lengthened, so that the fractional positive charge at the phosphorus atom is increased to a greater extent than in diphosphate, where an increase in positive fractional charge at the P atom is partly compensated for by withdrawing electron density from the bridging P—O—P bonds.

(ii) The surprising acceptor properties of iron–sulfur centers may be considered as due to the conjugated π-back-bonding system. Since the sulfur atoms are strongly bonded to the iron atoms by back donation, the fractional charge at each iron atom is low. Iron–sulfur proteins have very low redox potentials, in the range of -0.2 to -0.5 V. Hence, poor acceptor properties would be expected. However, a donor–acceptor interaction leads to a considerable charge-density rearrangement within the acceptor unit. The exercise of the acceptor function at $Fe_{(1)}$ initiates an electron transport process, by which any loss of fractional positive charge of the acceptor atom is overcompensated for by enhanced back donation toward the sulfur atoms: The spillover effect at the iron atom is considerable (see Sections 3.1 and 3.4). In this way the fractional positive charge at $Fe_{(1)}$ is increased and so is its acceptor strength. The electron transport takes place to other parts within the acceptor unit, such as to $Fe_{(2)}$. At this coordination center the final charge-density distribution will be different from that at $Fe_{(1)}$, which has attained a higher coordination number than $Fe_{(2)}$: The fractional positive charge at the latter will be smaller than at the former and hence there will be differences in chemical properties, such as redox behavior. The nonequivalence of the iron atoms may explain the observations by M. Gutmann *et al.* (1971), who found, while using beef-heart inner-mitochondrial membrane preparations, that under certain conditions one center was reoxidized by the respirating chain, whereas the other center remained reduced.

References

Aberdam, D., Baudoing, R., and Gaubert, C. (1975), *Surface Sci.* **52**, 125.
Aberdam, D., Baudoing, R., and Gaubert, C. (1976), *Surface Sci.* **57**, 715.
Achran, G. E., and Shore, S. G. (1965), *Inorg. Chem.* **4**, 125.
Adamson, A. W. (1958), *J. Am. Chem. Soc.* **80**, 3183.
Addison, C. C. (1967), *Endeavour* **26**, 91.
Agami, C., and Prévost, C. (1966), *C.R. Hebd. Séances Acad. Sci. Sér. C* **263**, 304.
Akimoto, S., and Syono, Y. (1969), *J. Geophys. Res.* **74**, 1653.
Akišin, P. A., and Rambidi, N. G. (1960), *Zhur. Neorg. Khim. SSSR* **5**, 23.
Alberghina, G., Arcoria, A., Fisichella, S., and Scarlata, G. (1973), *Gazz. Chim. Ital.* **103**, 319.
Alder, B. J., and Christian, R. H. (1961), *Phys. Rev. Lett.* **7**, 367.
Alder, B. J., Vaisnys, J. R., and Jura, G. (1959), *Phys. Chem. Solids* **11**, 182.
Alexander, R., Parker, A. J., Sharp, J. H., and Waghorne, W. E. (1972), *J. Am. Chem. Soc.* **94**, 1148.
Alexander, R., Owensby, D. A., Parker, A. J., and Waghorne, W. E. (1974), *Aust. J. Chem.* **127**, 933.
Alfenaar, M. (1975), *J. Phys. Chem.* **79**, 2200.
Alff, M. (1976), Ph.D. dissertation, Universität München.
Allen, A. D. (1971), *Adv. Chem. Ser. Amer. Chem. Soc.* **100**, 79.
Allerhand, A., and Schleyer, R. (1963), *J. Am. Chem. Soc.* **85**, 371.
Almhöf, J. (1973), *Chem. Scripta* **3**, 73.
Ames, D. P., and Sears, P. G. (1955), *J. Phys. Chem.* **59**, 16.
Amphlett, C. B., Mullinger, L. W., and Thomas, L. F. (1948), *Trans. Faraday Soc.* **44**, 927.
Anbar, M., and Hart, E. J. (1965), *J. Phys. Chem.* **69**, 973.
Andréŭ, P., and Madrid, J. (1973), *Acta Cientif. Venezol. Suppl. 2*, **24**, 169.
Andréŭ, P., Noller, H., and Paez, M. (1969), *Anal. Real. Soc. Espan. Fisica y Quimica, Ser. M. Quim.* **65**, 921.
Andrews, J. W., and Keefer, R. M. (1964), *Molecular Complexes in Organic Chemistry*, Holden–Day, Inc., San Francisco.
Angerman, N. S., and Jordan, R. B. (1969), *Inorg. Chem.* **8**, 2579.
Ashby, E. C., and Smith, M. B. (1964), *J. Am. Chem. Soc.* **86**, 4363.
Attig, R., and Williams, J. M. (1976), *Angew. Chem.* **88**, 507.
Aylett, B. J. (1968), *Adv. Inorg. Nucl. Chem.* **11**, 249.

Aylett, B. J., and Campbell, J. M. (1969), *J. Chem. Soc. A*, 1920.

Aylett, B. J., Emeléus, H. J., and Maddock, A. G. (1955), *J. Inorg. Nucl. Chem.* **1**, 187.

Baaz, M., Gutmann, V., and Kunze, O. (1962), *Mh. Chem.* **93**, 1142, 1162.

Baker, A. W., and Harris, G. H. (1960), *J. Am. Chem. Soc.* **82**, 1923.

Balasabramanian, A., and Rao, C. N. R. (1962), *Spectrochim. Acta* **18**, 1337.

Baniel, A., and Blumberg, R. (1968), in: *Fertilizer Science and Technology*, Vol. 1, Part II (A. V. Slack, ed.), Marcel Dekker, New York, p. 889.

Baraton, M. I., Gerbier, J., and Besnainou, S. (1973), *C.R. Acad. Sci. Paris B* **276**, 797.

Barnett, J. D., Vern, E. B., and Hall, H. T. (1966), *J. Appl. Phys.* **37**, 875.

Barraclough, C. C., Lewis, J., and Nyholm, R. S. (1961), *J. Chem. Soc.*, 2582.

Barthel, J. (1976), private communication.

Barthel, J., Schmithals, F., and Behret, H. (1970), *Z. Phys. Chem.* (*Frankfurt*) **71**, 115.

Basett, W. A., and Takahashi, T. (1965), *Amer. Mineral.* **50**, 1576.

Basolo, F., and Pearson, R. G. (1958), *Mechanisms of Inorganic Reactions*, Wiley, New York.

Basolo, F., Matoush, W. R., and Pearson, R. G. (1956), *J. Am. Chem. Soc.* **78**, 4883.

Basolo, F., Henry, P. M., and Pearson, R. G. (1957), *J. Am. Chem. Soc.* **79**, 5379, 5382.

Bauer, D., and Foucault, A. (1976), *J. Electroanal. Chem.* **67**, 19.

Bauer, E. (1958), *Z. Krist.* **110**, 372.

Bauer, W. H. (1971), *Amer. Mineral.* **56**, 1573.

Beattie, I. R., and Parrett, F. W. (1966), *J. Chem. Soc. A* 1784.

Beattie, I. R., Gilson, T., Webster, M., and McQuillan, G. P. (1964), *J. Chem. Soc.* 238.

Beck, M. T. (1970), *Chemistry of Complex Equilibria*, Van Nostrand–Reinhold, London.

Beck, W., and Lottes, K. (1964), *Z. Naturf.* **19b**, 987.

Beck, W., and Nitzschmann, R. E. (1962), *Z. Naturf.* **17b**, 577.

Bell, R. A., Christoph, G. G., Fronczek, T. R., and Marsh, E. R. (1975), *Science* **190**, 151.

Bellamy, L. J. (1958), *The Infrared Spectra of Complex Molecules* (2nd ed.), Methuen, London, p. 382.

Bellamy, L. J., and Pace, R. J. (1963), *Spectrochim. Acta* **19**, 1831.

Beltrame, P., Biate, G., Lloyd, D. J., Parker, A. J., Ruane, M., and Winstein, S. (1972), *J. Am. Chem. Soc.* **94**, 2240.

Bender, M. L. (1963), *Adv. Chem. Ser. Am. Chem. Soc.* **37**, 19.

Bene, J. D., and Pople, J. A. (1970), *J. Chem. Phys.* **52**, 4858.

Benesi, H. A., and Jones, A. C. (1959), *J. Phys. Chem.* **63**, 179.

Benkeser, R. A., Foley, K. M., Brutzner, J. B., and Smith, E. W. (1970), *J. Am. Chem. Soc.* **92**, 697.

Bennett, G. M., and Willis, G. H. (1929), *J. Chem. Soc.*, 256.

Bennetto, H. P., and Caldin, E. F. (1971), *J. Chem. Soc.* 2198.

Benson, G. C., Freeman, P. I., and Dempsey, E. (1961), *Adv. Chem. Ser. Am. Chem. Soc.* **33**, 26.

Berkowitz, J., and Chupka, W. A. (1958), *J. Chem. Phys.* **29**, 653, 1386.

Bernal, J. D. (1964), *Proc. Roy. Soc. London A* **280**, 299.

Bernal, J. D., and Fowler, R. H. (1933), *J. Chem. Phys.* **1**, 515.

Bernstein, S. C. (1970), *J. Am. Chem. Soc.* **92**, 699.

Berry, C. R. (1952), *Phys. Rev.* **88**, 596.

Bertini, I., Luchinat, C., and Scozzafava, A. (1976), *Biochim. Biophys. Acta* **452**, 239.

Bhattacharya, D. N., Lee, C. L., Smid, J., and Szwarc, M. (1963), J. Phys. Chem. **69**, 612.

Bikales, N. M., and Becker, E. I. (1962), *Can. J. Chem.* **41**, 1329.

Bjerrum, N. (1926), *Kgl. Danske Vidensk. Math. Fysike Medd.* **9**(7), 1.

Bjerrum, N., and Larsson, E. (1927), *Z. Physik. Chem.* **127**, 358.

Bobtelsky, M., Spiegler, K. S. (1949), *J. Chem. Soc.*, 143.

Boehm, H. P. (1966), *Angew. Chem.* **78**, 618.
Bolhuis, F. van, Koster, P. B., and Migchelsen, T. (1967), *Acta Cryst.* **23**, 90.
Bolton, W., and Perutz, M. F. (1970), *Nature* **228**, 557.
Booij, H. L. (1949), in: *Colloid Science* (H. R. Kruyt, ed.), Vol. 1, Elsevier, Amsterdam, p. 681.
Born, M. (1920), *Z. Physik* **1**, 45.
Borucka, J., and Kecki, Z. (1971), *Roczniki Chem.* **45**, 2133.
Borucka, J., Sadleyj, J., and Kecki, Z. (1973), *Adv. Mol. Relax. Processes* **5**, 253.
Boshard, H. H., and Zollinger, H. (1959), *Angew. Chem.* **71**, 375.
Boswell, F. W. C. (1951), *Proc. Phys. Soc.* **A64**, 465.
Bowmaker, G. A., and Hacobian, S. (1968), *J. Austral. Chem.* **21**, 551.
Bowmaker, G. A., and Hacobian, S. (1969), *J. Austral. Chem.* **22**, 2047.
Brackman, W. (1949), *Rec. Trav. Chim.* **68**, 147.
Brandon, J. K., and Brown, I. D. (1968), *Can. J. Chem.* **46**, 933.
Brennan, J. F., Schaeffer, R., Davison, A., and Wreford, S. S. (1973), *Chem. Comm.* 354.
Brice, V. T., and Shore, S. G. (1970), *Chem. Comm.*, 1312.
Brice, V. T., and Shore, S. G. (1973), *Inorg. Chem.* **12**, 309.
Brice, V. T., and Shore, S. G. (1975), *J. Chem. Soc. Dalton Trans.*, 334.
Briegleb, G. (1961), *Elektronen-Donator-Acceptor Komplexe*, Springer-Verlag, Berlin/Göttingen/Heidelberg.
Brønsted, J. N. (1923), *Rec. Trav. Chim. Pays-Bas* **42**, 718.
Brown, I. D., and Shannon, R. D. (1973), *Acta Cryst.* **A29**, 266.
Brubaker, G. L., Denniston, M. L., Shore, S. G., Carter, J. C., and Swicker, F. (1970), *J. Am. Chem. Soc.* **92**, 7216.
Brunauer, S. (1965), *Pure Appl. Chem.* **4**, 293.
Buchanan, B. B., and Arnon, D. I. (1970), *Adv. Enzymol.* **33**, 119.
Buckingham, D. A. (1973), in: *Inorganic Biochemistry* (G. L. Eichhorn, ed.), Elsevier, Amsterdam, London, New York, Chapter 1.
Bukowska, J., and Kecki, Z. (1975*a*), *J. Mol. Structure* **26**, 289.
Bukowska, J., and Kecki, Z. (1975*b*), *Roczniki Chem.* **49**, 11.
Bundy, F. P. (1962), *Science* **137**, 1055.
Burg, A. B., and Kuljian, E. S. (1950), *J. Am. Chem. Soc.* **72**, 3103.
Burger, K. and Fluck, E. (1974), *Inorg. Nucl. Chem. Lett.* **10**, 171.
Burgess, J. (1973), *J. Chem. Soc.*, 825.
Bürgi, H. B., Dunitz, J. D., and Shefter, E. (1973), *J. Am. Chem. Soc.* **95**, 5065.
Bürgi, H. B., Dunitz, J. D., and Shefter, E. (1974), *Acta Cryst.* **B30**, 1517.
Burtlich, J. M., and Petersen, R. B. (1970), *J. Organomet. Chem.* **24**, C 65.
Burton, R. E., and Daly, J. (1971), *Trans. Faraday Soc.* **67**, 1219.
Busmann, E. (1961), *Z. Anorg. Allg. Chem.* **313**, 90.
Caldin, E. F. (1964), *Fast Reactions in Solution*, Blackwell, London.
Campbell-Ferguson, J. J., and Ebsworth, E. A. V. (1965), *Chem. Ind.* (*London*) 301.
Campbell-Ferguson, J. J., and Ebsworth, E. A. V. (1967), *J. Chem. Soc. A*, 705.
Candlin, J. P., and Halpern, J. (1965), *Inorg. Chem.* **4**, 766.
Candlin, J. P., Halpern, J., and Trimm, D. L. (1964), *J. Am. Chem. Soc.* **86**, 1019.
Cannon, R. D., and Gardiner, J. (1970), *J. Am. Chem. Soc.* **92**, 3800.
Carreira, L. A., and Person, W. B. (1972), *J. Am. Chem. Soc.* **94**, 1485.
Carter, H. V., McClelland, B. J., and Warhurst, E. (1960), *Trans. Faraday Soc.* **56**, 455.
Carvajal, C., Tölle, K. J., Smid, J., and Szwarc, M. (1965), *J. Am. Chem. Soc.* **87**, 5548.
Casazza, J. J., and Cefola, M. (1967), *J. Inorg. Nucl. Chem.* **29**, 2595.
Chatt, J., Dilworth, R. L., Richards, R. L., and Sanders, J. R. (1969), *Nature* **224**, 1201.

Chatt, J., Dilworth, R. L., Leigh, G. J., and Richards, R. L. (1970), *Chem. Comm.*, 955.
Chatt, J., Fay, R. C., and Richards, R. L. (1971), *J. Chem. Soc. A*, 702.
Chatterji, A. C., and Tewari, P. (1955), *Kolloid Z.* **143**, 172.
Chen, J. M., and Papageorgeopopoulos, C. A. (1970), *Surface Sci.*, 377.
Chini, P. (1968), *Inorg. Chim. Acta Rev.* **2**, 48.
Clementi, E. (1967), *J. Chem. Phys.* **46**, 3851.
Clotman, D., Van Lerberghe, D., and Zeegers-Huyskens, T. (1970), *Spectrochim. Acta* **26A**, 1621.
Coetzee, J. F., McGuire, D. K., and Hendrick, J. L. (1963), *J. Am. Chem. Soc.* **67**, 1814.
Cogley, D. R., Butler, J. N., and Grunwald, E. (1971), *J. Phys. Chem.* **75**, 1477.
Coleman, J. E. (1973), in: *Inorganic Biochemistry* (G. L. Eichhorn, ed.), Elsevier, Amsterdam, London, New York, Chapter 16.
Collman, J. P., Gagne, R. R., Reed, C. A., Halbert, T. R., Lang, G., and Robinson, W. T. (1975), *J. Am. Chem. Soc.* **97**, 1427.
Conway, B. E. (1972), *Rev. Macromol. Chem.* **7**, 113.
Cook, D. (1958), *J. Am. Chem. Soc.* **80**, 49.
Cope, F. W. (1973), *Ann. N.Y. Acad. Sci.* **204**, 416.
Corbett, J. D. (1968), *Inorg. Chem.* **7**, 198.
Corey, J. Y., and West, R. (1963), *J. Am. Chem. Soc.* **85**, 4034.
Correa, A., Hunger, M., and Noller, H. (1968), *Z. Naturf.* **23b**, 275.
Costa, G. (1971), *Exp. Suppl.* **18**, 235.
Costa, G. (1972), *Pure Appl. Chem.* **30**, 335.
Costa, G., Mestroni, G., and Savorgnani, E. L. (1969), *Inorg. Chim. Acta* **3**, 323.
Costa, G., Puxeddu, A., and Reisenhofer, E. (1972), *J. Chem. Soc. Dalton Trans.*, 1519.
Cotton, F. A. (1964), *Inorg. Chem.* **3**, 702.
Cotton, F. A., and Kraihanzel, C. S. (1962), *J. Am. Chem. Soc.* **84**, 4432.
Cotton, F. A., and Wilkinson, G. (1967), *Advanced Inorganic Chemistry*, Interscience, New York.
Cotton, F. A., Faut, O. D., and Mague, J. T. (1964), *Inorg. Chem.* **3**, 17.
Cowan, D. O., and Mosher, H. S. (1962), *J. Org. Chem.* **27**, 1.
Cox, B. G., Hedwig, G. R., Parker, A. J., and Watts, D. W. (1974), *Austral. J. Chem.* **27**, 477–501.
Cram, D. J. (1965), *Fundamentals in Carbanion Chemistry*, Academic Press, New York, London.
Cram, D. J., Mateos, J. L., Hauck, F., Langemann, A., Kopecky, K. R., Nielsen, W. D., and Allinger, J. (1959), *J. Am. Chem. Soc.* **81**, 5774.
Creighton, J. A., and Thomas, K. M. (1972), *J. Chem. Soc. Dalton Trans.*, 403.
Cross, P. C., Burnham, J., and Leighton, P. A. (1937), *J. Am. Chem. Soc.* **59**, 1134.
Curtin, D. Y., and Engelman, J. H. (1972), *J. Org. Chem.* **37**, 3439.
Cyvin, S. J., Klaboe, P., Rytter, E., and Oye, H. A. (1970), *J. Chem. Phys.* **52**, 2276.
Dash, A. C., and Nanda, R. K. (1973), *Inorg. Chem.* **12**, 2024.
Davies, K. H., and Espenson, J. H. (1969), *J. Am. Chem. Soc.* **91**, 3093.
Davison, A., Traficante, D. D., and Wreford, S. S. (1972), *Chem. Comm.*, 1165.
Davison, A., Traficante, D. D., and Wreford, S. S. (1974), *J. Am. Chem. Soc.* **96**, 2862.
Delchar, T., and Tompkins, F. C. (1966), *Disc. Faraday Soc.* **41**, 72.
Dellwaulle, M. L. (1955), *Bull. Soc. Chim. France* **10**, 1204.
Delpuech, J. J. (1965), *Tetrahedron Lett.* **26**, 2111.
Demuth, J. E., Marcus, P. M., and Jepsen, D. W. (1975), *Phys. Rev.* **B11**, 1461.
Dennard, A. E., and Williams, R. J. P. (1966*a*), in: *Transition Metal Chemistry* (R. L. Carlin, ed.), Vol. 2, Marcel Dekker, New York.

Dennard, A. E., and Williams, R. J. P. (1966*b*), *J. Chem. Soc. A*, 812.
de Planta, T., Ghez, R., and Piuz, F. (1964), *Helv. Phys. Acta* **37**, 74.
Deryagin, B. V. (1970), *Sci. Am.* **223**, 52, 69.
Desai, A. G., Dodgen, H. W., and Hunt, J. P. (1970), *J. Am. Chem. Soc.* **92**, 798.
Desnoyers, J. E., and Jolicoeur, C. (1969), in: *Modern Aspects of Electrochemistry* (J. O'Bockris and B. E. Conway, eds.), Vol. 5, Plenum Press, New York, Chapter 1.
Dessy, R. E., Handler, G. S., Wotiz, J. H., and Hollingworth, C. A. (1957), *J. Am. Chem. Soc.* **79**, 3476.
Dickert, F., and Hoffmann, H. (1971), *Ber. Bunsenges. Phys. Chem.* **75**, 1320.
Dickert, F., Hoffmann, H., and Jaenicke, W. (1970), *Ber. Busenges, Phys. Chem.* **74**, 500.
Diebler, H., and Taube, H. (1965), *Inorg. Chem.* **4**, 1029.
Diercksen, G. H. F., and Kraemer, W. P. (1970), *Chem. Phys. Lett.* **5**, 570.
Diggle, J. W., and Parker, A. J. (1973), *Electrochim. Acta* **18**, 975.
Dimroth, K. (1953), *Marburger Sitzungsberichte* **76** (3), 3.
Dimroth, K., Reichhardt, C., Siepmann, T., and Bohlmann, F. (1963), *Ann. Chem.* **661**, 1.
Dixon, J. A., Gwinner, P. A., and Lini, D. C. (1965), *J. Am. Chem. Soc.* **87**, 3276.
Donoghue, J. T., and Drago, R. S. (1962), *Inorg. Chem.* **1**, 866.
Dostal, K., and Toučin, J. (1974), private communication.
Drabkin, O. L. (1950), *J. Biol. Chem.* **185**, 231.
Drago, R. S. and Wayland, B. B. (1965), *J. Am. Chem. Soc.* **87**, 3751.
Dragulescu, C., Petrovici, M., and Lupu, I. (1974), *Mh. Chem.* **105**, 1170, 1176, 1184.
Drickamer, H. G., and Frank, C. W. (1973), *Electronic Transitions and the High Pressure Chemistry and Physics of Solids*, Chapman and Hall, London.
Driessen, W. L., and Groeneveld, W. L. (1968), *Rev. Trav. Chim.* **87**, 786.
Driessen, W. L., and Groeneveld, W. L. (1969), *Rec. Trav. Chim.* **88**, 620.
Drost-Hansen, W. (1969), *Ind. Eng. Chem.* **61** (11), 10.
Duschek, O., and Gutmann, V. (1972), *Z. Anorg. Allg. Chem.* **394**, 243.
Duschek, O., and Gutmann, V. (1973*a*), *Mh. Chem.* **104**, 990.
Duschek, O., and Gutmann, V. (1973*b*), *Mh. Chem.* **104**, 1259.
Duschek, O., Gutmann, V., and Rechberger, P. (1974), *Mh. Chem.* **105**, 62.
Duterme, P., Clerbaux, T., Zeegers-Huyskens, T., and Huyskens, P. (1968), *J. Chim. Phys.* **65**, 1268.
Eagland, D. (1975), in: *Water, a Comprehensive Treatise* (F. Franks, ed.), Vol. 5, p. 1, Plenum Press, New York.
Eaton, G. R., and Lipscomb, W. N. (1969), *NMR-Studies of Boron Hydrides and Related Compounds*, Benjamin, New York.
Ebel, H. F., and Schneider, R. (1965), *Angew. Chem.* **77**, 914; *Int. Ed.* **4**, 878.
Ebsworth, E., and Weil, J. (1959), *J. Phys. Chem.* **63**, 1890.
Edgell, W. F., and Barbetta, A. (1974), *J. Am. Chem. Soc.* **96**, 415.
Edgell, W. F., and Lyford, J. (1971), *J. Am. Chem. Soc.* **93**, 6407.
Edgell, W. F., Watts, A. T., Lyford, J., and Risen, W. (1966), *J. Am. Chem. Soc.* **88**, 1815.
Edgell, W. F., Lyford, J., Wright, R., Risen, W., and Watts, A. T. (1970), *J. Am. Chem. Soc.* **92**, 2240.
Edgell, W. F., Lyford, J., Barbetta, A., and Lose, C. I. (1971), *J. Am. Chem. Soc.* **93**, 6403.
Eigen, M. (1963), *Pure Appl. Chem.* **6**, 97.
Eigen, M., and Wilkins, R. G. (1965), in: *Mechanisms in Inorganic Reactions* (J. Kleinberg, ed.), Advances in Chemistry Series, American Chemical Society, Vol. 49, p. 55.
Eisenberg, O., and Kauzmann, W. (1969), *The Structure and Properties of Water*, Clarendon, Oxford.

Elving, P. J., and Zemel, B. (1957), *J. Am. Chem. Soc.* **79**, 5855.

Endicott, J. F., and Taube, H. (1964), *J. Am. Chem. Soc.* **86**, 1686.

Englefeld, P., and Nemethy, G. (1973), *Ann. N.Y. Acad. Sci.* **204**, 77.

Erlich, R. H., and Popov, A. I. (1971), *J. Am. Chem. Soc.* **93**, 5620.

Erlich, R. H., Roach, E., and Popov, A. I. (1970), *J. Am. Chem. Soc.* **92**, 4989.

Erlich, R. H., Greenberg, M. S., and Popov, A. I. (1973), *Spectrochim. Acta* **29A**, 543.

Ertl, G. (1967), *Surface Sci.* **6**, 208.

Espenson, J. H., and Birk, J. P. (1965), *Inorg. Chem.* **4**, 527.

Fannin, A. A., King, L. A., and Seegmiller, D. W. (1972), *J. Electrochem. Soc.* **119**, 801.

Fauwarque, J., and Fauwarque, J. F. (1966), *Compt. Rend. Acad. Sci.* **263**, 488.

Fawcett, W. R., and Krygowski, T. M. (1975), *Austral. J. Chem.* **28**, 2115.

Fehrmann, K. R., and Garner, C. S. (1961), *J. Am. Chem. Soc.* **83**, 1276.

Felix, N., and Huyskens, P. (1975), *J. Phys. Chem.* **79**, 2316.

Felix, N., Neven, L., and Huyskens, P. (1975), in: *Symposium on Ions and Ion Pairs*, Leuven, Belgium.

Ferguson, B. J. (1956), *J. Chem. Phys.* **24**, 1263.

Ferraris, G., and Franchini-Angela, M. (1972), *Acta Cryst.* **B28**, 3572.

Finch, G. I., and Wilman, H. (1937), *Erg. Exakt. Naturwiss.* **16**, 418.

Firth, R. A., Hill, H. A. O., Mann, B. E., Pratt, J. M., and Thorp, R. G. (1967), *Chem. Comm.* 1013.

Fitzgerald, W. R., Parker, A. J., and Watts, D. W. (1968), *J. Am. Chem. Soc.* **90**, 5744.

Flood, H., and Förland, T. (1947), *Acta Chem. Scand.* **1**, 592, 781.

Flood, H., and Muan, A. (1950), *Acta Chem. Scand.* **4**, 364.

Flood, H., Förland, T., and Roald, B. (1947), *Acta Chem. Scand.* **1**, 790.

Foster, R. (1973), *Molecular Complexes*, Vol. 1, Elek Science, London.

Frank, H. S., and Quist, A. S. (1961), *J. Chem. Phys.* **34**, 605.

Frank, H. S., and Wen, W. Y. (1957), *Disc. Faraday Soc.* **24**, 133.

Frankel, L. S., Stengle, T. R., and Langford, C. H. (1965), *Chem. Comm.*, 393.

Frankel, L. S., Langford, C. H., and Stengle, T. R. (1970), *J. Phys. Chem.* **74**, 1376.

Franks, F. (1967), *Physicochemical Processes in Mixed Aqueous Solvents*, Heinemann, London.

Fratiello, A., and Cristie, E. G. (1964), *Trans. Faraday Soc.* **61**, 306.

Fratiello, A., and Douglass, D. C. (1963*a*), *J. Mol. Spectrosc.* **11**, 465.

Fratiello, A., and Douglass, D. C. (1963*b*), *J. Chem. Phys.* **39**, 2017.

Fratiello, A., and Miller, D. (1965), *J. Chem. Phys.* **42**, 796.

Fridborg, K., Kannan, K. K., Liljas, A., Lundin, J., Strandberg, B., Strandberg, R., Tilander, B., and Wrien, G. (1967), *J. Mol. Biol.* **25**, 505.

Gaines, D. F. (1973), *Accounts Chem. Res.* **6**, 416.

Gaines, D. F., and Iorns, T. V. (1967), *J. Am. Chem. Soc.* **89**, 4249.

Gaizer, F., and Beck, M. T. (1967), *J. Inorg. Nucl. Chem.* **29**, 21.

Galich, P. N., Golushenko, I. T., Gutyrya, V. S., and Il, V. G. (1965), *Dokl. Akad. Nauk SSSR, Ser. Khim.* **161** (**3**), 627.

Galin, C. A., Kiselev, A. V., and Lygin, V. I. (1962), *Russ. J. Phys. Chem.* (English translation) **36**, 951.

Garvie, R. C. (1966), *Material Res. Bull.* **1**, 161.

Garvie, R. C. (1967), *Material Res. Bull.* **2**, 897.

Gasvick, D., and Haim, A. (1971), *J. Am. Chem. Soc.* **93**, 7347.

Gažo, J., Bersuker, I. B., Garaj, J., Kabešová, M., Kohout, J., Langfelderova, H., Melnik, M., Serator, M., and Valach, F. (1976), *Coord. Chem. Revs.* **19**, 253.

Geier, G. (1965), *Ber. Bunsenges. Phys. Chem.* **69**, 617.

Gerlach, J. K., Gock, E. D., and Gosh, S. K. (1973), in: *International Symposium on Hydrometallurgy* (D. J. Evans and R. S. Shoemaker, eds.), University of Chicago Press, Chicago, p. 403.
Gilchrist, A., and Sutton, L. E. (1952), *J. Phys. Chem.* **56**, 319.
Gillespie, R. J., and Morton, M. J. (1971), *Quart. Revs.* **25**, 533.
Gillespie, R. J., and Passmore, J. (1975), in: *MTP International Review of Science—Inorganic Chemistry Series 2*, Vol. 3 (V. Gutmann, ed.), Butterworth, London, p. 121.
Gilman, H., and Harris, S. (1931), *Rev. Trav. Chim.* **50**, 1052.
Glaeser, H. H., Dodgen, H. W., and Hunt, J. P. (1965), *Inorg. Chem.* **4**, 1063.
Gold, V. (1956), *J. Chem. Soc.*, 4633.
Goldschmidt, V. M. (1958), *Geochemistry* (A. Muir, ed.), Clarendon Press, Oxford, p. 106.
Goodgame, M., Goodgame, D. M. L., and Cotton, F. A. (1962), *Inorg. Chem.* **1**, 233.
Gramstad, T., and Fuglerik, W. F. (1962), *Acta Chem. Scand.* **16**, 1369.
Greenberg, M. S., and Popov, A. I. (1975), *Spectrochim. Acta* **31A**, 697.
Greenwood, N. N., and Travers, N. F. (1966), *Inorg. Nucl. Chem. Lett.* **2**, 169.
Greenwood, N. N., and Travers, N. F. (1967), *J. Chem. Soc. A*, 880.
Greenwood, N. N., McGinnety, J. A., and Owen, J. D. (1972), *J. Chem. Soc. Dalton Trans.*, 989.
Grimley, T. B. (1967), *Proc. Phys. Soc.* **90**, 751.
Gritzner, G. (1975), *Mh. Chem.* **106**, 947.
Gritzner, G. (1976), *Mh. Chem.* **107**, 1499.
Gritzner, G. (1977*a*), *Inorg. Chim. Acta* **24**, 5.
Gritzner, G. (1977*b*), *Mh. Chem.* **108**, 271.
Gritzner, G., Gutmann, V., and Schmid, R. (1968), *Electrochim. Acta* **13**, 919.
Gritzner, G., Danksagmüller, K., and Gutmann, V. (1976*a*), *J. Electroanal. Chem.* **72**, 177.
Gritzner, G., Rechberger, P., and Gutmann, V. (1976*b*), *Mh. Chem.* **107**, 809.
Gritzner, G., Rechberger, P., and Gutmann, V. (1977), *J. Electroanal. Chem.* **75**, 739.
Grunwald, E., and Winstein, S. (1948), *J. Am. Chem. Soc.* **70**, 846.
Grunwald, E., Baughnan, G., and Kohnstamm, G. (1960), *J. Am. Chem. Soc.* **82**, 5801.
Guggenberger, L. J., and Rundle, R. E. (1964), *J. Am. Chem. Soc.* **86**, 5344.
Gut, R. (1960), *Helv. Chim. Acta* **43**, 830.
Gutmann, M., Singer, T. P., and Beinert, H. (1971), *Biochem. Biophys. Res. Commun.* **44**, 1572.
Gutmann, V. (1956), *Svensk Kem. Tidskr.* **68**, 1.
Gutmann, V. (1968), *Coordination Chemistry in Nonaqueous Solutions*, Springer-Verlag, Vienna, New York.
Gutmann, V. (1969), *Chimia* **23**, 265.
Gutmann, V. (1970), *Angew. Chem.* **81**, 858; *Int. Ed.* **9**, 843.
Gutmann, V. (1971*a*), *Chemische Funktionslehre*, Springer-Verlag, Vienna, New York, 1971.
Gutmann, V. (1971*b*), *Chem. in Britain* **7**, 102.
Gutmann, V. (1972), *Topics in Current Chem.* **27**, 59.
Gutmann, V. (1973), *Structure and Bonding* **15**, 141.
Gutmann, V. (1975), *Coord. Chem. Revs.* **15**, 207.
Gutmann, V. (1976*a*), *Coord. Chem. Revs.* **18**, 225.
Gutmann, V. (1976*b*), *Electrochim. Acta* **21**, 661.
Gutmann, V. (1976*c*), *Öst. Chem. Ztschr.* **4**, 1; **6**, 1.
Gutmann, V. (1977*a*), *Rev. Chim. Roum.* **22**, 679.
Gutmann, V. (1977*b*), *Mh. Chem.* **108**, 429.
Gutmann, V. (1977*c*), *Chimia* **31**, 1.
Gutmann, V., and Kunze, O. (1963), *Mh. Chem.* **94**, 786.

Gutmann, V., and Lindqvist, I. (1954), *Z. Physik. Chem.* (*Leipzig*) **203**, 250.
Gutmann, V., and Mayer, U. (1969), *Mh. Chem.* **100**, 2048.
Gutmann, V., and Mayer, U. (1971), *Structure and Bonding* **10**, 127.
Gutmann, V., and Mayer, H. (1976), *Structure and Bonding* **31**, 49.
Gutmann, V., and Noller, H. (1971), *Mh. Chem.* **102**, 20.
Gutmann, V., and Peychal-Heling, G. (1969), *Mh. Chem.* **100**, 1423.
Gutmann, V., and Schmid, R. (1969), *Mh. Chem.* **100**, 2113.
Gutmann, V., and Schmid, R. (1971), *Mh. Chem.* **102**, 1217.
Gutmann, V., and Schmid, R. (1974), *Coord. Chem. Revs.* **12**, 263.
Gutmann, V., and Schmidt, H. (1974), *Mh. Chem.* **105**, 653.
Gutmann, V., and Wegleitner, R. H. (1970), *Mh. Chem.* **101**, 1532.
Gutmann, V., and Weiss, A. (1969), *Mh. Chem.* **100**, 2104.
Gutmann, V., and Wychera, E. (1966), *Inorg. Nucl. Chem. Lett.* **2**, 257.
Gutmann, V., Weiss, A., and Kerber, W. (1969), *Mh. Chem.* **100**, 2096.
Gutmann, V., Beran, R., and Kerber, W. (1972), *Mh. Chem.* **103**, 764.
Gutmann, V., Danksagmüller, K., and Duschek, O. (1974*a*), *Z. Physik. Chem.* (*Frankfurt*) **92**, 199.
Gutmann, V., Mayer, U., and Krist, R. (1974*b*), *Synth. Inorg. Metalorg. Chem.* **4**, 523.
Gutmann, V., Gritzner, G., and Danksagmüller, K. (1976), *Inorg. Chim. Acta* **17**, 81.
Gutmann, V., Plattner, E., and Resch, G. (1977), *Chimia*, in press.
Haberfield, P., Clayman, L., and Cooper, J. S. (1969), *J. Am. Chem. Soc.* **91**, 787.
Haight, G. P., Jr., Richardson, D. C., and Coburn, N. H. (1964), *Inorg. Chem.* **3**, 1777.
Haim, A., and Sutin, N. (1976), *Inorg. Chem.* **15**, 476.
Hainton, J. F., and Amis, E. S. (1967), *Chem. Revs.* **67**, 367.
Hair, M. L. (1967), *Infrared Spectroscopy in Surface Chemistry*, Marcel Dekker, New York.
Hall, H. K. (1956), *J. Am. Chem. Soc.* **78**, 271, 2717.
Hall, T. H., Barnett, J. D., and Merril, L. (1963), *Science* **139**, 111.
Hall, D. O., Rao, K. K., and Cammack, R. (1975), *Sci. Progr. Oxford* **62**, 285.
Hamann, S. D., and Linton, M. (1975), *Austral. J. Chem.* **28**, 2567.
Hamilton, W. C., and Ibers, J. A. (1968), *Hydrogen Bonding in Solids* (1968), Benjamin, New York.
Hammes, G., and Schimmel, P. R. (1967), *J. Am. Chem. Soc.* **89**, 442.
Hammonds, C. N., and Day, M. C. (1969), *J. Phys. Chem.* **73**, 1151.
Hanania, G. I. H., and Irvine, D. H. (1962), *J. Chem. Soc.*, 2750.
Hankins, D., Moskowitz, J. W., and Stillinger, F. E. (1970), *J. Chem. Phys.* **53**, 4544.
Hare, E. F., and Zisman, W. A. (1960), *J. Phys. Chem.* **64**, 519.
Harkins, T. R., and Freiser, H. (1956), *J. Am. Chem. Soc.* **78**, 1143.
Harrison, D. F., Weissberger, E., and Taube, H. (1968), *Science* **159**, 320.
Harrison, W. D., Gill, J. B., and Goodall, D. C. (1976), in: *Proceedings of the Fifth International Conference on Nonaqueous Solutions* (abstracts), Leeds.
Hasegewa, M., and Noda, H. (1975), *Nature* **254**, 212.
Hassel, O., and Hoppe, N. (1960), *Acta Chem. Scand.* **14**, 341.
Hassel, O., and Hroslev, J. (1954), *Acta Chem. Scand.* **8**, 873.
Hassel, O., and Rømming, C. (1962), *Quart. Revs.* **16**, 1.
Hassel, O., and Strømme, K. O. (1958), *Acta Chem. Scand.* **12**, 1146.
Hassel, O., and Strømme, K. O. (1959*a*), *Acta Chem.* **13**, 1775.
Hassel, O., and Strømme, K. O. (1959*b*), *Acta Chem.* **13**, 275.
Haulait, M. Cl., and Huyskens, P. L. (1975), *J. Solution Chem.* **4**, 853.
Hauser, C. R., and Dunnavant, W. R. (1960), *J. Org. Chem.* **25**, 1296.
Hauser, C. R., and Puterbaugh, W. H. (1963), *J. Am. Chem. Soc.* **75**, 4756.

Hedvall, J. A. (1952), *Einführung in die Festkörperchemie*, Vieweg, Braunschweig.
Heininger, S. A., and Dazzi, J. (1951), *Chem. Eng. News* **35**, 87.
Heinzelmann, A., Letterer, R., and Noller, H. (1971), *Mh. Chem.* **102**, 1750.
Hengge, E., and Pretzer, K. (1963), *Chem. Ber.* **96**, 470.
Herasymenko, P. (1938), *Trans. Faraday Soc.* **34**, 1245.
Herasymenko, P. (1941), *Z. Elektrochem.* **47**, 588.
Herasymenko, P., and Speight, G. E. (1950), *J. Iron Steel Inst.* **166**, 169, 289.
Herlem, M., and Popov, A. I. (1972), *J. Am. Chem. Soc.* **94**, 1431.
Hermodsson, Y. (1969), *Arkiv Kemi* **31**, 218.
Hertz, H. G. (1970), *Angew. Chem. Int. Ed.* **9**, 124.
Hertz, H. G., and Raedle, C. (1973), *Ber. Bunsenges. Phys. Chem.* **77**, 521.
Herzberg, G. (1945), *Molecular Spectra and Molecular Structure*, Van Nostrand, New York.
Heyrovsky, J. (1941), *Polarography*, Springer-Verlag, Vienna/New York.
Heyrovsky, J., and Kůta, J. (1965), *Principles of Polarography*, Academic Press, New York.
Hieber, W., and Brendel, G. (1957), *Z. Anorg. Chem.* **289**, 338.
Hieber, W., and Schubert, E. H. (1965), *Z. Anorg. Chem.* **388**, 37.
Hieber, W., and Werner, R. (1957), *Chem. Ber.* **90**, 286, 1116.
Hieber, W., Sedlmaier, J., and Werner, R. (1957), *Chem. Ber.* **90**, 278.
Hill, H. A. O. (1973), in: *Inorganic Biochemistry* (G. L. Eichhorn, ed.), John Wiley and Sons, New York, Chap. 30.
Hill, H. A. O., Pratt, J. M., and Williams, R. J. P. (1969), *Disc. Faraday Soc.* **47**, 165.
Hinton, J. F., and Briggs, R. W. (1975), *J. Magnetic Res.* **19**, 393.
Hippel, P. H. von, and Wong, K. Y. (1965), *J. Biol. Chem.* **240**, 3409.
Hirsch, H., and Fuoss, R. M. (1960), *J. Am. Chem. Soc.* **82**, 1013, 1018.
Hirth, J. P., and Pound, G. M. (1963), *Condensation and Evaporation, Nucleation and Growth Kinetics*, Pergamon, Oxford.
Hoffmann, H. (1975), *Pure Appl. Chem.* **41**, 327.
Hoffmann, H., Janjic, T., and Sperati, R. (1974), *Ber. Bunsenges. Phys. Chem.* **78**, 223.
Hogen-Esch, T. E., and Smid, J. (1966), *J. Am. Chem. Soc.* **88**, 307, 318.
Hogen-Esch, T. E., and Smid, J. (1967), *J. Am. Chem. Soc.* **89**, 2764.
Hoheisel, C., and Kutzelnigg, W. (1975), *J. Am. Chem. Soc.* **97**, 6970.
Holm, T. (1966), *Acta Chem. Scand.* **20**, 1141.
Honigmann, B. (1958), *Gleichgewichts- und Wachstumsformen von Kristallen*, D. Steinkopf, Darmstadt.
Hopf, G., and Paetzold, R. (1972), *Z. Physik. Chem. (Leipzig)* **251**, 273.
Horill, W., and Noller, H. (1976), *Z. Physik. Chem. (Frankfurt)* **100**, 155–163.
Hübner, G., Jung, K., and Winkler, E. (1970), *Die Rolle des Wassers in biologischen Systemen*, Akademie Verlag, Berlin; Vieweg, Braunschweig.
Hughes, W. N. (1972), *The Inorganic Chemistry of Biological Processes*, John Wiley and Sons, New York.
Ichiba, S., Sakai, H., Negita, H., and Maeda, Y. (1971), *J. Chem. Phys.* **54**, 1627.
Ignatiev, A., Jona, F. P., and Shih, H. D. (1975), *Phys. Rev.* **B11**, 4780.
Iler, R. K. (1955), *The Colloid Chemistry of Silica and Silicates*, Cornell University Press, Ithaca, New York.
Ingold, C. K. (1933), *J. Chem. Soc.*, 1120.
Ingold, C. K. (1934), *Chem. Revs.* **15**, 225.
Ingold, C. K. (1953), *Structure and Mechanism in Organic Chemistry*, Cornell University Press, Ithaca, New York, Chapter VII.
Jackl, C., and Sundermeyer, W. (1973), *Chem. Ber.* **106**, 1752.
Jackson, S. E., and Symons, M. C. R. (1976), *Chem. Phys. Lett.* **37**, 551.

Jacobson, B. (1955), *J. Am. Chem. Soc.* **77**, 2919.
Jamieson, J. C. (1963*a*), *Science* **139**, 762.
Jamieson, J. C. (1963*b*), *Science* **139**, 1291.
Janz, G. J. (1967), *Molten Salts Handbook*, Academic Press, New York.
Jaworiwsky, I. S. (1975), Ph.D. Thesis, Ohio State University.
Jette, E. R., and Foote, F. (1935), *J. Chem. Phys.* **3**, 605.
Jones, T. P., Harris, W. E., and Wallace, W. J. (1961), *Can. J. Chem.* **39**, 2371.
Jørgensen, C. K. (1959), *Mol. Phys.* **2**, 309.
Jørgensen, C. K. (1962), *J. Inorg. Nucl. Chem.* **24**, 1587.
Jørgensen, C. K. (1964), *Inorg. Chem.* **3**, 1201.
Justice, J. C., and Justice, M. C. (1976), in: *Abstracts Symposium on Ions and Ion Pairs in Nonaqueous Media*, Leuven, p. 143.
Kagiya, T., Sumida, Y., and Inoe, T. (1968), *Bull. Soc. Chim. Japan* **41**, 767.
Kamlet, M. J., and Taft, R. W. (1976), *J. Am. Chem. Soc.* **98**, 377.
Karioris, F. G., Woyci, F. F., and Buckrey, R. R. (1967), in: *Advances in X-Ray Analysis* (G. R. Mallet *et al.*, eds.), Vol. 10, Plenum Press, New York, p. 250.
Karpfen, A., and Schuster, P. (1976), *Chem. Phys. Lett.* **44**, 459.
Kecki, Z., and Gulik-Krzywicki, T. (1964), *Roczniki Chem.* **38**, 277.
Kendrew, J. C., Watson, H. C., Strandberg, B. E., Dickerson, R. E., Phillips, D. C., and Share, V. C. (1960), *Nature* **185**, 422.
Kerridge, D. H. (1975), *Pure Appl. Chem.* **41**, 365.
Kettle, S. F. A., and Paul, I. (1972), *Adv. Organimet. Chem.* **10**, 199.
Kietaibl, H., Völlenkle, H., and Wittmann, A. (1972), *Mh. Chem.* **103**, 1360.
Kleber, W. (1967), *Kristall und Technik* **2**, 13.
Kleber, W., and Weis, J. (1958), *Z. Krist.* **110**, 1.
Kleiman, G. G., and Burkstrand, J. M. (1975), *Surface Sci.* **50**, 493.
Klement, W., Jr., and Jayaraman, A. (1967), *Progr. Solid State Chem.* **3**, 289.
Knobler, C., Baker, C., Hope, H., and McCullough, J. D. M. (1971), *Inorg. Chem.* **10**, 697.
Koepp, H. M., Wendt, H., and Strehlow, H. (1960), *Z. Elektrochem.* **64**, 483.
Kolthoff, I. M., and Chantooni, Jr., M. K. (1971), *J. Am. Chem. Soc.* **93**, 7104.
Kortüm, G., and Weller, H. (1950), *Z. Naturf.* **5a**, 451, 590.
Kortüm, G., Vogel, J., and Braun, W. (1958), *Angew. Chem.* **70**, 651.
Kosower, E. M. (1956), *J. Am. Chem. Soc.* **78**, 5700.
Kosower, E. M. (1958), *J. Am. Chem. Soc.* **80**, 3253, 3261, 3267.
Kosower, E. M., and Klinedinst, E. P. (1956), *J. Am. Chem. Soc.* **78**, 3483.
Kotz, J. C., and Turnipseed, C. O. (1970), *Chem. Comm.*, 41.
Koutecký, J. (1964), *Angew. Chem.* **75**, 364.
Krakow, K., and Ast, D. C. (1976), *Surface Sci.* **58**, 485.
Krebs, H. (1968), *Grundzüge der anorganischen Kristallchemie*, Enke, Stuttgart.
Kriegsmann, H. (1959), *Z. Anorg. Allg. Chem.* **299**, 149.
Krindel, P. and Elizier, I. (1971), *Coord. Chem. Revs.* **6**, 217.
Krygowski, T. M., and Fawcett, W. R. (1975), *J. Am. Chem. Soc.* **97**, 2143.
Kuchitsu, K. (1968), *J. Chem. Phys.* **49**, 4456.
LaGrange, J., Leroy, L., and Louterman-Leloup, G. (1977), private communication.
Lander, J. J. (1964), *Surface Sci.* **1**, 125.
Lander, J. J., and Morrison, J. (1962), *J. Chem. Phys.* **37**, 792.
Lander, J. J., and Morrison, J. (1963), *J. Appl. Phys.* **34**, 1403.
Lander, J. J., and Morrison, J. (1964), *Surface Sci.* **2**, 553.
Langford, C. H., and Sastri, V. S. (1972), in: *MTP International Review of Science—Inorganic Chemistry*, Series 1, Vol. 9 (M. L. Tobe, ed.), Butterworth, London, p. 203.

Lanir, A., and Navon, G. (1974), *Biochim. Biophys. Acta* **341**, 65.
Lappert, M. F., and Smith, J. K. (1961), *J. Chem. Soc.* 3224.
Laramore, G. E. (1974), *Phys. Rev.* **B9**, 1204.
Laramore, G. E., and Duke, C. B. (1972), *Phys. Rev.* **B5**, 267.
Laramore, G. E., and Switendick, A. C. (1973), *Phys. Rev.* **B7**, 3615.
Larkindale, J. P., and Simkin, D. J. (1971), *J. Chem. Phys.* **55**, 5048.
Larsen, D. W. (1966), *Inorg. Chem.* **5**, 1109.
Larsen, E. M., Mayer, J. W., Gil-Arnao, F., and Camp, M. J. (1974), *Inorg. Chem.* **13**, 574.
Lederer, M., and Mazzei, M. (1968), *J. Chromatogr.* **35**, 201.
Leftin, H. P., and Hobson, M. C., Jr. (1963), *Adv. Catal. Rel. Subj.* **14**, 150.
Lehninger, A. (1965), *Bioenergetics*, Benjamin, New York.
Lenhert, P. G. (1968), *Proc. Roy. Soc. London A* **303**, 45.
Lennard-Jones, E. J. (1928), *Proc. Roy. Soc. London A* **121**, 247.
Leroy, G., and Louterman-Leloup, G. (1975), *J. Mol. Structure* **28**, 33.
Leroy, G., Louterman-Leloup, G., Gaultier, J., and Schroer, M. (1975), *J. Mol. Structure* **25**, 205.
Letterer, K., and Noller, H. (1969), *Z. Physik. Chem.* (*Frankfurt*) **67**, 317.
Lewis, G. N. (1923), *Valence and Structures of Atoms and Molecules*, The Chemical Catalog Co., New York.
Lewis, H. L., and Brown, T. L. (1969), *J. Am. Chem. Soc.* **92**, 4664.
Linck, R. G. (1972), in: *MTP International Review of Science—Inorganic Chemistry Series 1*, Vol. 9 (M. L. Tobe, ed.), Butterworth, London, p. 303.
Lindberg, J. J., Kenttaman, J., and Nissema, A. (1961), *Suomen Kem.* **34B**, 98.
Lindqvist, I. (1963), *Inorganic Adduct Molecules of Oxo-Compounds*, Springer-Verlag, Berlin/Göttingen/Heidelberg.
Lindqvist, I., and Zackrisson, M. (1960), *Acta Chem. Scand.* **14**, 453.
Liotta, C. L., and Harris, H. P. (1974), *J. Am. Chem. Soc.* **96**, 2250.
Little, L. H. (1966), *Infrared Spectra of Adsorbed Species*, Academic Press, London/New York.
Lloyd, D. J., and Parker, A. J. (1970), *Tetrahedron Lett.* **57**, 5029.
Long, J. R. (1973), Ph.D. Thesis, Ohio State University.
Long, L. H. (1972), *Progr. Inorg. Chem.* **15**, 1.
Long, L. H., and Dollimore, D. (1954), *J. Chem. Soc.*, 4457.
Lowry, T. M. (1923), *Trans. Faraday Soc.* **18**, 285.
Lubinsky, A. R., and Duke, C. B. (1976), *J. Vac. Sci. Tech.* **13**, 189.
Luck, W. P. A. (1974), *Fortschr. Chem. Forschung* **4**, 653.
Luck, W. P. A. (1976*a*), *Topics in Current Chem.* **64**, 115.
Luck, W. P. A. (1976*b*), *The Hydrogen Bond—Recent Developments in Theory and Experiment* (P. Schuster *et al.*, eds.), North-Holland Publ. Co., Amsterdam, Chap. 28.
Luder, W. F., and Zuffanti, S. (1946), *The Electronic Theory of Acids and Bases*, Wiley, New York.
Lüdemann, H. D., and Franck, E. U. (1967), *Ber. Bunsenges. Phys. Chem.* **71**, 455.
Lüdemann, H. D., and Franck, E. U. (1968), *Ber. Bunsenges. Phys. Chem.* **72**, 514.
Lundgren, J. O. (1970), *Acta Cryst.* **B26**, 1893.
Lundgren, J. O., and Tellgren, R. (1974), *Acta Cryst.* **B30**, 1937.
Lundgren, J. O., and Olavsson, I. (1976), in: *The Hydrogen Bond—Recent Developments in Theory and Experiment* (P. Schuster *et al.*, eds.), North-Holland Publ. Co., Amsterdam, Chap. 10.
Lux, H. (1939), *Z. Elektrochem.* **45**, 303.
Lux, H. (1942), *Z. Anorg. Allg. Chem.* **250**, 150.

Lux, H. (1948), *Z. Elektrochem.* **52**, 220.
Lux, H. (1949), *Z. Elektrochem.* **53**, 41.
Luz, Z., and Meiboom, S. (1964), *J. Chem. Phys.* **40**, 1058.
Maciel, G. E., and Ruben, G. C. (1963), *J. Am. Chem. Soc.* **85**, 3903.
Maciel, G. E., and Hancock, J. K., Lafferty, L. F., Mueller, P. A., and Musker, W. K. (1966), *Inorg. Chem.* **5**, 554.
MacKellar, W. J., and Rorabacher, D. B. (1971), *J. Am. Chem. Soc.* **93**, 4379.
Malin, J. M., and Swinehart, J. H. (1969), *Inorg. Chem.* **8**, 1407.
Malkin, C. R., and Bearden, A. J. (1971), *Proc. Nat. Acad. Sci. USA* **68**, 16.
Malmström, B. G., and Neilands, J. B. (1964), *Ann. Rev. Biochem.* **33**, 331.
Marcus, P. M., Demuth, J. E., and Jepsen, D. W. (1975), *Surface Sci.* **53**, 501.
Marcus, Y., and Kertes, A. S. (1969), *Ion Exchange and Solvent Extraction of Metal Complexes*, Wiley–Interscience, London/New York/Toronto.
Markl, P. (1972), *Extraktion und Extraktionschromatographie in der anorganischen Analytik*, Akademie Verlagsgesellschaft, Frankfurt.
Martin, M. R., and Somorjai, G. A. (1973), *Phys. Rev.* **B7**, 3607.
Matage, N., and Kubota, T. (1970), *Molecular Interactions and Electronic Spectra*, Marcel Dekker, New York.
Mathias, A., and Warhurst, E. (1962), *Trans. Faraday Soc.* **58**, 948.
Matthews, B. A., and Watts, D. (1974), *Inorg. Chim. Acta* **11**, 127.
Matthews, B. A., Turner, J. F., and Watts, D. W. (1976), *Austral. J. Chem.* **29**, 551.
Matts, T. C., and Moore, P. (1969), *J. Chem. Soc.*, 219.
Maxey, B. W., and Popov, A. I. (1967), *J. Am. Chem. Soc.* **89**, 2230.
Maxey, B. W., and Popov, A. I. (1968), *J. Am. Chem. Soc.* **90**, 4470.
Maxey, B. W., and Popov, A. I. (1969), *J. Am. Chem. Soc.* **91**, 20.
May, W. (1970), *Adv. Catal.* **21**, 151.
Mayburg, P. C., and Koski, W. S. (1953), *J. Chem. Phys.* **21**, 742.
Mayer, U. (1975), *Pure Appl. Chem.* **41**, 291.
Mayer, U. (1976), *Coord. Chem. Revs.* **21**, 159.
Mayer, U. (1977), *Mh. Chem.* in press.
Mayer, U., and Gutmann, V. (1970), *Mh. Chem.* **101**, 912.
Mayer, U., and Gutmann, V. (1972), *Structure and Bonding* **12**, 113.
Mayer, U., and Gutmann, V. (1975), *Adv. Inorg. Chem. Radiochem.* **17**, 189.
Mayer, U., Gutmann, V., and Lodzinska, A. (1973), *Mh. Chem.* **104**, 1045.
Mayer, U., Gutmann, V., and Gerger, W. (1975), *Mh. Chem.* **106**, 1235.
Mayer, U., Kösters, K., and Gutmann, V. (1976), *Mh. Chem.* **107**, 845.
Mayer, U., Kotocova, A., Gerger, W., and Gutmann, V. (1977*a*), *Mh. Chem.*, in press.
Mayer, U., Gerger, W., and Gutmann, V. (1977*b*), *Mh. Chem.* **108**, 489.
Mayer, U., Gutmann, V., and Weihs, P. (1977*c*), unpublished.
Mays, C. W., Vernaak, J. S., and Kuhlmann-Wilsdorf, D. (1968), *Surface Sci.* **12**, 134.
McClelland, B. J. (1964), *Chem. Revs.* **64**, 301.
McKinney, W. J., and Popov, A. I. (1970), *J. Phys. Chem.* **74**, 535.
Middaugh, R. L., Drago, R. S., and Niedzielski, R. J. (1964), *J. Am. Chem. Soc.* **86**, 388.
Miessner, H., Labes, D., and Heckner, K. H. (1976), *Z. Chem.* **16**, 198.
Migchelsen, T., and Vos, A. (1967), *Acta Cryst.* **23**, 796.
Miller, J., Gregoriou, G., and Mosher, H. S. (1961), *J. Am. Chem. Soc.* **83**, 3966.
Mochida, I., Kato, A., and Seiyama, T. (1971), *J. Catal.* **22**, 23.
Moesta, H. (1968), *Chemisorption und Ionisation in Metall–Metall Systemen*, Springer-Verlag, Berlin/New York, p. 47.
Moore, P., and Wilkins, R. G. (1964), *J. Chem. Soc.*, 3454.

Morf, W. E., and Simon, W. (1971), *Helv. Chim. Acta* **54**, 794.
Morimoto, H., Lehmann, H., and Perutz, J. F. (1971), *Nature* **232**, 408.
Moritz, W. (1976), Ph.D. dissertation, University of Munich, Germany.
Movius, W. G., and Linck, R. G. (1970), *J. Am. Chem. Soc.* **92**, 2677.
Muetterties, E. L. (1960), *J. Inorg. Nucl. Chem.* **15**, 182.
Muetterties, E. L. (1967), *The Chemistry of Boron and Its Compounds*, Wiley, New York.
Muller, J. P., Vergruysse, G., and Zeegers-Huyskens, Th. (1972), *J. Chim. Phys.* 1439.
Mulliken, R. S. (1950), *J. Am. Chem. Soc.* **72**, 600.
Mulliken, R. S. (1951), *J. Chem. Phys.* **19**, 514.
Mulliken, R. S. (1952*a*), *J. Am. Chem. Soc.* **74**, 811.
Mulliken, R. S. (1952*b*), *J. Phys. Chem.* **56**, 801.
Mulliken, R. S., and Person, W. B. (1962), *Ann. Rev. Phys. Chem.* **13**, 107.
Murray-Rust, D. M., Hadow, H. J., and Hartley, Sir Harold (1931), *J. Chem. Soc.*, 215.
Narten, A. H., Vaslow, F., and Levy, H. A. (1973), *J. Chem. Phys.* **58**, 5017.
Nelson, N. J., Kime, N. E., and Shriver, D. F. (1969), *J. Am. Chem. Soc.* **91**, 5173.
Nemethy, G. (1974), in: *Structure of Water and Aqueous Solutions* (W. P. A. Luck, ed.), Verlag Chemie, Weinheim, p. 74.
Neuhaus, A. (1964), *Chimia* **18**, 93.
Newton, I. (1687), *Philosophiae Naturalis Principia Mathematica*, S. Pepys, London, p. 402.
Niedermayer, R. (1970), *Kristall Tech.* **5**, 263.
Niendorf, K., and Paetzold, R. (1973), *J. Mol. Structure* **19**, 693.
Nobbs, C. L., Watson, H. C., and Kendrew, J. C. (1966), *Nature* **209**, 339.
Noll, W. (1963), *Angew. Chem.* **75**, 123.
Noller, H., and Kladnig, W. (1976), *Catal. Rev. Sci. Eng.* **13**, 149.
Noller, H., and Ostermeier, K. (1956), *Z. Elektrochem. Ber. Bunsenges Phys. Chem.* **60**, 921.
Noller, H., and Wolff, H. (1959), *Mitteilungsblatt der Chemischen Gesellschaft der DDR, Sonderheft Katalyse*, Berlin, p. 232.
Noller, H., Andréŭ, P., Schmitz, E., Zahlout, A., and Ballesteros, R. (1966), *Z. Physik. Chem. (Frankfurt)* **49**, 299.
Nordman, C. E. (1957), *Acta Cryst.* **10**, 777.
Nordman, C. E. (1962), *Acta Cryst.* **15**, 18.
Nordman, C. E., and Reimann, C. (1959), *J. Am. Chem. Soc.* **81**, 3538.
Normant, H. (1967), *Angew. Chem.* **79**, 1029.
Ohkaku, N., and Nakamoto, K. (1973), *Inorg. Chem.* **12**, 2440.
Olah, G. A. (1974), *Carbocations and Electrophilic Reactions*, Verlag Chemie, Weinheim.
Olah, G. A., and Lukas, J. (1967), *J. Am. Chem. Soc.* **89**, 4739.
Olah, G. A., Baker, E. B., Evans, J. C., Tolgyesi, W. S., McIntyre, J. S., and Bastien, I. J. (1964), *J. Am. Chem. Soc.* **86**, 1360.
Olavsson, I. (1968), *J. Chem. Phys.* **49**, 1063.
Olsen, T. (1970), *Acta Chem. Scand.* **24**, 3081.
O'Sullivan, W. I., Swanner, F. W., Humphlett, W. J., and Hauser, C. R. (1961), *J. Org. Chem.* **26**, 2306.
Overbeck, J. Th. G. (1972), *Chem. in Britain* **8**, 370.
Owensby, D. A., Parker, A. J., and Diggle, J. W. (1974), *J. Am. Chem. Soc.* **96**, 2682.
Oye, H. A., and Gruen, D. M. (1964), *Inorg. Chem.* **3**, 836.
Paetzold, R. (1968), *Spectrochim. Acta* **24A**, 717.
Paetzold, R. (1975*a*), *Z. Chem.* **15**, 377.
Paetzold, R. (1975*b*), in: *MTP International Review of Science—Inorganic Chemistry Series 2*, Vol. 3 (V. Gutmann, ed.), Butterworth, London, p. 201ff.
Paetzold, R., and Niendorf, K. (1974), *Z. Anorg. Allg. Chem.* **405**, 129.

Paetzold, R., and Niendorf, K. (1975), *Z. Physik. Chem.* (*Leipzig*) **256**, 361.
Paoloni, L., Patti, A., and Mangano, F. (1975), *J. Mol. Structure* **27**, 123.
Parker, A. J. (1962), *Quart. Revs.*, 163
Parker, A. J. (1967), in: *Advances in Physical Organic Chemistry* (V. Gold, ed.), Academic Press, London, New York.
Parker, A. J. (1969), *Chem. Revs.* **69**, 1.
Parker, A. J., and Alexander, R. (1968), *J. Am. Chem. Soc.* **90**, 3313.
Parry, R. W., and Shore, S. G. (1958), *J. Am. Chem. Soc.* **80**, 15.
Parshall, G. W. (1964), *J. Am. Chem. Soc.* **86**, 361.
Pauling, L. (1929), *J. Am. Chem. Soc.* **51**, 1010.
Pauling, L. (1959), in: *Hydrogen Bonding* (D. Hadzi, ed.), Pergamon, London.
Pauling, L. (1960), *The Nature of the Chemical Bond*, Cornell University Press, Ithaca, New York.
Pearson, R. G., and Ellgen, P. (1967), *Inorg. Chem.* **6**, 1379.
Pedersen, B. (1974), *Acta Cryst.* **B30**, 289.
Perutz, M. F. (1970), *Nature* **228**, 557.
Perutz, M. F., and Lehmann, H. (1968), *Nature* **219**, 902.
Perutz, M. F., Muirhead, H., Cox, J. M., and Goaman, L. G. (1968), *Nature* **219**, 29, 131.
Peters, C. R., and Nordman, C. E. (1960), *J. Am. Chem. Soc.* **82**, 5758.
Peters, K. (1962), in: *Symposium Zerkleinern*, Verlag Chemie, Weinheim, p. 18.
Peters, K., and Pajakoff, S. (1962), *Microchim. Acta*, 314.
Petersen, R. B., Stezowski, J. J., Wan, C., Burtlich, J. M., and Hughes, R. E. (1971), *J. Am. Chem. Soc.* **93**, 3532.
Peterson, S. W., Taylor, M., and Lin, S. C. (1976), *Chem. Zeit.* **100**, 199.
Pfeiffer, P. (1922), *Organische Molekülverbindungen* Enke, Stuttgart, p. 263.
Pignolet, L. H., and Horrocks, W. DeW., Jr. (1968), *J. Am. Chem. Soc.* **90**, 922.
Pines, H., and Manassen, J. (1966), *Adv. Catal.* **16**, 49.
Pirson, D. J., and Huyskens, P. L. (1974), *J. Sol. Chem.* **3**, 305.
Pitochelli, A. R., Ettinger, R., Dupont, J. A., and Hawthorne, M. F. (1962), *J. Am. Chem. Soc.* **84**, 1057.
Pleskov, V. A. (1947), *Usp. Khim.* **16**, 254.
Pople, J. A. (1951), *Proc. Roy. Soc. London* **A205**, 163.
Popov, A. I. (1975), *Pure Appl. Chem.* **41**, 275.
Pritchard, J. (1963), *Trans. Faraday Soc.* **59**, 437.
Prout, V. K., and Wright, J. D. (1968), *Angew. Chem.* **80**, 688.
Puterbaugh, W. H., and Hauser, C. R. (1959), *J. Org. Chem.* **24**, 416.
Racker, E. (1965), *Mechanisms in Bioenergetics*, Academic Press, New York.
Rahmi, A. K., and Popov, A. I. (1976), *Inorg. Nucl. Chem. Lett.* **12**, 703.
Reichhardt, Ch. (1969), *Lösungsmitteleffekte in der organischen Chemie*, Verlag Chemie, Weinheim.
Remmel, R. J., and Shore, S. G. (1977), unpublished.
Remmel, R. J., Johnson II, H. D., Jaworiwsky, I. S., and Shore, S. G. (1975), *J. Am. Chem. Soc.* **97**, 5395.
Rhodin, T. N. (1953), *Adv. Catal.* **5**, 39.
Rice, S. A. (1975), *Topics in Current Chem.* **60**, 109.
Riepe, R., and Wang, J. H. (1967), *J. Am. Chem. Soc.* **89**, 4229.
Ringwood, A. E. (1967), *Earth Planet Sci. Lett.* **2**, 255.
Ritchie, C. D., and Unschold, R. E. (1967), *J. Am. Chem. Soc.* **89**, 2752.
Roberts, I., and Hammett, L. P. (1937), *J. Am. Chem. Soc.* **59**, 1063.
Robinson, D. R., and Grant, M. E. (1966), *J. Biol. Chem.* **241**, 4030.
Rode, B. M. (1975), *Chem. Phys. Lett.* **32**, 38.

Rode, B. M. (1976), private communication.
Rode, B. M., Breuss, M., and Schuster, P. (1975), *Chem. Phys. Lett.* **32**, 34.
Rogers, T. E., Swinehart, J. H., and Taube, H. (1965), *J. Phys. Chem.* **69**, 134.
Rose, J. (1967), *Molecular Complexes*, Pergamon, London.
Rosseinski, D. R. (1965), *Chem. Revs.* **65**, 467.
Rosseinski, D. R. (1971), *Ann. Rep. Chem. Soc. A*, **63**, 92.
Rossotti, F. J. C., Rossotti, H. S., and Whewell, R. J. (1972), *J. Inorg. Nucl. Chem.* **33**, 2051.
Russegger, P., Lischka, H., and Schuster, P. (1972), *Theor. Chim. Acta* **24**, 191.
Ruzicka, S. J., and Merbach, A. E. (1976), *Inorg. Chim. Acta* **20**, 221.
Ryschkewitsch, G. E., and Zutshi, K. (1970), *Inorg. Chem.* **9**, 411.
Ryschkewitsch, G. E., Mathur, M. A., and Sullivan, T. E. (1970), *Chem. Comm.*, 117.
Saito, K., and Nishizawa, M. (1977), private communication.
Sakurai, H., Okada, A., Kira, M., and Yonezawa, K. (1971), *Tetrahedron Lett.* **19**, 1511.
Sapunov, V. N., and Schmid, R. (1977), to be published.
Schaefer, W. P., and Marsh, R. E. (1966), *J. Am. Chem. Soc.* **88**, 178.
Schaschel, E., and Day, M. C. (1968), *J. Am. Chem. Soc.* **90**, 503.
Schläfer, H. L., and Schaffernicht, W. (1960), *Angew. Chem.* **72**, 618.
Schleich, T., and Hippel, P. H. von (1969), *Biopolymers* **7**, 861.
Schlosser, M. (1973), *Struktur und Reaktivität polarer Organometalle*, Springer, Berlin/Heidelberg/New York.
Schmid, R., Sapunov, V. N., and Gutmann, V. (1976*a*), *Ber. Bunsenges. Phys. Chem.* **80**, 456.
Schmid, R., Sapunov, V. N., and Gutmann, V. (1976*b*), *Ber. Bunsenges. Phys. Chem.* **80**, 1302.
Schmid, R., Sapunov, V. N., Krist, R., and Gutmann, V. (1977), *Inorg. Chim. Acta* **24**, 25.
Schnieder, H., and Schulz, H. (1976), in: *Proceedings of the Fifth International Conference on Nonaqueous Solutions*, Leeds, lecture no. B2.18.
Schnieder, H., and Strehlow, H. von (1965), *Ber. Bunsenges. Phys. Chem.* **69**, 674.
Schnell, E. (1961), *Mh. Chem.* **92**, 1055.
Schrauzer, G. N., and Deutsch, E. (1969), *J. Am. Chem. Soc.* **91**, 3341.
Schroer, D., and Nininger, R. C., Jr. (1967), *Phys. Rev. Lett.* **19**, 632.
Schroer, D., Marzke, R. F., Erickson, D. J., Marshall, S. W., and Wilenzick, R. M. (1970), *Phys. Rev.* **11**, 4414.
Schuller, D. (1973), *Naturwissenschaften* **60**, 145.
Schulz, L. G. (1951), *Acta Cryst.* **4**, 487.
Schuster, P. (1969), *Int. J. Quantum Chem.* **3**, 851.
Schuster, P. (1970), *Theor. Chim. Acta* **19**, 212.
Schuster, P. (1973), *Z. Chem.* **13**, 41.
Schuster, P. (1977), The fine structure of the hydrogen bond, in: *Intermolecular Interactions from Diatomics to Biopolymers* (B. Pullman, ed.), John Wiley, New York, p. 363ff.
Schuster, P., and Preuss, H. W. (1971), *Chem. Phys. Lett.* **11**, 35.
Schuster, P., Jakubetz, W., and Marius, W. (1975), *Topics in Current Chem.* **60**, 1.
Schuster, P., Lischka, H., and Bayer, A. (1977), *Progr. Theor. Org. Chem.* **2**, 89.
Schwab, G. M., Block, J., Müller, W., and Schütze, D. (1957), *Naturwissenschaften* **44**, 582.
Schwarzenbach, G., and Girgls, K. (1975), *Helv. Chim. Acta* **58**, 2391.
Schwarzenbach, G., and Schellenberg, M. (1965), *Helv. Chim. Acta* **48**, 28.
Sears, P. G., Wilhoit, E. D., and Dawson, L. R. (1955), *J. Phys. Chem.* **59**, 16.
Seebach, D., and Enders, D. (1975), *Angew. Chem.* **87**, 1.
Seewald, D., Sutin, N., and Watkins, K. O. (1969), *J. Am. Chem. Soc.* **91**, 7307.
Seifert, H. (1956), *Naturwissenschaften* **7**, 156.
Shapiro, I., and Weiss, H. G. (1953), *J. Phys. Chem.* **57**, 219.
Shih, H. D., and Jona, F. (1976), *J. Phys.* **9**, 1405.

Shin-Piaw, C., Loong Seng, W., and Yoke Seng, L. (1966), *Nature* **209**, 1300.
Shu, F. R., and Rorabacher, D. B. (1972), *Inorg. Chem.* **11**, 1496.
Siddiqui, M. M., and Tompkins, F. C. (1962), *Proc. Roy. Soc.* **A268**, 452.
Sidgwick, N. V. (1927), *The Electronic Theory of Valency*, Clarendon Press, Oxford.
Sigel, H., and McCormick, D. B. (1970), *Accounts Chem. Res.* **3**, 201.
Sillén, L. G., and Martell, A. E. (1971), *Stability Constants of Metal Ion Complexes*, Supplement No. 1, The Chemical Society, London Special Publ. No. 25.
Smith, J. H., and Brill, T. B. (1976), *Inorg. Chim. Acta* **18**, 225.
Smith, P. L., and Martin, F. E. (1965), *Phys. Lett.* **19**, 541.
Smith, S., Fainberg, A., and Winstein, S. (1961), *J. Am. Chem. Soc.* **83**, 618.
Sneddon, L. G., Hoffmann, J. C., Schaeffer, R. O., and Streib, W. E. (1972), *Chem. Comm.*, 474.
Sobol, H., Garfias, J., and Keller, J. (1976), *J. Phys. Chem.* **80**, 1941.
Spaziante, P. M., and Gutmann, V. (1971), *Inorg. Chim. Acta* **5**, 273.
Spiegel, K. (1967), *Surface Sci.* **7**, 125.
Spiro, T. G. (1973), in: *Inorganic Biochemistry* (G. L. Eichhorn, ed.), Elsevier, Amsterdam, London/New York, Chap. 17.
Stackelberg, M. von (1949), *Naturwissenschaften* **36**, 327, 359.
Stackelberg, M. von, and Müller, H. R. (1954), *Z. Elkektrochem.* **58**, 251.
Steinhaus, R. K., and Margerum, D. W. (1966), *J. Am. Chem. Soc.* **88**, 441.
Stephens, J. S., and Cruickshank, D. W. J. (1970), *Acta Cryst.* **B26**, 222.
Stishov, S. M., and Popova, S. V. (1961), *Geochemistry* **6**, 923.
Strandberg, B., and Liljas, A. (1971), cited in: *Structure and Bonding* **8**, 153.
Strehlow, H. von (1966), in: *The Chemistry of Non-aqueous Solvents* (J. J. Lagowski, ed.), Academic Press, New York, p. 129.
Strehlow, H. von, and Koepp, H. M. (1958), *Z. Elektrochem.* **62**, 373.
Strehlow, H. von, and Schneider, H. (1969), *J. Chim. Phys.* **66**, 118.
Strehlow, H. von, and Schneider, H. (1971), *Pure Appl. Chem.* **25**, 327.
Strohmeier, W. (1968), *Structure and Bonding* **5**, 96.
Strohmeier, W., and Onada, T. (1968), *Z. Naturf.* **23b**, 1527.
Strom, E. T., Orr, W. L., Snowden, B. S., and Woessner, D. E. (1967), *J. Phys. Chem.* **71**, 4017.
Struss, A. W., and Corbett, J. D. (1970), *Inorg. Chem.* **9**, 1373.
Suhrmann, R., Wedler, G., Wilke, H. G., and Reusmann, G. (1960), *Z. Phys. Chem. (Frankfurt)* **26**, 85.
Sujishi, S., and Witz, S. (1954), *J. Am. Chem. Soc.* **76**, 4631.
Sujishi, S., and Witz, S. (1957), *J. Am. Chem. Soc.* **79**, 2447.
Sumner, G. G., Klug, H. P., and Alexander, L. E. (1964), *Acta Cryst.* **17**, 732.
Sundermeyer, W. (1961), *Z. Anorg. Allg. Chem.* **313**, 290.
Sundermeyer, W. (1963), *Chem. Ber.* **96**, 1293.
Sundermeyer, W. (1965), *Angew. Chem. Int. Ed.* **4**, 222.
Sutin, N. (1968), *Accounts Chem. Res.* **1**, 225.
Swain, C. G. (1948), *J. Am. Chem. Soc.* **70**, 1119.
Swart, E. R., and Le Roux, G. J. (1957), *J. Chem. Soc.*, 406.
Swift, T. J. (1964), *Inorg. Chem.* **3**, 526.
Swift, T. J., and Connick, R. J. (1962), *J. Chem. Phys.* **37**, 307.
Symons, M. C. R. (1976), private communication.
Szabó, Z. G. (1962), *Kinetics and Mechanisms of Inorganic Reactions in Solutions*, The Chemical Society, Special Publ. No. 16, p. 113.
Szabó, Z. G., and Bercés, T. (1968), *Z. Phys. Chem. (Frankfurt)* **57**, 113.

Szabó, Z. G., and Solymosi, F. (1961), *Actes du Dixième Congrès de Catalyse, Paris*, Editions Technip, Paris, p. 1627.
Szabó, Z. G., and Thege, I. K. (1975), *Acta Chim. Sci. Hungary* **86**, 127.
Szmant, H. H. (1972), in: *Dimethylsulfoxide* (S. W. Jakob *et al.*, eds.), Marcel Dekker, Inc., New York.
Szwarc, M. (1972), *Ions and Ion Pairs in Organic Reactions*, Wiley and Sons, New York.
Taft, R. W., and Kamlet, M. F. (1976), *J. Am. Chem. Soc.* **98**, 2886.
Tamres, M., and Searles, S. (1962), *J. Phys. Chem.* **66**, 1099.
Taube, H. (1965), *Adv. Chem. Ser. Amer. Chem. Soc.* **49**, 107.
Taube, H., and Posey, F. A. (1953), *J. Am. Chem. Soc.* **75**, 1463.
Taube, H. and Myers, H. (1954), *J. Am. Chem. Soc.* **76**, 2103.
Thomas, G., and Holba, V. (1969), *J. Inorg. Nucl. Chem.* **31**, 1749.
Thomke, K., and Noller, H. (1972), *Z. Naturf.* **27b**, 1462.
Tobe, H. (1965), *Adv. Chem. Ser. Amer. Chem. Soc.* **49**, 7.
Torsi, G., and Mamantov, G. (1972), *Inorg. Chem.* **11**, 1439.
Townes, C. H., and Dailey, B. P. (1952), *J. Chem. Phys.* **20**, 35.
Tracy, J. C., and Palmberg, P. W. (1969), *J. Chem. Phys.* **51**, 4852.
Treitel, I. M., Flood, M. T., Marsh, R. E., and Gray, H. B. (1969), *J. Am. Chem. Soc.* **91**, 6512.
Tschebull, W. (1974), Ph.D. Thesis, Technical University of Vienna.
Tschebull, W., Gutmann, V., and Mayer, U. (1975), *Z. Anorg. Chem.* **416**, 323.
Ukaji, T., and Kuchitsa, K. (1966), *Bull. Chem. Soc. Japan* **39**, 2153.
Unertl, W. N., and Thapliyal, H. V. (1976), *J. Vac. Sci. Tech.* **12**, 263.
Ussanovich, M. (1939), *Zhur. Obshch. Khim. USSR* **9**, 182.
Van Hove, M. A., and Tong, S. Y. (1976), *Surface Sci.* **54**, 91.
Venkataraman, H. S., and Hinshelwood, C. N. (1960), *J. Chem. Soc.* 4977.
Verwey, E. J. (1941), *Rec. Trav. Chim.* **60**, 687.
Vinek, H., and Noller, H. (1976), *Z. Phys. Chem.* (*Frankfurt*) **102**, 255.
Volpin, M. E., and Kolomnikov, I. S. (1975), in: *Organometallic Reactions* (E. I. Becker, M. Tsutsui, eds.), Vol. 5, Wiley–Interscience, New York, p. 313.
Vonk, C. G., and Wiebenga, E. H. (1957), *Acta Cryst.* **10**, 378.
Waack, R., and Doran, M. A. (1962), *Chem. Ind.* 1290.
Waack, R., and Doran, M. A. (1963), *J. Phys. Chem.* **67**, 148.
Waack, R., and Doran, M. A. (1964), *J. Phys. Chem.* **68**, 11.
Waack, R., Doran, M. A., and Stevenson, P. E. (1966), *J. Am. Chem. Soc.* **88**, 2101.
Waack, R., McKeerer, L. D., and Doran, M. A. (1969), *Chem. Comm.* 127.
Wagner, C. (1975), *Metallurg. Trans.* **6B**, 405.
Wannagat, U., and Schwarz, R. (1954), *Z. Anorg. Chem.* **277**, 73.
Wannagat, U., and Vielberg, F. (1957), *Z. Anorg. Chem.* **291**, 310.
Watkins, K. O., and Jones, M. M. (1964), *J. Inorg. Nucl. Chem.* **26**, 469.
Weidemann, E. G., and Zundel, G. (1970), *Z. Naturf.* **25a**, 627.
Weidemann, E. G., and Zundel, G. (1972), *J. Am. Chem. Soc.* **94**, 2387.
Weidemann, E. G., and Zundel, G. (1973), *J. Chem. Soc. Faraday II* **69**, 305.
Weiser, H. B. (1950), *A Textbook of Colloid Chemistry*, John Wiley, New York.
Weiss, A. (1975), private communication.
Weiss, E. (1964), *J. Organomet, Chem.* **2**, 314.
Weiss, E., and Lucken, E. C. A. (1964), *J. Organomet. Chem.* **2**, 197.
Weitz, E., Franck, H., and Schuchard, M. (1950), *Chem. Ztg.* **74**, 256.
Wells, A. F. (1962), *Structural Inorganic Chemistry*, 3rd ed., Clarendon, Oxford.
Wells, C. H. J. (1966), *Spectrochim. Acta* **22**, 2125.

Werner, A. (1913), *Neuere Anschauungen auf dem Gebiete der Anorganischen Chemie*, Vieweg und Sohn, Braunschweig.
Weyl, W. A. (1951), Mineral Industries Experimental Station Bulletin No. 57, cited in Brunauer (1965).
Wiberg, E., and Krüerke, U. (1953), *Z. Naturf.* **8b**, 610.
Widom, J. M., Philippe, R. J., and Hobbs, M. E. (1957), *J. Am. Chem. Soc.* **79**, 1383.
Williams, E. A., Cargioli, J., and Larochelle, R. W. (1976), *J. Organomet. Chem.* **108**, 153.
Winstein, S., and Robinson, G. C. (1958), *J. Am. Chem. Soc.* **80**, 169.
Winstein, S., Clippinger, E., Fainberg, A. H., and Robinson, G. C. (1954), *J. Am. Chem. Soc.* **76**, 2597.
Witschonke, C. R., and Kraus, C. A. (1947), *J. Am. Chem. Soc.* **69**, 2472.
Wittig, G. (1958), *Experientia* **14**, 389.
Wolf, D. (1972), Ph.D. dissertation, University of Munich, Germany.
Wong, M. K., McKinney, W. J., and Popov, A. I. (1971), *J. Phys. Chem.* **75**, 56.
Woolf, A. A., and Emeléus, H. J. (1949), *J. Chem. Soc.* 2865.
Wooster, C. B., and Ryan, J. F. (1932), *J. Am. Chem. Soc.* **54**, 2419.
Wright, C. P., Murray-Rust, D. M., and Hartley, Sir Harold (1931), *J. Chem. Soc.* 199.
Wuepper, J. L., and Popov, A. I. (1969), *J. Am. Chem. Soc.* **91**, 4352.
Wuepper, J. L., and Popov, A. I. (1970), *J. Am. Chem. Soc.* **92**, 1493.
Wyckoff, R. W. G. (1965), *Crystal Structures*, 2nd ed., Vol. 1, Wiley, New York/London/Sydney.
Yamada, M., Saruyama, H., and Aida, K. (1972), *Spectrochim. Acta* **28A**, 439.
Yarwood, J. (1967), *Chem. Comm.* 809.
Yarwood, J. (1970), *Spectrochim. Acta* **26A**, 2099.
Yates, J. T., Jr. (1974), *Chem. Eng. News*, August 26.
Yates, J. T., Jr., and King, D. A. (1972), *Surface Sci.* **30**, 601.
Yates, J. T., Jr., and Lucchesi, P. J. (1961), *J. Chem. Phys.* **35**, 243.
Yates, J. T., Jr., Madey, T. E., and Erickson, N. S. (1974), *Surface Sci.* **43**, 257.
Yeagle, P. L., Lochmüller, C. H. and Henkens, R. W. (1975), *Proc. Nat. Acad. Sci. USA* **72**, 454.
Yokoyama, T., Taft, R. W., and Kamlet, M. J. (1976), *J. Am. Chem. Soc.* **98**, 3233.
Zandstra, P. J., and Weissman, S. I. (1962), *J. Am. Chem. Soc.* **84**, 4408.
Zoltai, T., and Buerger, M. J. (1959), *Z. Krist.* **111**, 129.
Zundel, G. (1976), in: *The Hydrogen Bond—Recent Developments in Theory and Experiment* (P. Schuster *et al.*, eds.), North-Holland Publ. Co., Amsterdam, Chap. 15.

Index